CRYSTALLIZATION AND MATERIAL SCIENCE OF MODERN ARTIFICIAL AND NATURAL CRYSTALS

Edited by **Elena Borisenko** and **Nikolai Kolesnikov**

Crystallization and Materials Science of Modern Artificial and Natural Crystals
Edited by Elena Borisenko and Nikolai Kolesnikov

Published by InTech

Janeza Trdine 9, 51000 Rijeka, Croatia

Publishing Process Manager Sasa Leporic

Technical Editor Teodora Smiljanic

Cover Designer InTech Design Team

Additional hard copies can be obtained from orders@intechopen.com

Crystallization and Materials Science of Modern Artificial and Natural Crystals,
Edited by Elena Borisenko and Nikolai Kolesnikov
p. cm.
ISBN 978-953-307-608-9

For EU product safety concerns:
IN TECH d.o.o., Prolaz Marije Krucifikse Kozulić 3, 51000 Rijeka, Croatia,
info@intechopen.com or visit our website at intechopen.com.

Meet the editors

Dr. Elena Borisenko is currently a Senior Researcher at the Institute of Solid State Physics, at the Russian Academy of Sciences. She graduated from Moscow Institute of Steel and Alloys as a Master of Science in Metal Physics, in 1985 and she did her PhD study, on condensed matter physics, at ISSP RAS, in 1992. Dr. Borisenko is experienced in studying properties and microstructure of advanced materials, especially, inorganic semiconductor compounds and alkali halide crystals. Her research includes phase transitions, physical-chemical aspects of crystal growth, grain growth and textures, structural defects, and testing of mechanical properties of solids.

Nikolai N. Kolesnikov graduated from the Physical-Chemical Department of the Moscow Chemical-Technological Institute in 1982. From 1982 to 1993, he worked at the Institute of Solid State Physics (ISSP), Russian Academy of Sciences, and has experienced the journey from Engineer to Senior Researcher. He obtained his PhD degree in Condensed Matter Physics in 1989. Since 1993, Dr Kolesnikov has been Chief of Laboratory of the Physical-Chemical Basis of Crystallization of ISSP. Nikolai N. Kolesnikov is successfully working in crystal growth from the melt and related fields. His research interests are crystal growth of II-VI compounds (ZnS, ZnSe, ZnTe, CdS, CdSe, CdTe, Cd1-xZnxTe) by zone melting, the Bridgman technique, growth of the nanocrystals of the wide-gap semiconductors, investigation of properties of the refractory compound melts, and the growth of Tl-based HTSC single crystals. Dr. Kolesnikov has published over 100 papers in scientific journals and has received 38 patents.

Contents

Preface

Crystal growth is a fundamental phenomenon, which provides the basis for a wide variety of naturally occurring processes, and an astonishing range of diverse human activities. This book includes a representative collection of contemporary studies on crystal growth across various aspects, including; problems concerning crystal growth in high technology, medicine, pharmacy, and mineralization of marine organisms; modern techniques developed for crystal growth of inorganic and organic crystals from melt, solutions, via chemical synthesis, and atmospheric plasma spraying; phase transitions in solid and liquid states; a particular problem of vitrification and formation of glass-crystalline composites in complex systems of oxides; interaction between organic and inorganic materials leading to formation of the advanced class of inorganic/organic composites; macromolecular crystallization, biomineralization, and biomimetic crystallization; modern investigative techniques, such as x-ray diffraction (XRD), scanning electron microscopy (SEM), electron back scattering diffraction (EBSD), energy dispersive x-ray spectroscopy (EDS), x-ray electron probe microanalysis (EPMA), IR spectroscopy (IRS), Raman light spectroscopy (RS), photoluminescence (PL), optical microscopy, thermogravimetric (TG) and differential thermal analysis (DTA), used to study phase transitions, crystal structure, morphology, microstructure and properties of the grown materials.

The first chapter *Structure and Properties of Advanced Inorganic Materials* contains the studies of thermal barrier ceramic coatings, in particular; yttrium-partially stabilized zirconium (PSZ) produced by atmospheric plasma spraying; the role of oxide formation in crack nucleation and growth in PSZ; optical microscopy, SEM, EBSD, used to study morphology of grain structure, distribution of crystal orientations (crystallographic texture) and crystal symmetry of the coatings. EDS was applied to find a correlation between formation of Cr and Al oxides, and crack propagation in the coating layers.

The second chapter considers the nucleation and crystal growth in phase separated lithium silicate glasses, and studies the dependences of crystal growth rate in the lithium silicate systems on SiO_2 content, and heating temperature.. It has been found that, - comtrary to the perception -phase separation does not facilitate the conditions for the formation and growth of crystals.On the contrary, in some cases even retards the crystal growth process. As a result of the studies, optimum compositions and annealing regimes are recommended for formation of the lithium silicate glass-ceramic materials for practical applications.

Third chapter deals with Cu-SiO_2-GeO_2 metamaterials consisting of glass and crystalline phase, grown by the floating zone technique in the pseudo-ternary system; Eutectic crystallization, phase composition, microstructure, and optical properties of the grown materials are studied, and self-organization of a structure is examined, depending on growth parameters and chemical composition, to find optimal fiber sizes and filling factor. Crystal state of the materials is studied by x-ray diffraction, EBSD and Fourier transformation techniques. These results in combination with EPMA data and analysis of diffusion of the components are used to revise the pseudo-ternary diagram. A phase boundary of vitrification for a silica glass containing Cu is demonstrated in graphic form. Shape and crystallography of the etched crystalline matrix and glass fibers are studied, and increase in a difference of refraction indexes in the self-organized structure and optical band gap opening caused by fiber etching are analyzed.

Within the second section of the book *Inorganic/Organic Materials for High Technology*, the effect of amino acid additives on crystal growth of ammonium dihydrogen phosphate (ADP) is studied; a metastable zone width of pure and doped ADP is measured; a dependence of induction period of nucleation on supersaturation is investigated; the effect of amino acid additives on crystal growth conditions, including crystal growth rate is discussed. Powder x-ray diffraction (XRD) experiments confirm that pure and doped crystals are single phased. High resolution XRD (HRXRD) proves high perfection of the doped crystals. Some optical, electric and piezoelectric properties of the pure and doped ADP are studied. These materials are regarded as promising for application in non-linear optic, and they are used as piezoelectric crystals.

The following chapter is a study of topochemical conversion from inorganic-organic hybrid precursors to inorganic nanoplates and nanofilms. Oriented films were produced from $WO_3 \cdot H_2O$ and WO_3 tungsten oxide nanoplates through dropping–evaporation process. Topochemical conversion of tungstate-based inorganic-organic hybrid belts to W_2N nanoplates is considered. Average particle size calculated from XRD patterns is 3.2 nm. TEM images indicate that the nanoplates are composed of nanocrystals. Application fields of the nanoplates are shown; WO_3 nanoplates are used as a photocatalyst. An amount of O_2 generated when using WO_3 nanoplates for visible-light-induced water splitting is greater than that of commercial WO_3 powders by an order of magnitude. The enhanced photocatalytic properties should be attributed to the developed specific surface area and high crystal perfection of the synthesized WO_3 crystal nanoplates. WO_3 (α-MoO_3) gas sensors were fabricated and tested.

The third chapter of this section describes open-framework structures of metal phosphates synthesized under hydrothermal or solvothermal conditions in the presence of linear polyamines used as the structure-directing agents (SDA).

Diethylenetriamine (DETA), triethylenetetramine (TETA) and tetraethylenepentamine (TEPA) were used to produce $CoZnPO_4$-IV, $CoZnPO_4$-V and $CoZnPO_4$-VI cobalt-zinc phosphates. Microporous cobalt-zinc phosphates $CoZnPO_4$-I, $CoZnPO_4$-II, and $CoZnPO_4$-III were synthesized under hydrothermal conditions from gel at the temperature of 200°C with ammonium, ethylenediaminum, and diethylenetriaminum involved in the crystallization process. The effect of crystallization temperature and crystallization time on transformations of the cobalt-zinc phosphates is studied; decomposition of linear polyamines, template molecules, and its effect on crystallization of zinc-cobalt phosphates is investigated. They have a variety of potential applications in catalysis, separation processes, and as photo-luminescent materials.

The third section of this book, *Biomineralization,* discusses the mineralization of calcareous organisms. The first chapter considers biomineralization of marine organisms, various mollusks and oysters. Interaction of calcareous organisms with metal in seawater is studied by x-ray diffraction (phase analysis) and x-ray spectroscopy (elemental composition). Ability of the living organisms, in particular, freshwater and subterranean snails to accumulate dolomite is discussed, and genetic aspects of biomineralization are tackled.

The following chapter in this section is concerned with calcium-based biominerals; biomineralization of different families of proteins present in Pacific red abalone (*Haliotis rufescens*) and Japanese pearl oyster (*Pinctada fucata*); The role of biomacromolecules in the mineralization of calcium carbonate and calcium oxalate; biomimetic crystallization, which mimics some aspects of biomineralization; role of natural proteins and peptides in mineralization of calcium carbonate; sources of amorphous calcium carbonate (ACC) and its role in biomineralization; formation of ACC in closed pores. Synthetic polymers and peptides are used in mimicked biomineralization, and calcium oxalate crystals in kidney stones are displayed.

Section four, *Crystal Growth for Health protection,* includes growth of pharmaceutical crystals, cryoprotection of living cells, and various techniques of protein crystal growth.

Within this first chapter, studies focused on the cryoprotective effect of disaccharide molecules on living cells are discussed. Trehalose and sucrose, which consist of fructose and glucose rings connected by a glycosidic bond, are natural disaccharide compounds found in cryoprotectants for living cells. Cryoprotective mechanism of these substances, considered as controlled by hydrogen bonding between a lipid bilayer head group and trehalose, is confirmed by Raman spectroscopy. Another mechanism of protection is inhibition of ice crystal growth in extracellular space. Cryoprotective mechanism of disaccharide is controlled by the hydration of disaccharide molecules in water solution. Kinetics of ice crystallization depending on concentration of trehalose solution and cooling rate of the solution is analyzed. Electron-microscopic studies of replicas and powder X-ray diffraction analysis were used to find structure and phase composition of disaccharide and water molecules in the frozen state. The disaccharide molecules are considered to act as the cell-impermeable cryoprotectants.

The second chapter deals with growth from melt of pharmaceutical crystals, such as erythritol, atenolol, pindolol, which are studied by differential scanning calorimetry, polarized light thermal microscopy, infrared spectroscopy, x-ray diffractometry. The problems of nucleation, phase transitions caused by solidification of melt, are studied. The conditions of vitrification and crystallization are found in view of binodal and spinodal decomposition of a compound in a liquid state. Morphology and polymorphism of the crystals and mechanisms of crystallization depending on cooling or heating conditions are considered. The effect of hydrogen bonding on nucleation and a crystalline structure is analysed by IR spectroscopy, and relationships between molecular and crystal structures are discussed.

The third chapter is a consistent, representative, and well-illustrated scientific description of modern alternative techniques used in macromolecular crystallization of proteins. It considers general principles of macromolecular crystallization and finds correlations between the nucleation and growth kinetics, and the phase diagram. Crystallization of a protein in presence of a precipitant is considered on a basis of two-dimensional solubility diagram. There is a difference between a classical molecular crystal and a protein crystal; in the former, all the atoms can be described in terms of a regular lattice, while in the latter a crystal coexists with a fraction of material in a liquid state. The ways of lowering an energy barrier to crystal nucleation from a solution are discussed. A role of selective excipients, which might form the self-assembly in protein crystallization is analyzed. Four well-known crystallization methods: vapor diffusion, free interface diffusion, batch, and dialysis are compared, and their advantages and disadvantages are considered. Production of a fine-sized crystal is regarded as a main drawback of the vapor diffusion technique. The drawback of free interface diffusion is that it is protein-consuming, since the solution must be highly concentrated for crystallization. Batch method is more expensive than the others. Dialysis is not a universal technique, as it does not work with PEG solutions. Alternative crystallization techniques involve additives, which affect crystallization. Additive screening, use of cofactors, ligands and heavy-atom ligands are considered. Salting out is described as the precipitation process in which solute–solute interactions play an important role. The effect of pH of additives on crystallization kinetics of proteins is considered.

The final section of this book, *Molecular Crystals,* is closely connected to the previous book section in regard to specific features of molecular crystalline structure. However, in this section the focus is on physical properties, which are of interest for technical applications. General problem of statistics of the spontaneous breakage of chiral symmetry in crystallization processes and distribution of left (L) and right-handed (D) forms in molecules or crystals are studied.

In the first chapter optical properties of prototypical molecular crystals: 1,1,4,4-tetraphenyl-1,3-butadiene (TPB), dibenzo[d,d′]thieno[3,2-b;4,5-b′]dithiophene (DBTDT), N-pyrrole end-capped thiophene/phenyl co-oligomer (TPP) are discussed. Some methods of crystal growth are considered, as well as the light emission of these molecular crystals. The general focus is on the UV-visible absorption properties considered in the framework of the exciton theory of molecular materials. Electronic and excitonic transitions, exciton polarization, exciton-phonon coupling, the Davydov splitting of exciton bands of these molecular crystals are discussed. The role of polymorphism, intermolecular interactions in the transitions is considered using the experimental data on TPB and DBTDT crystals. TPP is studied to analyze directional dispersion of the absorption spectra.

The second chapter presents statistics of distributions of enantiomeric excess of laevorotatory (L) and dextrorotatory (D) $NaBrO_3$ crystals grown from aqueous solution. A character of the distribution depending on growth conditions is studied and a conclusion is drawn on the effect of temperature, supersaturation, crystal growth rate, a type of crystallizer, handedness of seeds on a relative percentage of L and D-crystals, and racemates grown from unstirred and stirred solutions. Cloning of the crystals in stirred solution is studied.

Acknowledgements

It was a pleasure working on this project with the Publishing Process Managers, Ms. Tajana Jevtic and Ms. Sasa Leporic. I appreciate their competent and timely assistance.

Elena B. Borisenko and Nikolai Kolesnikov
Institute of Solid State Physics, The Russian Academy of Sciences
Russia

Part 1

Structure and Properties of Advanced Inorganic Materials

1

Crystallographic Observation and Delamination Damage Analyses for Thermal Barrier Coatings Under Thermal Exposure

Kazunari Fujiyama
Meijo University,
Japan

1. Introduction

Recently, as the service condition of high temperature components are becoming severer and almost beyond its ultimate performance, protective coatings are expected to be effective solutions to keep the reliability and durability of high efficiency apparatus. Thermal barrier coatings(TBCs) have been applied to cool down the metal temperature and to protect damage under thermal exposure of hot gas path components in gas turbines[1]. As TBCs are used in very severe conditions, thermally induced damage and material degradation are inevitably induced during service and delamination of coating layer may occur finally as the result of coalescing multiple lateral cracks and some vertical cracks after the evolution of oxide layer between top coat and bond coat schematically shown in Fig.1.

However, the mechanism of damage and degradation is still not clear enough because there are so many factors affecting the delamination life[2]. Therefore, one objective of this paper is to focus on how the EBSD observation of thermal exposure samples of TBC top coatings can be applied to identify the particle morphologies after the plasma spraying and another is to evaluate the damage process until the delamination of top coatings.

Optical microscope observation was conducted on laboratory test samples of TBC system after thermal exposure using electric furnace and for measuring the pore fraction amounts during the process. SEM observation was also conducted to measure cracks induced by thermal exposure. It should be noted that the detailed microstructural features of TBC top coat have not been clearly observed by the conventional optical microscope or scanning electron microscope (SEM) because those measures cannot reveal the detailed proper boundaries of top coat splat particles. Electron backscatter diffraction (EBSD) method[3][4] is expected to be an effective tool for observing the morphologies of such particles, but the application of EBSD to the TBCs has not been popularized enough due to the difficulties in preparing the observation surface of TBCs and in identifying the exact crystal systems. We demonstrate the current status for visualizing the splat morphologies in top coat by EBSD and depict some problems in applying the technique. EDS(Energy Dispersive Spectroscopy) analyses and indentation tests are also used as the tools for investigating the sintering of top coatings and the evolution of TGO(Thermally Grown Oxide) layers. Finally this article presents some evaluation diagram for the top coat delamination based on the obtained experimental results.

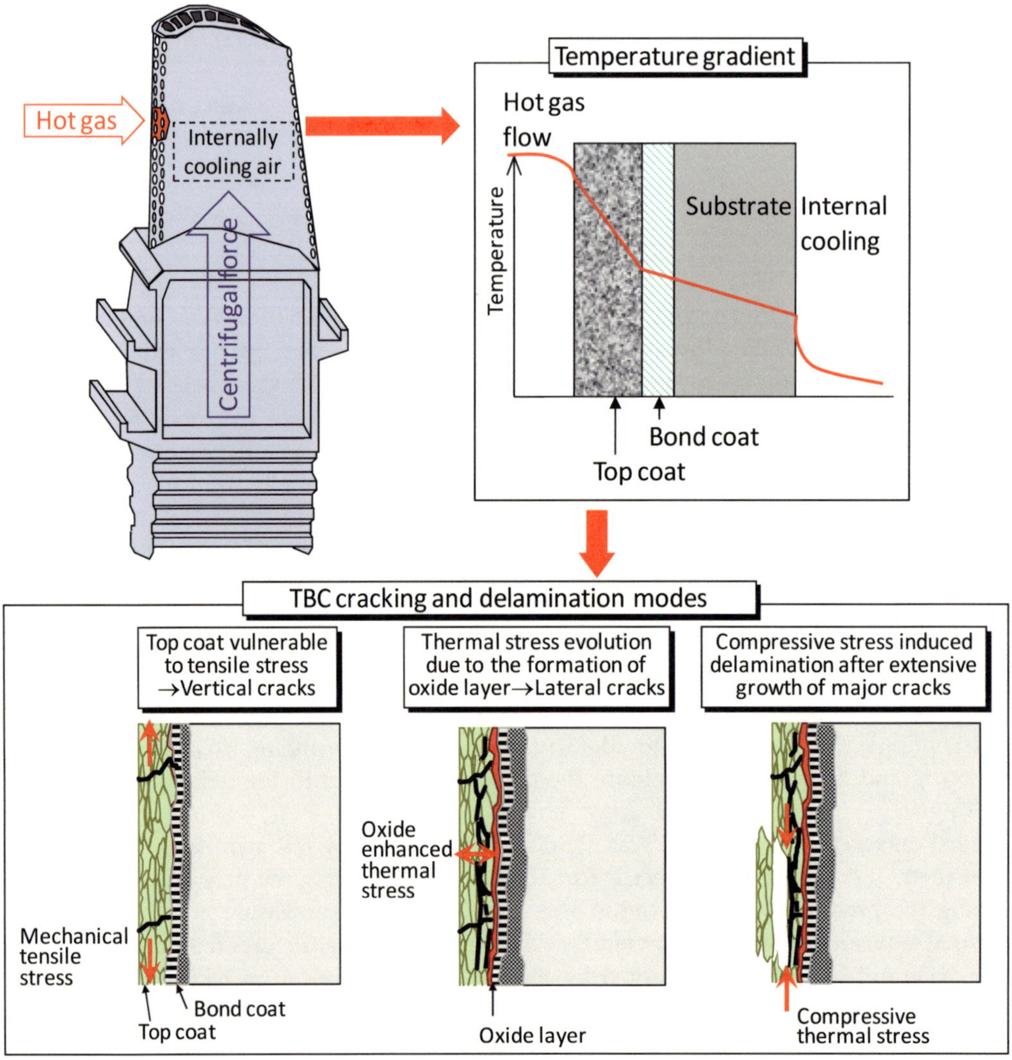

Fig. 1. Schematic illustration for the delamination of TBC in gas turbine blade under temperature gradient.

2. Specimen preparation [5]

Atmospheric plasma sprayed ceramic coatings are widely used as the thermal barrier coating (TBC) in high temperature components typically in gas turbine hot gas path sections. The TBC samples tested here are consisted with three layers: top coat, bond coat and substrate as shown in Fig.2. The top coat is consisted with Yttria- Partially Stabilized Zirconia(PSZ). The thickness of top coat region is 1mm, considerably thicker compared with the commercial TBC system in actual gas turbines.

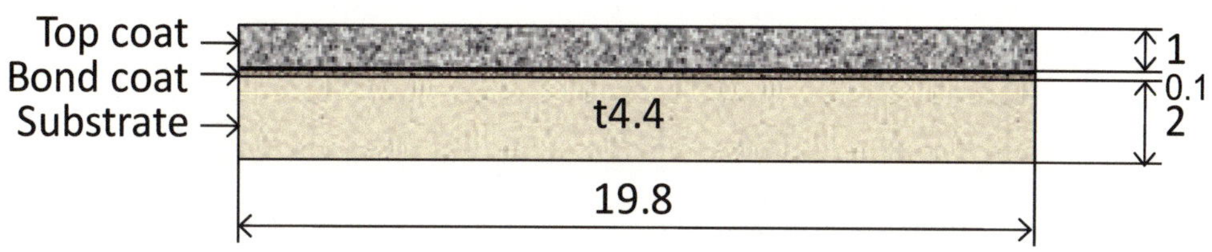

Top coat : 8wt%Y_2O_3-ZrO_2(1mm thick); APS(Atmospheric Plasma Spraying)
Bond coat : CoNiCrAlY(100μm thick); LPPS(Low Pressure Plasma Spraying)
Substrate : MA263(2mm thick)

Fig. 2. TBC specimen geometry.

Thermal exposure tests were conducted up to 1000 hours under constant temperature conditions at 900°C and at 1000°C using an electric furnace up to 1000hrs. The specimens were cut into observation samples with the surface finished with colloidal alumina with particle diameters as 0.1 to 3 micron meters.

3. Optical microscope observation and measurement of pores and cracks [5]

Figure 3 shows the delamination process exposed at 900°C. Macro cracks grow laterally in the top coat just above the bond coat. The surface macro crack is located at the interface between top coating and bond coat besides at the mid section crack is located above the bond coat within top coat. Fig.4 shows the cross section exposed for 50hours at1000°C. The delamination was clearly found in this case.

Figure 5 shows optical microscope observation of top coat at the region near the bond coat. Reduction of the area by pores was observed for samples after thermal exposure compared with as-sprayed samples despite of the non-monotonic trend with exposure time. Fig.6 shows the traced pore image for image processing based on optical microscope photos. Area fractions of pores were obtained from the area ratio of pores (black area) to the observed area.

Figure 7 shows the trend of area fraction of pores against thermal exposure time. Reduction in area fraction of pores was observed at the initial stage of heating but the decreasing trend was not monotonic. The area fraction of pores showed similar levels exposed at 900°C for 500h and at 1000°C for 75h.

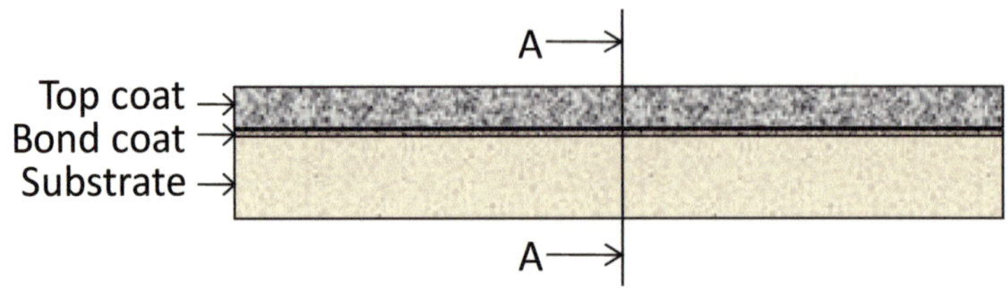

(A-A : Observed portion)

(a)As sprayed

(b)900°C 100h

(c)900°C 300h

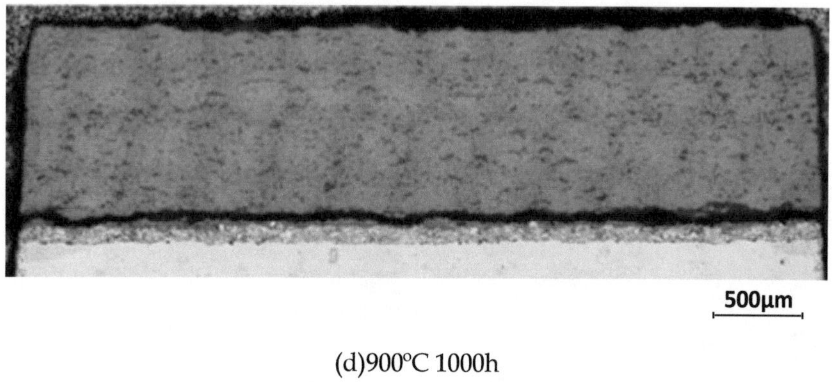

Fig. 3. Optical microscope observation of the cross section at 900 °C exposure tests.

Fig. 4. Optical microscope observation of the cross section at 1000 °C-50h exposure test.

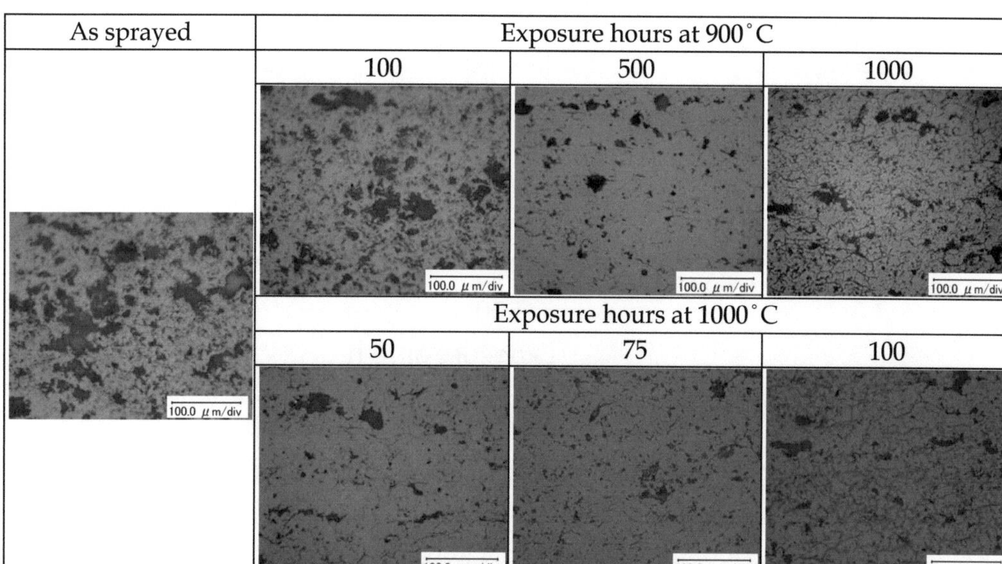

Fig. 5. Optical microscope observation of top coat.

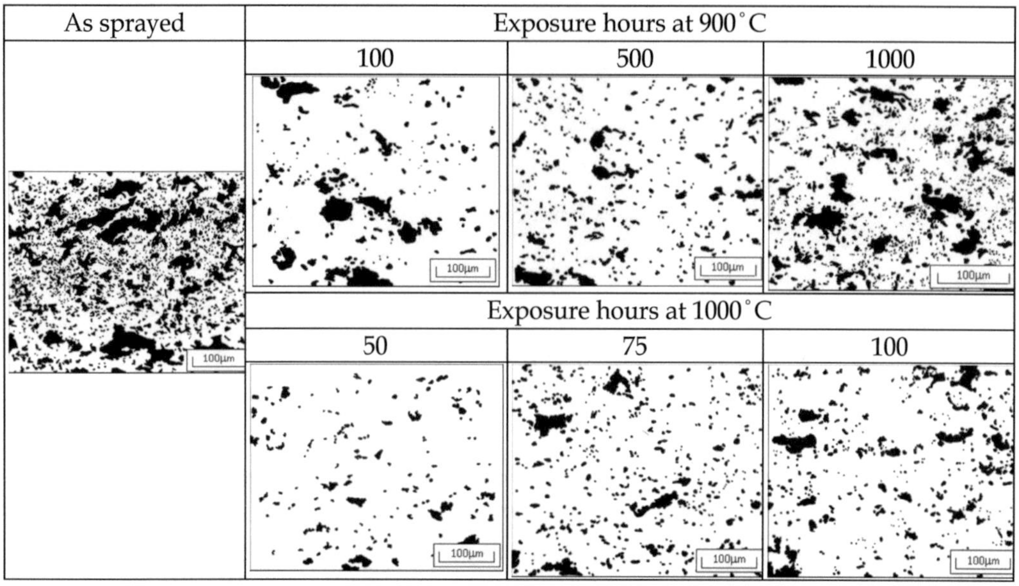

Fig. 6. Pore area trace of optical microscope image in top coat.

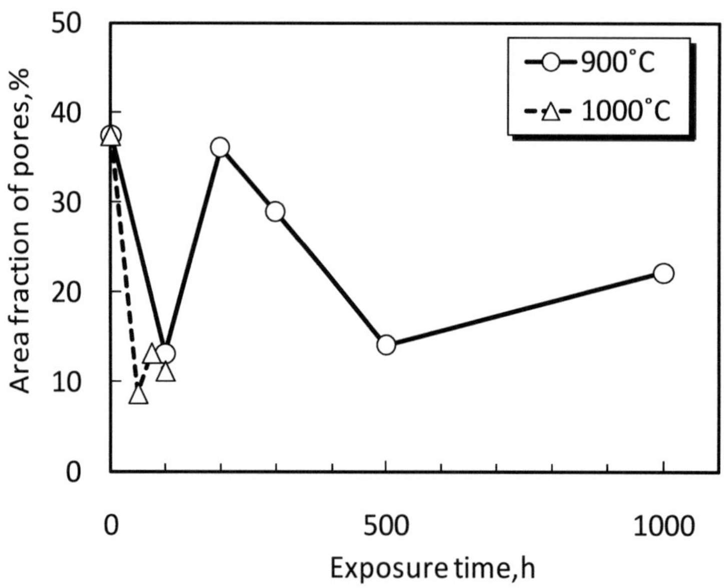

Fig. 7. The trend of area fraction of pores in top coat against exposure time.

4. SEM/EBSD observation and measurement of crack growth trend [5]

SEM observation was conducted using the thermal field-emission scanning electron microscope mainly used for investigating crack morphologies. Observed cracks were traced manually and then processed by image processing software to measure crack length. TGO layer was investigated using EDS system of SEM to identify the elements of oxides.

EBSD observation was conducted using Tex SEM Laboratory OIM 4.6 system attached to the SEM. IPF(Inverse Pole Figure) maps were obtained from the position adjacent to the bond coat within top coat(bottom region), middle of top coat thickness(middle region) and near surface region of top coat(top region). The tentative crystal system for EBSD observation was ZrO_2 cubic system because of the easiness of observation of particle morphologies though PSZ has commonly tetragonal system. It should be noted that the EBSD equipment has the limitations to characterizing the tetragonal system from the cubic system for the subject top coating PSZ, requiring further development of the technique to identify the tetragonal system clearly and easily.

Figure 8 shows the matching of SEM image and EBSD IPF maps near bond coat for as-sprayed samples. The splat particle morphologies were clearly observed by IPF maps, which cannot be obtained from the SEM image. The splat morphologies are classified into two typical groups. One is large granular type particles which might not be melted completely at the spraying process, and the other is the cluster of small columnar particles which might be formed by crystallization from completely melted particles. Cracks are found to be affected by splat microstructures after crystallization completed.

Figure 9 shows the matching of SEM image and IPF maps from bottom to top region of top coat for samples exposed at 900°C for 500h. Though there is no significant difference in crystallographic features and crack morphologies in test samples and locations, subsequent crack growth and reduction of pores can be seen by comparing with as-sprayed sample shown in Fig.8.

Figure 10 shows the matching of SEM image and IPF maps near delamination portion samples for 1000°C/500h exposed sample. Larger cracks can be found by comparing with Figs.8 and 9.

Figure 11 shows IPF maps at higher magnification for typical crack morphologies. There are three major cracking patterns. The first is the interface cracking between large granular particles and the cluster of small columnar particles, the second is the interface cracking along larger granular particles which is often perpendicular to thickness direction of coating and the third is the transgranular cracking across the cluster of small columnar particles. The cracking orientation seems almost perpendicular to crystal growth direction at the columnar small particle regions. Those cracks are thought to be introduced during cooling process after crystallization was completed. For heated samples cracks are thought to grow from initially introduced cracks during spraying process and increasing in number at successive exposure test. There is no apparent dependence of crack morphologies on the position toward the thickness direction of top coat.

Figure 12 shows the comparison of IPF maps before and after indentation tests for as-sprayed sample. Indentation tests were conducted by 500mN load. Cracks or pores were emanated from the corner of the diamond shaped indentation and showed the apparent tendency that cracks developed along the intergranular path along relatively larger splat particles and the extensive drop out occurred at the small particle zones. This result suggested the very low resistance at small particle (or amorphous) zones and particle boundaries but relatively higher resistance at larger particles. Fig.13 shows the local zoomed up IPF maps with SEM image of green circle region in Fig.12 before and after indentation test. This map clearly indicated the transgranular cracking path from the indentation corner and coalesced with the pre-existing crack across relatively large particles.

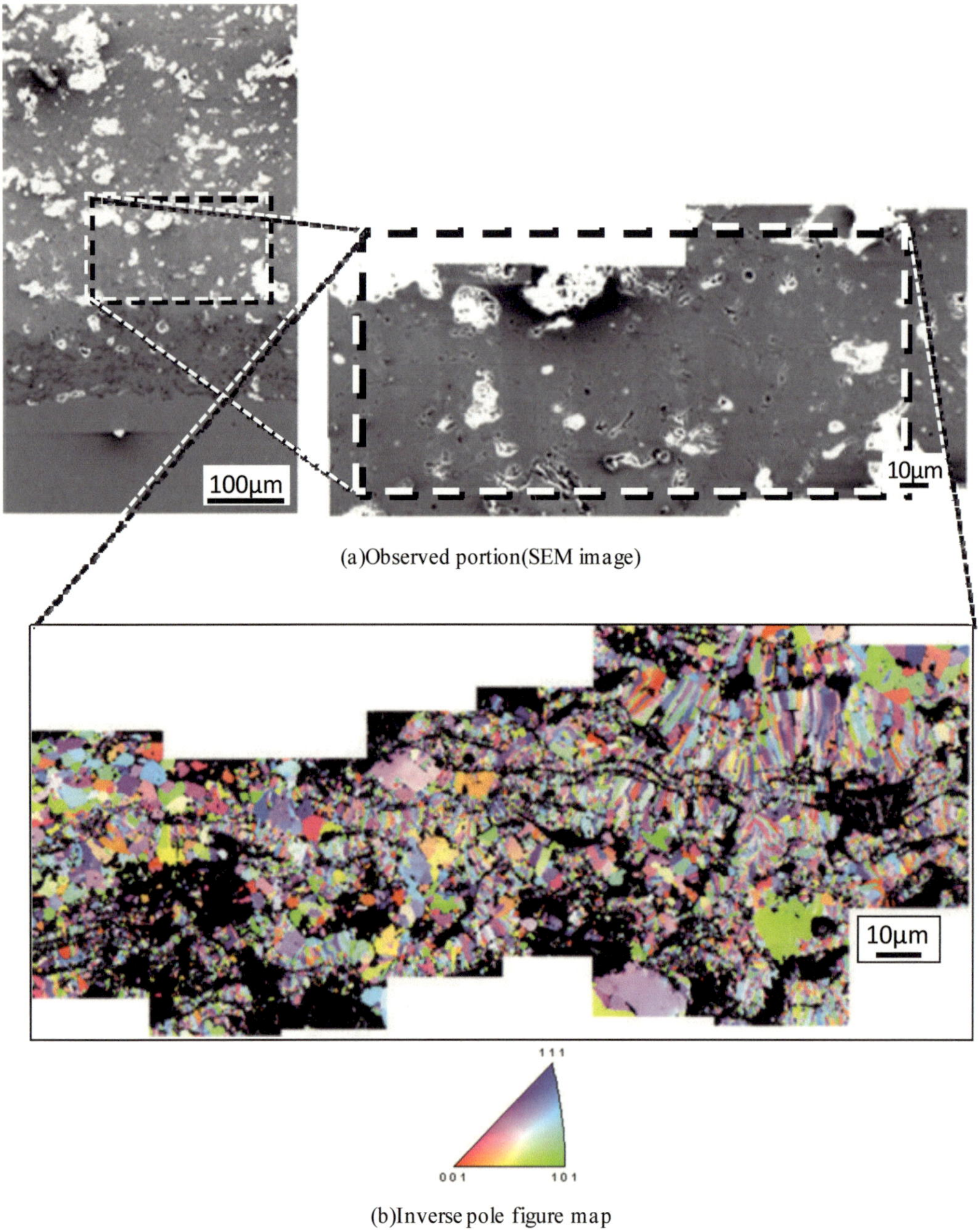

Fig. 8. Matching of SEM observations and IPF map of as-sprayed sample.

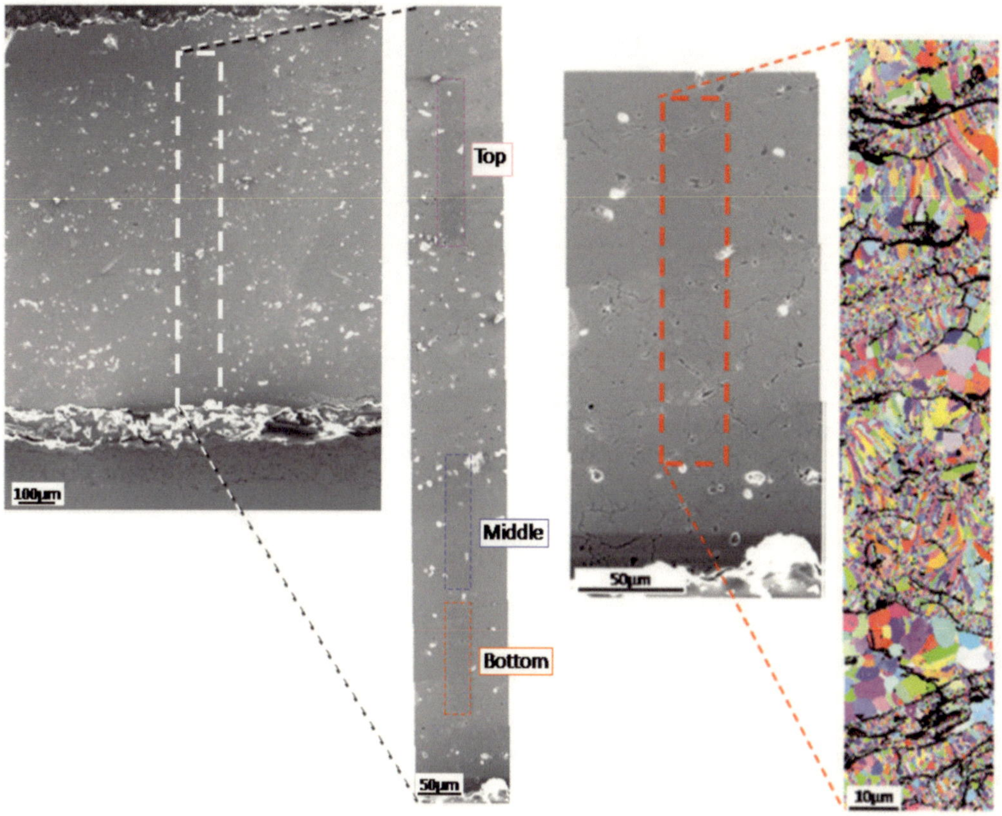

(a) Location of observation (b) Bottom portion

Fig. 9. Matching of IPF maps and SEM image for the sample exposed at 900°C for 500h.

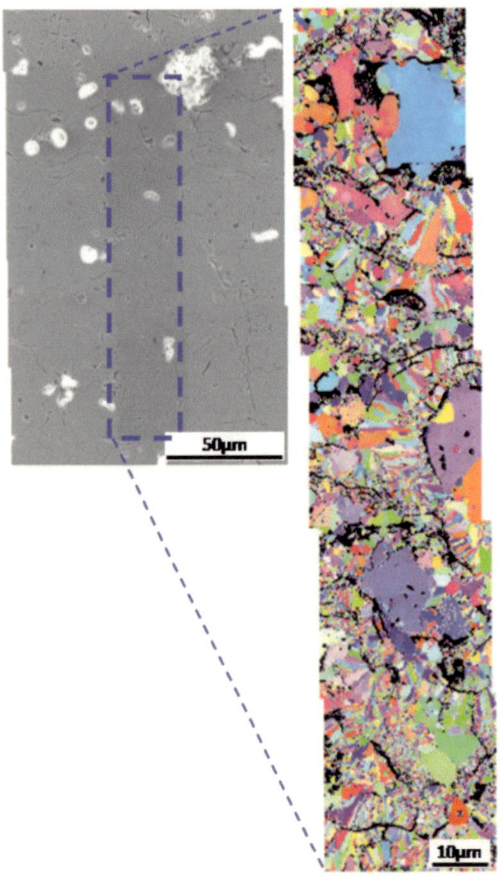

(c) Middle portion

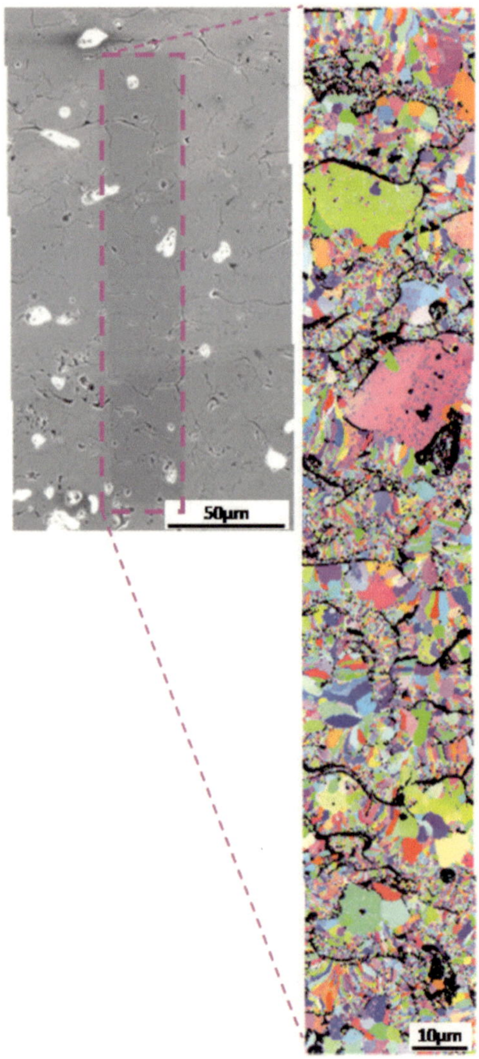

(d) Top portion

Fig. 9. Matching of IPF maps and SEM image for the sample exposed at 900°C for 500h (continued).

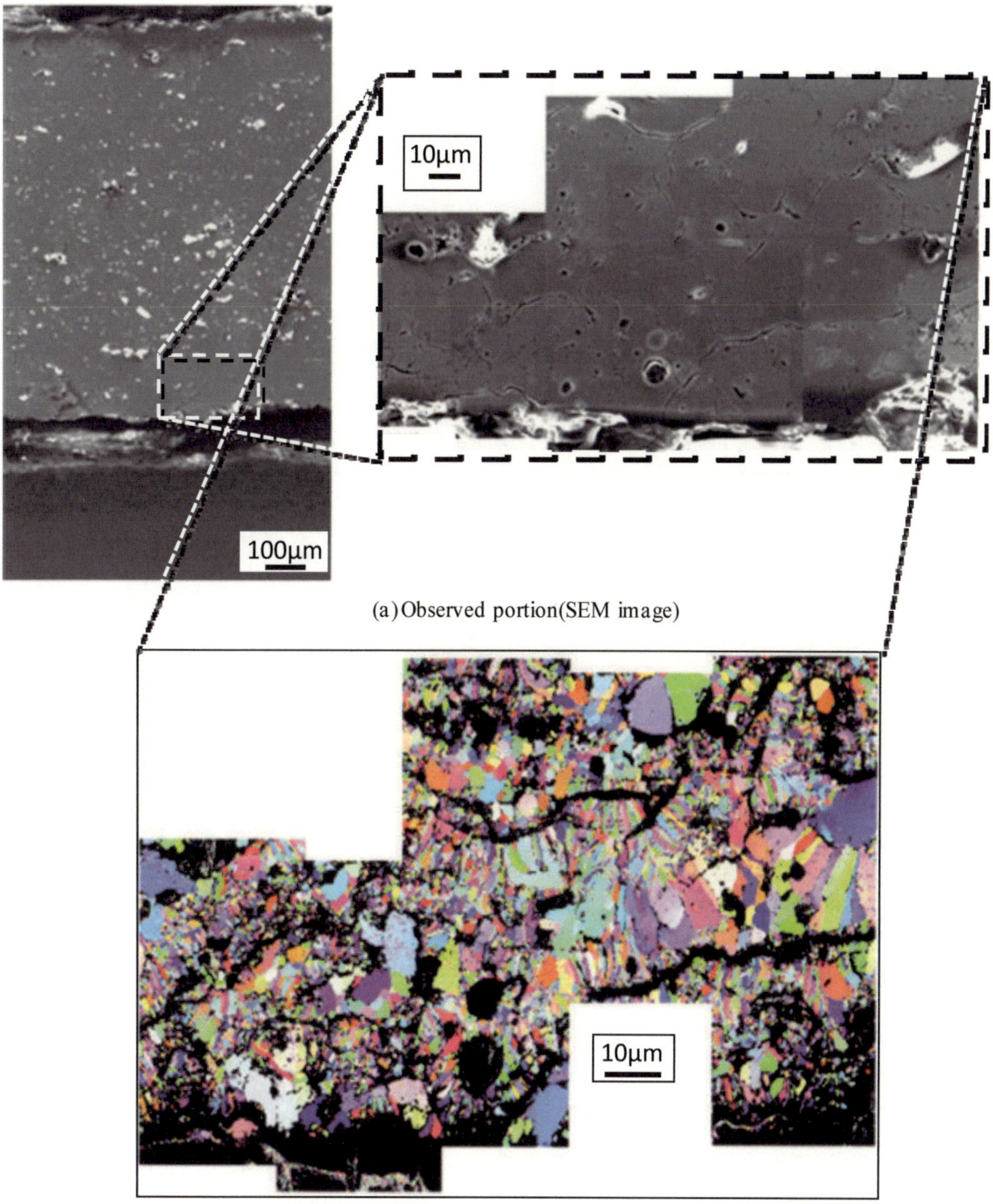

Fig. 10. Matching of SEM observations and IPF map of 1000°C-50h aged sample.

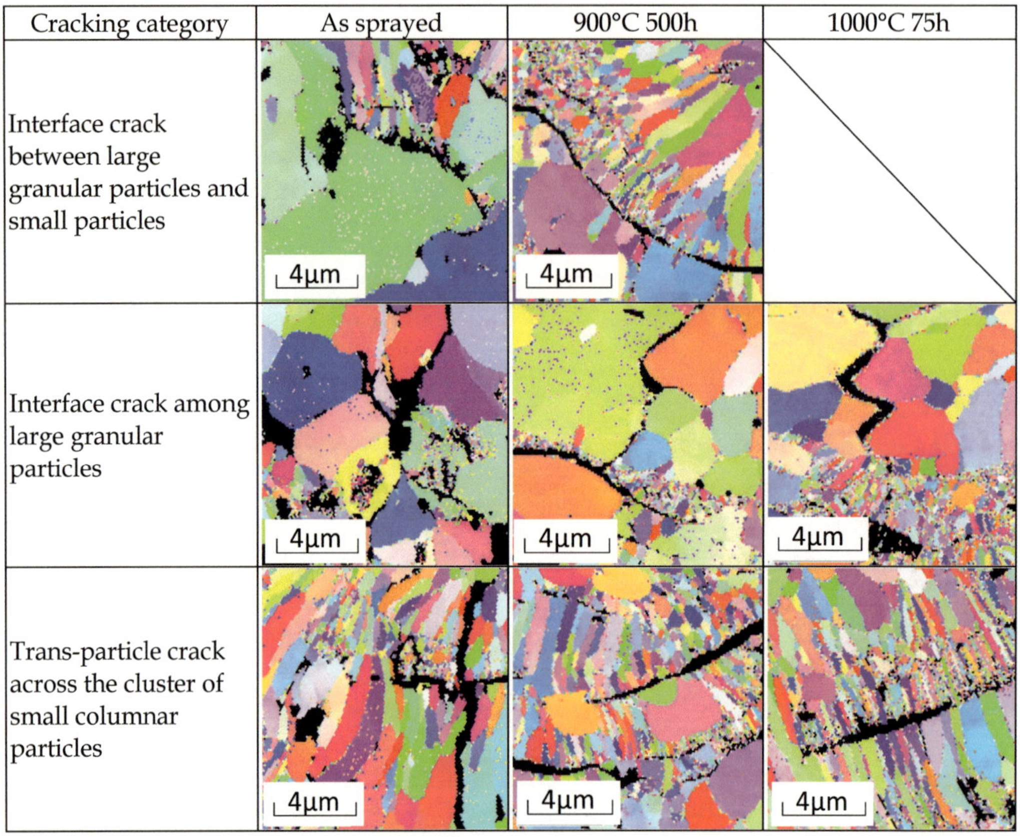

Fig. 11. Crack morphologies observed by IPF maps at top coat.

Figure 13 shows the IPF maps of heated sample at 1000°C/75hours before and after indentation test. The features of cracking showed no significant difference in the cracking morphologies with as sprayed samples shown in Fig.12. Fig.14 shows another view of IPF map with SEM image after indentation test for the same sample in Fig.13. The tendency of intergranular cracking in columnar particles is apparent but some cracks propagate across the particles. As a crack emanated from the bottom corner of the indentation is arrested at the splat boundary, the higher resistance for transgranular cracking than intergranular cracking is strongly suggested.

As demonstrated here, EBSD observation is proved to be a very powerful tool for identifying the crack morphologies and studying the resistance for cracking with respect to splat morphologies though further study is required to study precise crystal system and orientations.

Figure 16 shows the trace examples of micro cracks based on SEM observation in the top coat near bond coat region. The crack orientation is relatively random according to the splat morphologies as shown above. Crack growth may occur due to the coalescence of micro cracks which becomes more frequently at the later stage of exposure time. Crack length density is obtained as the ratio of total sum of crack length and the observed area.

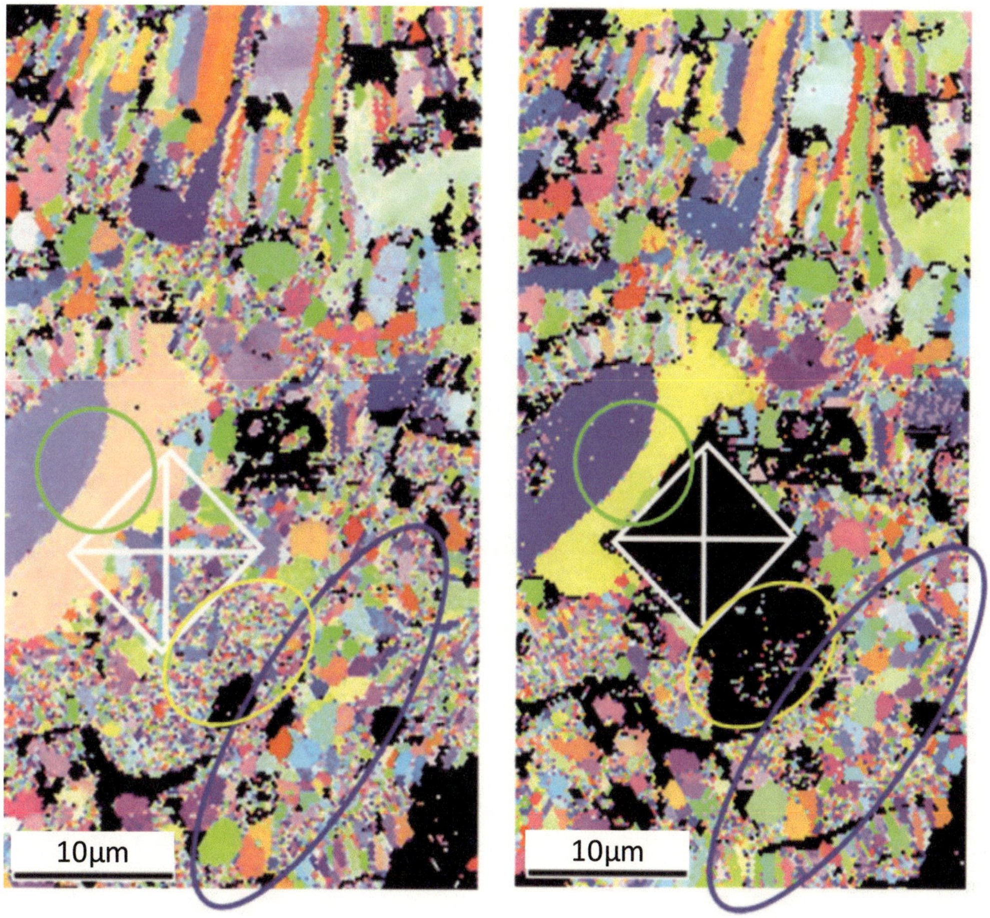

(a) Before indentation test (b) After indentation test

Fig. 12. IPF maps for as sprayed top coat sample before and after the indentation test.

Figure 17 shows the trend of crack length density against exposure time. Significant increase in crack length density was observed at the final stage of delamination for both exposure temperatures.

EDS analysis showed Al oxides and Cr oxides at TGO as shown in Fig.18. Cr oxides bulged toward top coat from TGO film consisted of Al oxides and enhancing top coat cracking. The initial significant increase and successive constant trend of TGO thickness was observed. The time to attain around 5% area percent of Cr oxide in TGO total area was corresponding to the intensive cracking in top coat. Thus, the Cr oxide growth enhanced the cracking of top coat strongly.

Figure 19 shows the trend of average TGO thickness against exposure time. The initial significant increase and successive constant trend of TGO thickness was observed for both exposure temperatures. Fig.20 shows the area fraction (%) of Al oxide and Cr oxide to total (Al + Cr) oxide area against exposure time. For Al oxide, the growth trend is almost similar in both exposure temperatures but for Cr oxide the growth trend is quite different. The time to exceed 30% area fraction might corresponding to the onset of intensive cracking in top coat. Thus, the Cr oxide growth was thought to be an enhancing factor of cracking in top coat.

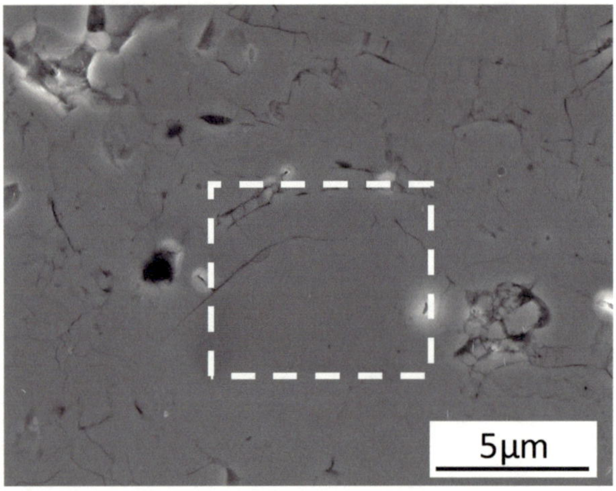

(a) Observation area in SEM image

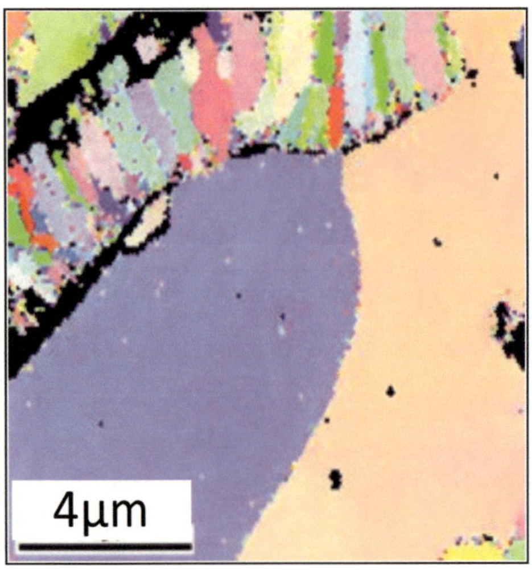

(b) IPF map before indentation test

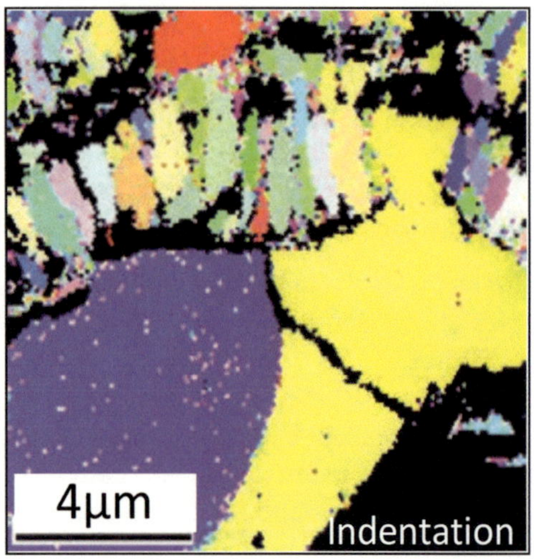

(c) IPF map after indentation test

Fig. 13. IPF maps for as sprayed top coat sample before and after the indentation test in higher magnification of Fig. 12.

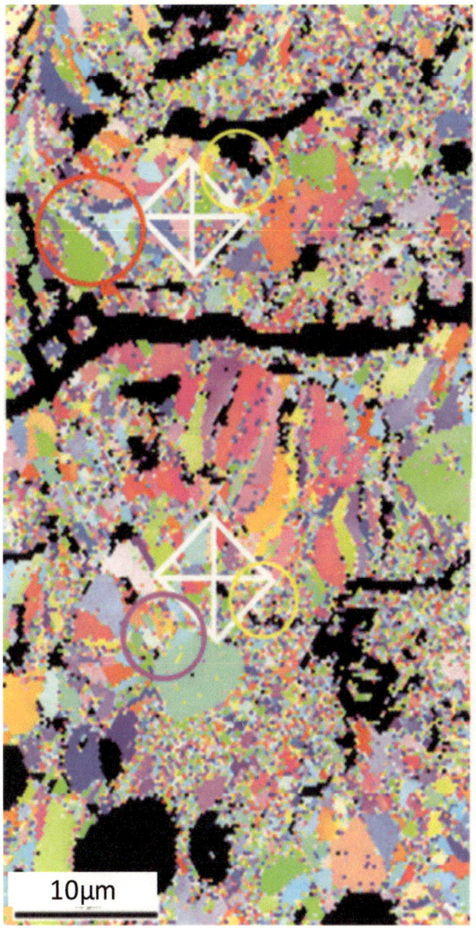

(a) Before indentation test

(b) After indentation test

Fig. 14. IPF maps for 1000 °C -75h heated top coat samples before and after the indentation test.

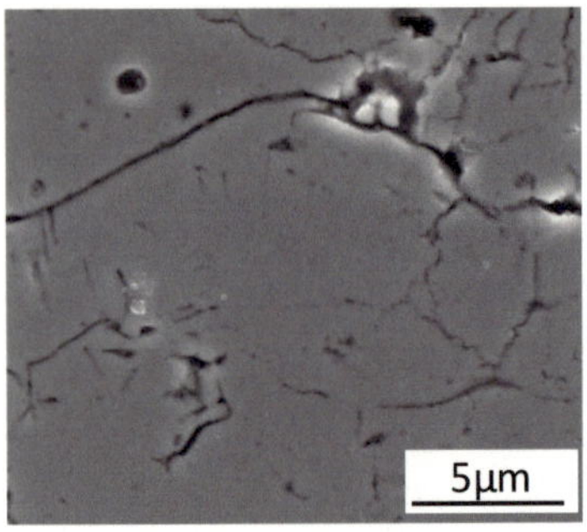

(a) SEM image before indentation testing

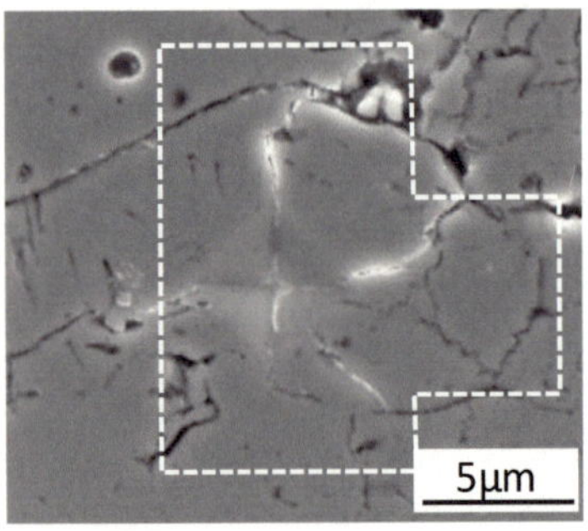

(b) SEM image after indentation testing

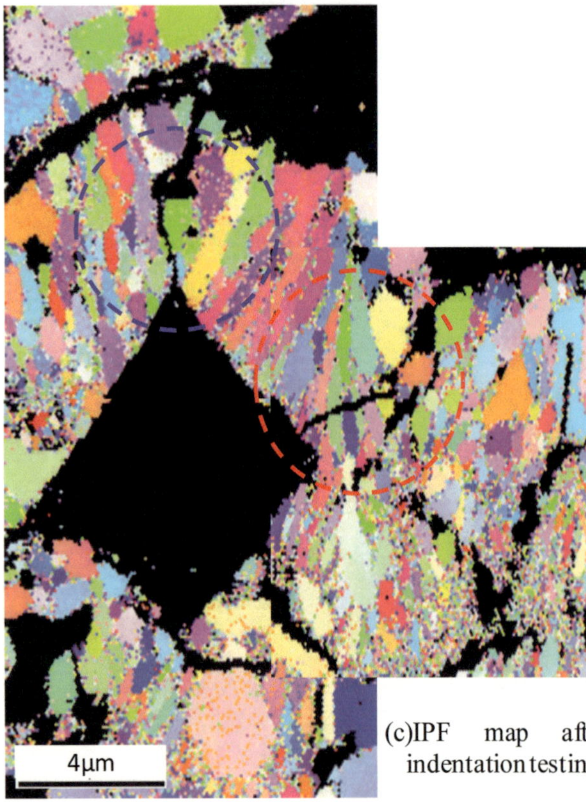

(c)IPF map after indentation testing

Fig. 15. Matching of SEM image and IPF map for 1000 °C -75h heated top coat samples before and after the indentation test.

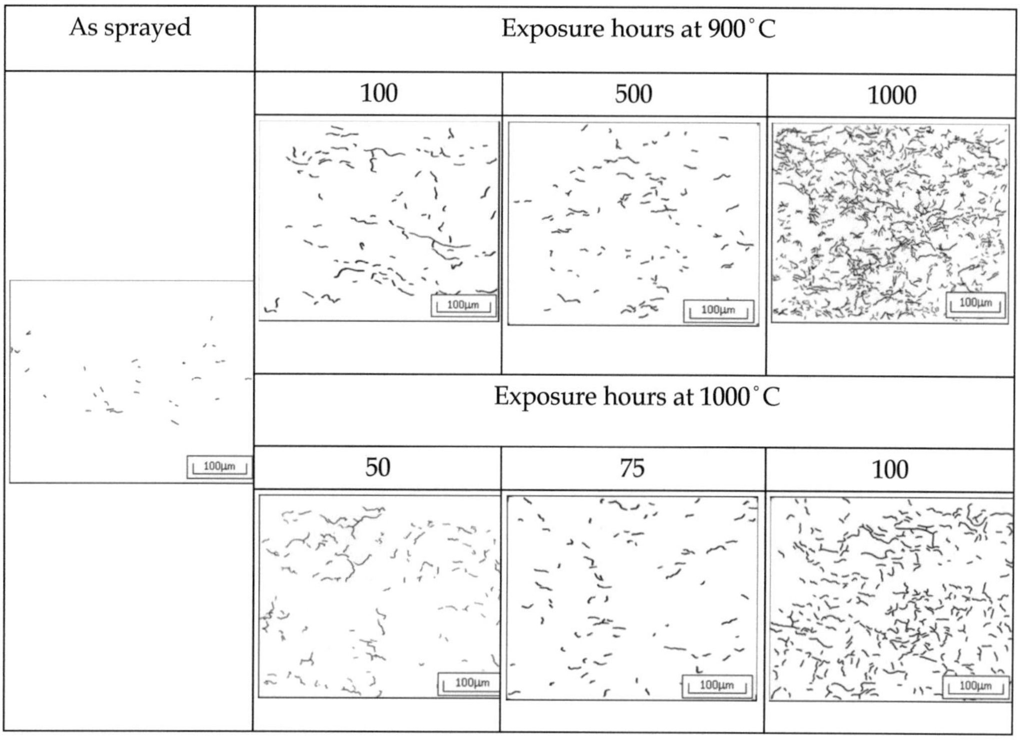

Fig. 16. Trace of cracks in top coat.

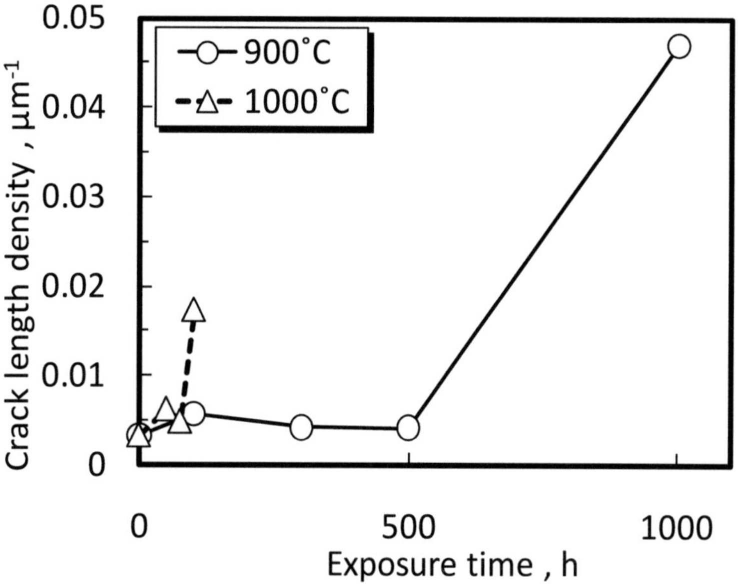

Fig. 17. Crack length density in top coat against exposure time.

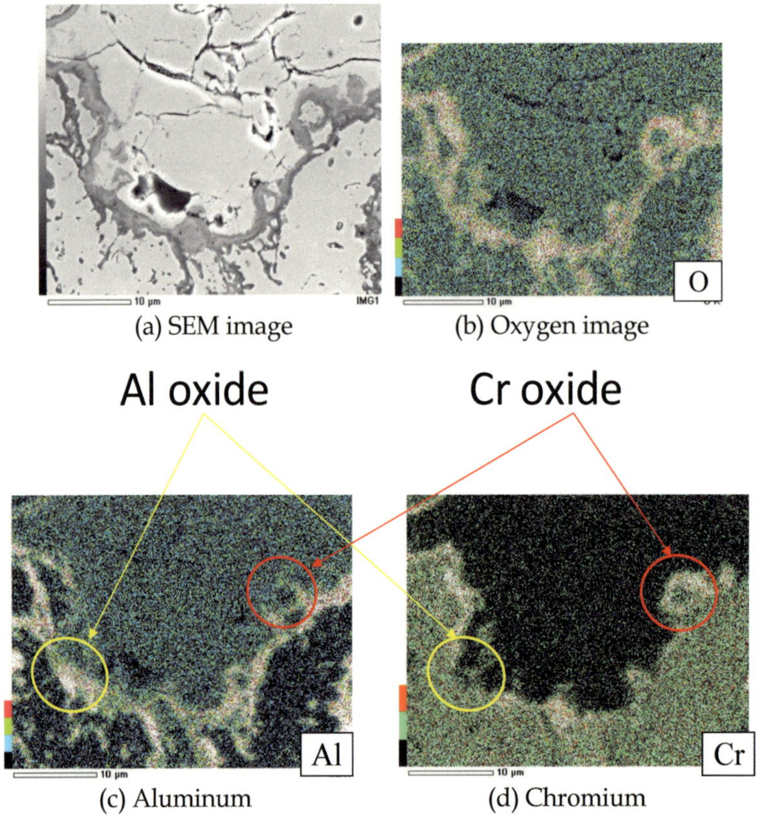

Fig. 18. Observation and element analysis of TGO.

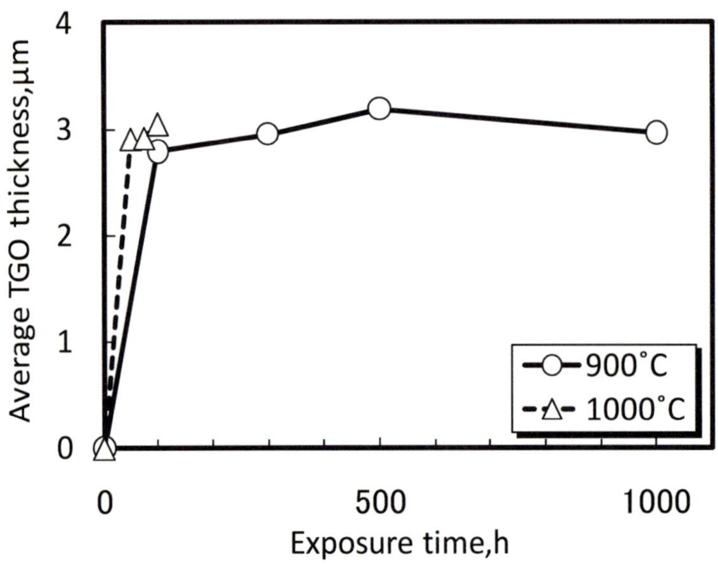

Fig. 19. Trend of average TGO thickness against exposure time.

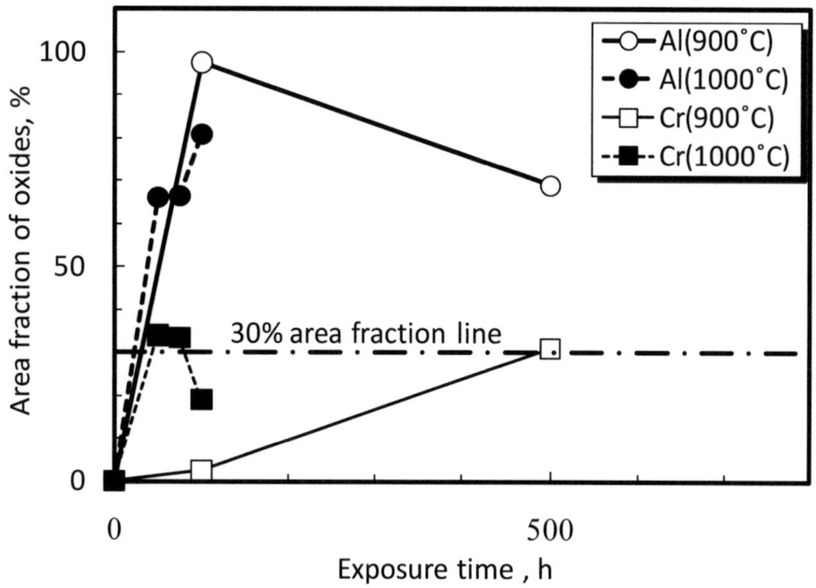

Fig. 20. Relationship between the area fraction of oxides and exposure time.

5. Summary

Microscopic and crystallographic observation techniques are very useful for investigating the damage and degradation process for structural materials. This article presented several results on the microstructural and crystallographic features during the heating process of TBC as follows.

1. Reduction of pores was observed by optical microscope observation due to thermal exposure and possibly attributed to the sintering of ceramics in top coat, though the reduction trend was not completely monotonic.
2. Micro cracking morphologies of top coat were divided into three typical patterns through EBSD observation: intergranular type along large splats, of interfacial type between large splats and the cluster of small columnar splats and of transgranular type across the cluster of small columnar splats.
3. The cracking were supposed to occur during cooling process after crystallization because the crack paths were strongly affected by particle morphologies and across the small columnar particles.
4. By indentation testing coupled with EBSD observation, the vulnerable spot of top coat to cracking and falling out were proved to be intergranular paths between relatively large splat particles and very small particle zones rsspctively.
5. The transition from small distributed cracking to large macro cracking was closely related to the Cr oxide formation toward top coat.

Further investigation should be conducted to reveal the effect of crystallographic characteristics on micro cracking in TBCs by upgrading EBSD analysis techniques for more precise crystalline system's identification and damage evaluation.

6. References

[1] Okazaki, M., The Potential for the Improvement of High Performance Thermal Barrier Coatings (Review), Material Science Research International, Vol.9, No.1 (2003), pp.3-8.

[2] Annigeri, R., DiMascio, P.S., Orenstein, R.M., Zuiker, J.R., Thompson, A.M., Lorraine, P.W. and Dubois, M., Life Assessment of Thermal Barrier Coatings for Gas Turbine Applications, ASME 200-GT-580, (2000), pp.1-8.

[3] Wilkinson, A.J., Meaden, G. and Dingley, D.J., High Resolution Mapping of Strains and Rotations using Electron Backscatter Diffraction, Materials Science and Technology, Vol.22, No.11, (2006), pp.1271-1278.

[4] Schwartz, A.J., Kumar, M., Adams, L.B., Field, D.P., Electron Backscatter Diffraction in Materials Science, Second Edition, Springer, (2009).

[5] K. Fujiyama, H. Nakaseko, Y. Kato and H. Kimachi, EBSD Observation of Micro Crack Morphologies after Thermal Exposure in Thermal Barrier Coatings, J. Solid Mechanics and Material Engineering, JSME, Vol.4, No.2, (2010), pp.178-188.

2

Nucleation and Crystal Growth in Phase Separated Glasses in the Lithium Silicate System

G. A. Sycheva
*Grebenshchikov Institute of Silicate Chemistry,
Russian Academy of Sciences,
St. Petersburg,
Russia*

1. Introduction

The glass forming systems may be regarded as model systems for technical glass ceramics. Many papers on investigation of nucleation in oxide glasses are known. Table 1 shows the compound glass-forming systems, for which the temperature dependence of the steady and unsteady crystal nucleation in the bulk nucleation were studied.

As can be seen from Table 1, the greatest attention was paid by researchers to the system Li_2O–SiO_2 [1-68] because of its practical importance. A specific feature of the system Li_2O–SiO_2 is that, in the region of compositions from 0 to 32 mol % Li_2O, the glasses have a tendency towards the phase separation. Many authors [1, 2, 3, 4] have believed that the process of phase separation of glasses into two vitreous phases favors a homogeneous crystallization of a glass with the formation of small crystals. As follows from [5, 6] the role of the phase separation is reduced to the appearance of phase boundaries that serve as a particular catalyst of the crystallization. According to [93], the crystallization of phase separated glasses cannot be referred to as catalytic. It is necessary to speak about the influence of the phase separation on a further crystallization process rather than about catalysis. Actually, a review of a number of experimental works [7, 8, 9, 10, 81, 11] demonstrates that the liquid–liquid phase separation before crystallization affects the crystal structure of glass-ceramic materials through at least three different ways.

1. The boundary between phases plays a role of an initiating agent.
2. Since, after the phase separation, the composition of a matrix or dispersed phase becomes closer to that of a crystallizing phase, the formation and growth of crystals progress more easily.
3. In one of the phases, the metastable phase precipitates and plays a role of a catalyst for the main crystalline phase.
4. The purpose of this work is to investigate the phase separation and its effect on the crystallization process in lithium silicate glasses that serve as a basis for the preparation of glass ceramic materials.

Nucleated compound	System	№ References
The system $Li_2O–SiO_2$ [12, 13, 14, 15, 16, 17, 18, 19, 20, 21, 22, 23, 24, 25, 26, 27, 28, 29, 30, 31, 32, 33, 34, 35, 36, 37, 38, 39, 40, 41, 42, 43, 44, 45, 46, 47, 48, 49, 50, 51, 52, 53, 54, 55, 56, 57, 58, 59, 60, 61, 62, 63, 64, 65, 66, 67, 68, 69, 70, 71, 72, 73, 74, 75, 76, 77, 78]		
Glasses of the stoichiometric composition		
$Li_2O·2SiO_2$	$Li_2O–SiO_2$	[15, 16, 17, 20, 21, 22, 23, 24, 27, 28, 32, 35, 39, 46, 52, 79]
$Li_2O·SiO_2$	$Li_2O–SiO_2$	[40]
$Na_2O·2CaO·3SiO_2$	$Na_2O–CaO–SiO_2$	[80, 81]
$2Na_2O·CaO·3SiO_2$	$Na_2O–CaO–SiO_2$	[82, 83]
$BaO·2SiO_2$	$BaO–SiO_2$	[84, 85, 86]
$CaO·SiO_2$	$CaO–SiO_2$	[87]
Glasses of the compositions close to stoichiometric ones		
$Na_2O·SiO_2$	$Na_2O–SiO_2$	[88, 89, 90, 91]
$Na_2O·2SiO_2$	$Na_2O–SiO_2–Cr_2O_3$	[56]
$BaO·2SiO_2$	$Na_2O–BaO–SiO_2$	[92]
$Na_2O·2CaO·3SiO_2$	$Na_2O–CaO–SiO_2$	[93, 94, 95]
Glasses of the non-stoichiometric compositions		
$Li_2O·SiO_2$	$Li_2O–Al_2O_3–SiO_2$	[96]
$MgCr_2O_4$	$CaO–MgO–Al_2O_3–SiO_2$	[97]
$CaO·MgO·2SiO_2$	$CaO–MgO–Al_2O_3–SiO_2$	[86]
$Na_2O·Al_2O_3·6SiO_2$	$Na_2O–Al_2O_3–SiO_2$	[98]
$2SnO·P_2O_5$	$SnO–SnO_2–ZnO–P_2O_5$	[99]
$Na_2O·ZnO·P_2O_5$	$Na_2O–ZnO–P_2O_5$	[100]
$BaO·B_2O_3·P_2O_5$	$BaO–P_2O_5–B_2O_3$	[101]

Table 1. Compound of glass-forming systems, for which the temperature dependences of the steady and unsteady crystal nucleation in the bulk nucleation were studied.

2. Sample preparation and experimental technique

The rates of nucleation and growth of crystals and phase separated inhomogeneities were studied in glasses of the analyzed compositions $xLi_2O·(100 - x)SiO_2$ (where x = 23.4, 26.0, 29.1 and 33.5 mol % Li_2O). Since it was established earlier that the crystal nucleation parameters are strongly affected by water [35] and bubbles formed in the course of glass synthesis [102], the glasses were homogenized to a maximum extent under specially established temperature–time conditions of synthesis. As was previously shown, bubbles can play a role of initiators of nucleation of lithium disilicate crystals and increase a number of crystals to 20% [103]. As the water content increases, the stationary nucleation rate increases and the position of the maximum shifts toward lower temperatures [35]. In this respect, all cares were taken in order to prepare glasses homogenized to a maximum extent. Water was removed by bubbling glass melts with argon for 20 min [49]. Bubbles were removed by performing the synthesis at a high temperature. The glasses were synthesized according to the technique described in [103]. A batch was prepared from lithium carbonate (analytical grade) and dehydrated silicon dioxide (analytical grade) by mixing in a drum. The glasses were synthesized in a platinum crucible (volume, 200 ml) in a Globar-heater furnace with silicon carbide heaters (synthesis temperature, 1550 °C; synthesis time, 5 h). Heat treatments were performed in SShOL electric shaft furnaces and a gradient furnace designed at a laboratory. The temperature was maintained accurate to within ±1°C. The results of the chemical analysis of the synthesized glasses are presented in Table 2.

Glass no.	Li_2O	SiO_2
1	23.4	76.6
2	26.0	74.0
3	29.1	70.9
4	33.5	66.5

Table 2. Chemical compositions under investigation (mol %)

X-ray powder diffraction analysis was performed on a DRON-2 diffractometer (CuK_α radiation; operating voltage 30 kV; current 20 mA; detector rotation rate, 2 deg/min). Differential thermal analysis (DTA) was carried out on a MOM derivatograph (heating rate, 10 K/min; sample weight, 1 g; galvanometer sensitivity, 1/5; reference sample, Al_2O_3; platinum crucible). Optical microscopy investigations in transmitted and reflected light were performed on Carl Zeiss Jenaval and Neophot 32 microscopes (Germany). Electron microscopy studies were carried out on an EM-125 transmission electron microscope (accelerating voltage, 75 kV). Samples were prepared using the method of celluloid–carbon replicas. The viscosity was measured by the bending method on a Klyuev viscometer. The temperature dependence data on the viscosity was processed by the least squares technique with the conventional computer program for determining the coefficients A, B, and T_0 in the approximation according to the Vogel–Fulcher–Tammann equation [104]. The nucleation and growth rates of phase separated inhomogeneities and crystals were determined by the development method: preliminary heat treatment of the glass at a low temperature T, followed by the development at a higher temperature $T_{dev} > T$ bringing to sizes that can be fixed in an optical microscope. The crystal growth rate was measured by quenching the samples. The samples in the form of small glass pieces were held at a specified temperature for different times and were used to prepare polished plane parallel disks 0.5 mm thick. Crystals with a maximum size were found in the bulk of a sample, and their radius R_{max} was measured using the Jenaval optical microscope. The radius R_{max} of the phase separated droplet or crystal was taken to be equal to the average value over ten maximum radii of crystals or phase separated droplets. This requires the explanation. If some number of phase separated droplets or crystals in the shape of spheres with the same radius R are randomly distributed in the glass bulk (their number per unit volume is designated as N), certain crystals or phase separated droplets will be cut by the plane when preparing the cleavage or cross section of the sample. This holds true only for crystals or phase separated droplets with the centers located at a distance that is not larger than R from the cut plane. Therefore, all crystals or phase separated droplets with the centers located in the layer of thickness $2R$ leave a trace on the cut plane. It is clear that the maximum radius of the trace of the crystal or phase separated droplet will correspond to the real radius of these particles.

3. Temperature dependence of the nucleation and growth rate of phase separated in homogeneities in lithium silicate glasses of the compositions $23.4Li_2O{\cdot}76.6SiO_2$ (1), $26Li_2O{\cdot}74SiO_2$ (2), and $29.1Li_2O{\cdot}70.9SiO_2$ (3)

In the $33.5Li_2O{\cdot}66.5SiO_2$ glass with the composition closest to the stoichiometric composition of the lithium disilicate, excess SiO_2 is absent and, therefore, no phase separation is observed. The compositions of the glasses containing 23.4 mol % Li_2O (no. 1), $26Li_2O$ (no. 2), and $29.1Li_2O$ (no. 3) lie in the metastable phase separation region of the lithium-silicate system. The samples of glasses no. 1, no. 2, no. 3 and no. 4 were subjected to preliminary heat treatment at temperatures in the range 370–560 °C for different times. Then, they were held at a development temperature of 600 °C for 10 min. If glass no. 1, no. 2 or 3 is held at the temperature T = 600°C for a time in the range 0–10 h, it remains transparent without visible opalescence in visual examination. If the glass is heat treated for t = 2 h 40 min at temperatures of 400–560°C, it remains visually transparent and does not become opalescent. After additional heat treatment at 600 °C for 10 min, the glass acquires bright blue (yellow in transmission) going to milky opalescence. The electron microscopic images of the glasses preliminarily heat treated at temperatures in the range 400–560 °C and developed at 600 °C for 10 min are displayed in Figs. 1*a*–1*c*.

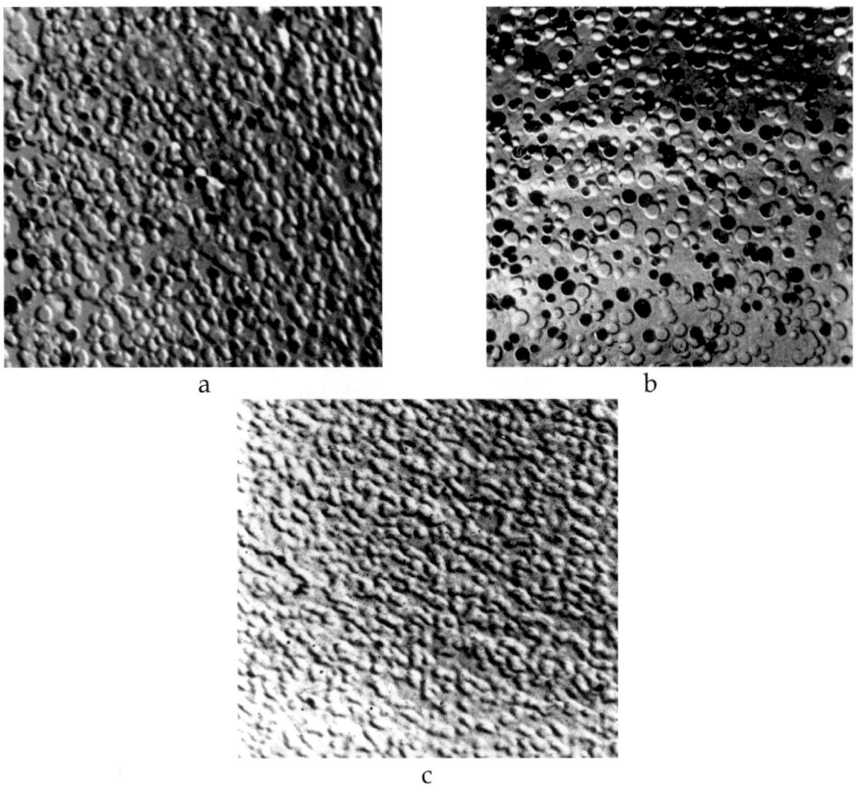

Fig. 1. Electron microscopic images of glasses (*a*) no. 1, (*b*) no. 2 and (*c*) no. 3 preliminarily heat treated at temperature 460 °C for 2 h 40 min and developed at 600 °C for 10 min (magnification, 28800).

We counted the number of traces of particles per unit area in the electron microscopic image N_S/S_0 and determined the sizes of the maximum traces of particles R. The corresponding results for glass no. 1 are listed in Table 3.

T, °C	N_S, μm^{-2}	R, Å
370	90	103
390	108	150
400	112	160
420	120	190
460	134	250
464	131	260
490	118	300
520	94	335
540	74	375
560	60	408
580	48	440

Table 3. Numbers of traces of particles per unit area in electron microscopic images N_S/S_0 and their sizes in glass no. 3 after heat treatment at T for 160 min and development at 600 °C for 10 min.

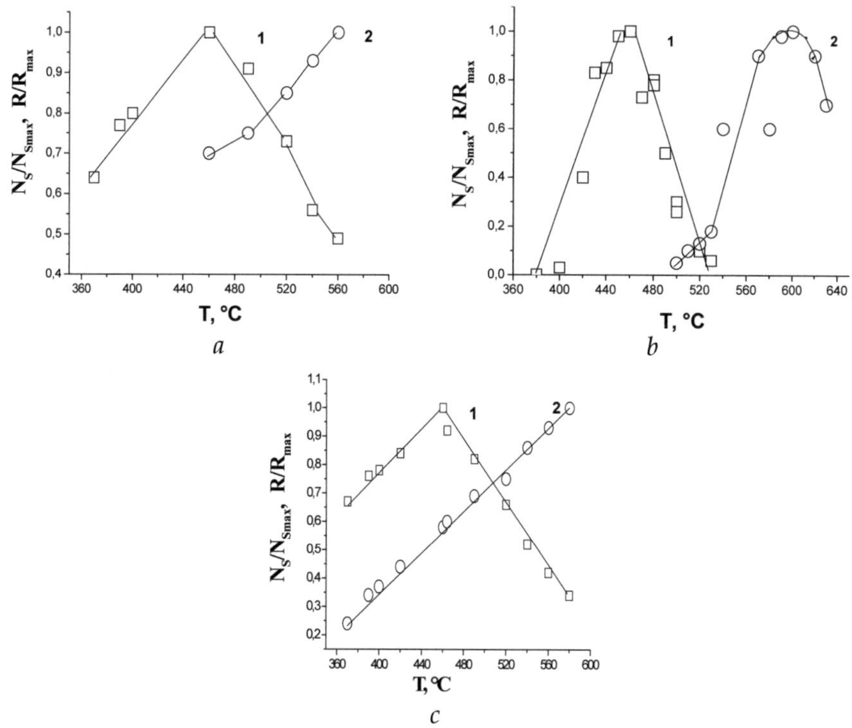

Fig. 2. Quantitative characteristics of the phase separation structure in glasses (*a*) no. 1, (*b*) no. 2 and (*c*) no. 3 as a function of the temperature: (*1*) ratio of the number of traces to the maximum number of traces N_S/N_{Smax} and (*2*) ratio of the radius of the particle trace to the maximum radius R/R_{max}.

Figure 2 presents the quantitative characteristics of the phase separation structure in glasses no. 1 and no. 2 as a function of the temperature, namely, the ratio of the number of traces to the maximum number of traces $N_S/N_{S\max}$ and the ratio of the radius of the particle trace to the maximum radius $R/R_{\max}$.

Here, we assume that the dependence of the number of traces of phase separated droplets per unit area of the sample surface on the heat treatment temperature (at a constant heat treatment time) obeys the same function law as the temperature dependences of the stationary nucleation rate of droplets. Strictly speaking, in order to determine the stationary nucleation rate $I(T)$, initially, it is necessary to determine the number of droplets per unit volume of the glass by the number of traces per unit area with the use of stereological formulas. Then, it is required to determine a variation in the number of droplets per unit volume as a function of the heat treatment time. Thereafter, it is necessary to check that the nucleation rate is stationary and to determine its value. This is not a simple problem, which is complicated by the fact that the droplets in the volume of the glass can have different sizes. The obtained dependence of the number of traces on the temperature for a fixed heat treatment time can differ substantially from the dependence of the stationary nucleation rate of droplets. However, we used a simpler method to approximately evaluate the temperature dependence of the nucleation rate of droplets. Instead of the dependence $I(T)$, we constructed the dependence $N_S(T)$ of the number of traces of particles developed for the same specified time of development t_{dev} at T_{dev} after preliminary heat treatment for the same specified time t at all temperatures T. Since the quantities I and N_S are approximately proportional to each other, the curves $I(T)/I_{\max}$ and $N_S(T)/N_{S\max}$ are very close to each other and, in order to evaluate the temperature of the maximum nucleation rate, it is sufficient to measure the dependence of the number traces of crystals N_S without regarding for their shape and sizes [77, 80]. It can be seen from Fig. 2 that these dependences exhibit maxima. The maximum values of the number of traces and sizes correspond to temperatures of 460 and 600 °C, respectively. On the whole, the results obtained can be explained as follows. The nucleation rate of a new phase (phase separated or crystalline) I has the form [68]

$$I = A\exp[-(\Phi_A + \Phi^*)/kT], \tag{1}$$

where Φ^* is the increment of the thermodynamic potential of the glass due to the appearance of a critical (in composition and size) inhomogeneity nucleus, Φ_A is the thermodynamic potential of diffusion jumps and transformations during the phase separation, and A is the factor that, like the quantities Φ_A and Φ^*, depends on the glass composition and temperature more weakly than an exponential function. If the glass composition c_0 belongs to the bimodal region at a given temperature, the thermodynamic potential Φ^* is relatively high because the inhomogeneities that do not reach some limiting concentration c_1 appear to be energetically unfavorable irrespectively of their sizes. When the glass composition belongs to the spinodal region, inhomogeneities with an arbitrarily small difference $|c - c_0|$ and, correspondingly, with negligible Φ^* can be energetically favorable if only their size exceeds the critical size. As a consequence, there is an increase in the nucleation rate of inhomogeneities I (expression (1)). By using the notion of the concentration wave length λ introduced by Cahn, in the spinodal region, the value $|c - c_0|$ most rapidly increases in amplitude and, hence, the inhomogeneities with the size λ_m that is larger than the critical size λ_{cr} by a factor of $\sqrt{2}$ nucleate most rapidly. The wave amplitude λ_m increases as $\exp(t/\tau_m)$; in this case, the magnitude $1/\tau_m$ can be considered not only as the measure of the nucleation rate of the amplitude but also as the measure of the nucleation rate of inhomogeneities by taking $I = B/\tau_m$. As a result, we have (2)

$$I = \frac{B}{\tau_m} = c(T_S - T)^2 \exp\left(-\frac{\Phi_A}{kT}\right) \tag{2}$$

where B and c are temperature independent constants and T_S is the spinodal temperature. From the condition $dI/dT = 0$, we find that the maximum of the rate is located at the temperature $T_{\max}$ satisfying the equation (3).

$$T_S - T_m = \frac{2RT_m^{\ 2}}{E_A(T_m)}, \quad E_A(T) = \Phi_A(T) - T\frac{\partial \Phi_A(T)}{\partial T} \tag{3}$$

By assuming that the thermodynamic potential Φ_A does not depend on the temperature T, i.e., $\Phi_A = E_A$, and setting $\Phi_A = 83$ kcal/mol (activation energy of oxygen diffusion), at $T_m \cong 973K$, Eq. (3) takes the form

$$T_S - T_m \cong T_m(2RT_m / \Phi_A) \cong 25K \tag{4}$$

Under the assumption that the thermodynamic potential Φ_A is equal to the activation energy of lithium self diffusion, we obtain

$$T_S - T_m \cong 60K \tag{5}$$

Therefore, the maximum of the ratio $N_S/N_{S\max}$ in Fig. 2 reflects the maximum of the nucleation rate of droplets with a diameter λ=200–400 Å. This reflection is rather indirect because the number of inhomogeneities n' nucleated for the time t' $n' = \int I(t)dt$ can differ significantly (and differs) from the number n of large particles growing in the course of the development. The reason is that, after the increase in the temperature to 600 °C, the inhomogeneities with the composition and size that reach critical values at 600 °C are predominantly retained and will develop. The inhomogeneities with the composition that approaches the binodal composition at the corresponding temperature during low temperature heat treatment go beyond the binodal with the increase in temperature to 600 °C and should dissolve in the glass located between inhomogeneities. A number of inhomogeneities can merge together when reaching the critical sizes for a temperature of 600 °C. This dissolution and merging of small inhomogeneities can lead to the formation of larger and stable inhomogeneities with a composition close to the binodal composition at the temperature of 600 °C.

Although the relation between the quantities n and n' is complex, the maximum of the ratio $N_S/N_{S\max}$ in Fig. 2 reflects the rate of nucleation and development of phase separated inhomogeneities. Then, the temperature of its left boundary T_S' can be identified with the spinodal temperature T_S and the difference between this temperature and the temperature of the maximum T_{max} is appropriately identified by the difference $T_S^{'}$ - T_{max}, which was estimated above as 25–60 °C. It is difficult to accurately determine the spinodal temperature. In our experiment, we determined this temperature from the appearance of pronounced phase separated droplets in the electron microscopic images. Experimental data on the determination of the spinodal are absent in the literature.

Haller et al. [105] calculated the spinodal for glasses in the lithium silicate system. A comparison of our data with those obtained in [105,105] leads to the following results. According to [105], the spinodal temperature T_S amounts to approximately 600 °C for glass no. 1 (23.4 mol % Li_2O) and is in the vicinity of 500 °C for glass no. 2 (26.0 mol % Li_2O). Below 500 °C, data in [105] are absent. However, if the curve is extrapolated to 400 °C, the spinodal temperature T_S for glass no. 2 appears to be 400 °C. In our work, the maximum of the ratio N_S/N_{Smax} in Fig. 2 lies in the temperature range 350–570°C for glass no. 1 and ≈ 400–650°C for glass no. 2. Therefore, the experimental and calculated data for glass no. 2 (26.0 mol % Li_2O) are in good agreement, whereas, for glass no. 1 (23.4 mol % Li_2O), the calculated value is by 200 °C higher than the experimental. For glasses no. 1 and no. 2, the difference $T_S^{'}$ -T_{max} ≅ 40–50 °C actually lies in this range. The maximum of the rate of nucleation and development of inhomogeneities is located at the glass transition temperature T_g.

In order to compare more rigorously the theory with the experiment, it is necessary to know the spinodal temperature T_S. This temperature was evaluated as follows. Quenched glasses no. 1 and no. 2 (without low temperature heat treatments) were held in the gradient furnace at temperatures in the range 300–800 °C for 2 h. This led to the appearance of the opalescent band along the sample length, which was stable with respect to a further increase in the holding time. This band had a higher intensity in the central part and weaker intensities toward higher and lower temperatures up to its complete visual disappearance. The temperature of the upper edge of the visible opalescence (770 °C for glass no. 1) coincided with the phase separation temperature T_l (770 °C for glass no. 1) determined as the temperature of the disappearance of the visible opalescence after the opalescent part of the rod was displaced in the range of temperatures a priori higher than T_l. The analysis of the electron microscopic images demonstrates that, in going from the high temperature edge of the band to the low temperature edge, the number of particle traces per unit area (or the relative area occupied by particle traces) increases linearly. Initially, this process goes slowly and then accelerates. It is clear because the nucleation rate is minimum in the vicinity of the phase separation temperature T_l and increases as the spinodal temperature T_S is approached as a result of the decrease in the thermodynamic potential Φ^* in expression (1). When the traces of inhomogeneities begin to cover a larger part of the replica area and to merge together, the increase in the relative area S/S_0 occupied by particles at T_S ' is retarded sharply. The temperature of this kink in the curve S/S_0 is naturally identified by the spinodal temperature T_S corresponding to the beginning of fast nucleation of inhomogeneities occupying the volume of the glass. This temperature almost absolutely coincide with the low temperature edge of the visible opalescence. The temperature width of the opalescent band ΔT_{op} corresponds to the difference $T_l - T_S \cong \Delta T_{op}$.

3.1 Effect of displacement of the glass composition from the stoichiometric lithium disilicate composition toward an Increase in the SiO_2 content on the nucleation of lithium disilicate crystals

In order to accelerate the search for the temperature range of crystal nucleation, we used the DTA curves and the temperature dependences of the viscosity, because it was previously shown [68] that the maximum of the crystal nucleation rate is located in the range of the endothermic effect in the DTA curve at the temperature corresponding to a viscosity of 10^{13} P. The temperature dependences of the viscosity for glasses no. 1–no. 3 are plotted in Figs. 3*a*–3*c*. The inset to Fig. 3 presents the coefficients *A*, *B*, and T_0 in the approximation according to the Vogel–Fulcher–Tammann equation [104]. It can be seen from Fig. 3 that the glass transition temperature determined as the temperature corresponding to a viscosity of 10^{13} P is equal to 453 °C for glass no. 1, 458 °C for glass no. 2, 457 °C for glass no. 3 and 450 °C for glass no. 4. The slope of the temperature dependence of the natural logarithm of the viscosity on the reciprocal of the temperature (Fig. 3*d*) was used to determine the activation energies of the viscous flow $E_\eta = R d\ln\eta / d(1/T)$, which were equal to 148, 149, 150 and 152 kcal/mol for glasses no. 1–no. 4, respectively.

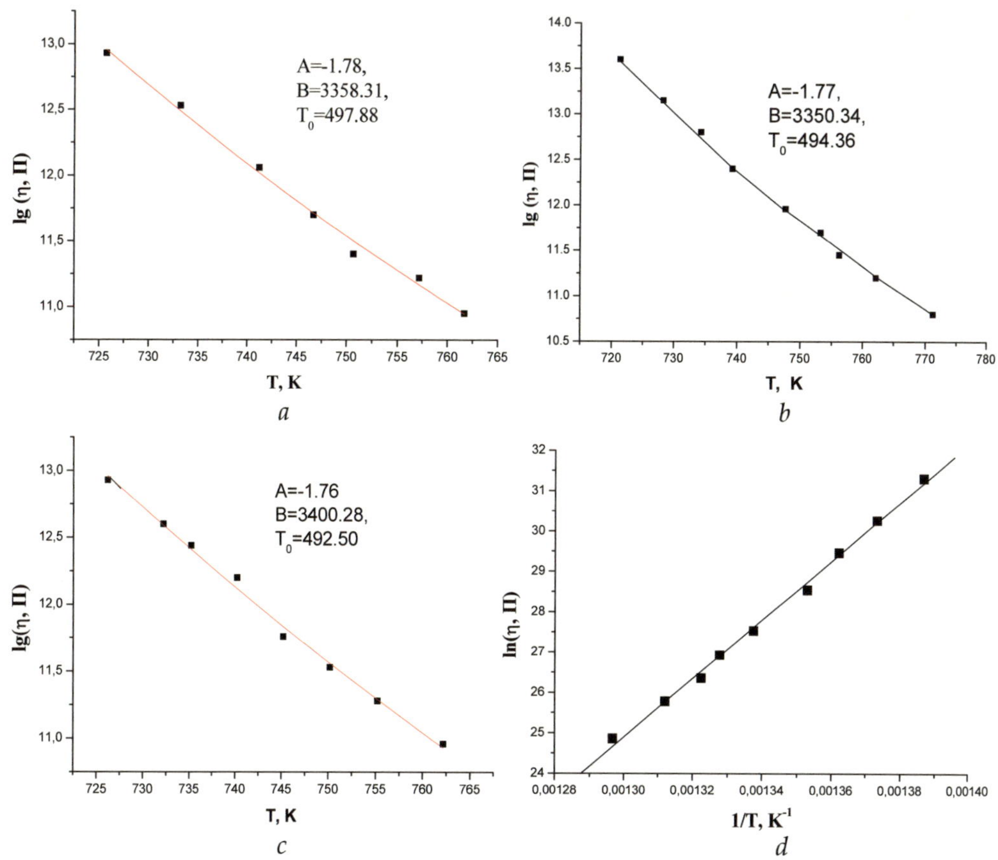

Fig. 3. Temperature dependences of the decimal logarithm of the viscosity for glasses (*a*) no. 1, (*b*) no. 2, and (*c*) no. 3 and (*d*) temperature dependence of the natural logarithm of the viscosity for glass no. 2 (used for determining the activation energy of viscous flow in the glass).

The X-ray powder diffraction data (Fig. 4) demonstrate that the initial glasses are X-ray amorphous. The lithium disilicate crystals nucleated at temperatures of preliminary heat treatments in the range 420–520 °C for all times cannot be revealed using X-ray diffraction analysis. This is associated with the fact that the lithium disilicate nuclei are extremely small and their total weight is insignificant and, hence, insufficient for identifying by X-ray diffraction analysis, because its sensitivity is equal to a few weight percent of a compound. After heat treatments, lithium disilicate Li_2O $2SiO_2$ (PCPDFWIN no. 80-1470; orthorhombic structure; unit cell parameters a = 5.687 Å, b = 4.784 Å, and c = 14.640 Å) crystallizes in glasses no. 1–no. 3 at a development temperature of 600 °C for 10 min.

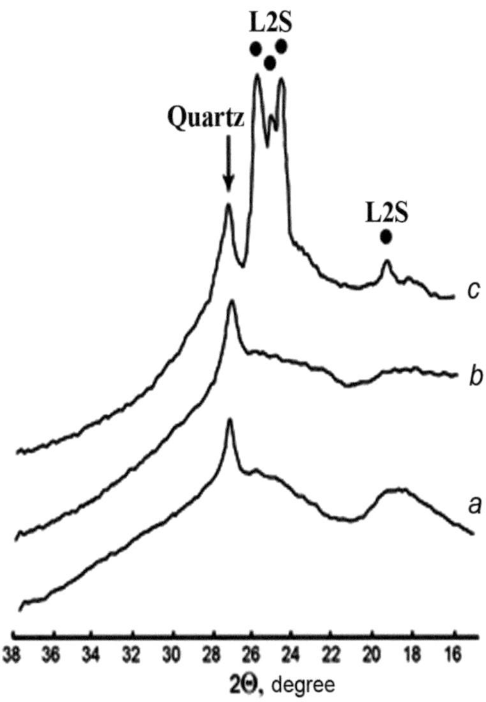

Intensity, arb. units

Fig. 4. X-ray diffraction patterns of initial and heat treated glasses no. 2: (*a*) initial glass, (*b*) glass heat treated at 450°C for 5 h, and (*c*) glass preliminarily heat treated at 450°C for 5 h and developed at 600°C for 10 min.

It can be seen from the images obtained using the optical microscope in reflected and transmitted light (Fig. 5) that the lithium disilicate grows in the form of ellipsoids of revolution with an axial ratio of 1.0 : 1.6 in glass no. 4 (with the composition closest to the stoichiometric lithium disilicate composition), with an axial ratio of 1.0 : 2.0 in glass no. 3 and in the shape of regular spheres in glass no. 2. The number of crystals increases with an increase in the holding time during preliminary heat treatment.

In order to nucleate lithium disilicate crystals, the samples of initial glasses no. 1–no. 4 were held at temperatures in the range 420–520 °C (range of the endothermic effect in the DTA curves). Then, the nucleated crystals were developed (to sizes visible in the optical microscope) at 600 °C for 10 min, i.e., at the temperature corresponding to the ascending branch of the exothermic peak in the DTA curve. The kinetics of homogeneous crystal nucleation was described using the stationary nucleation rate I_{st}, nonstationary nucleation time τ, and temperature dependences of these quantities.

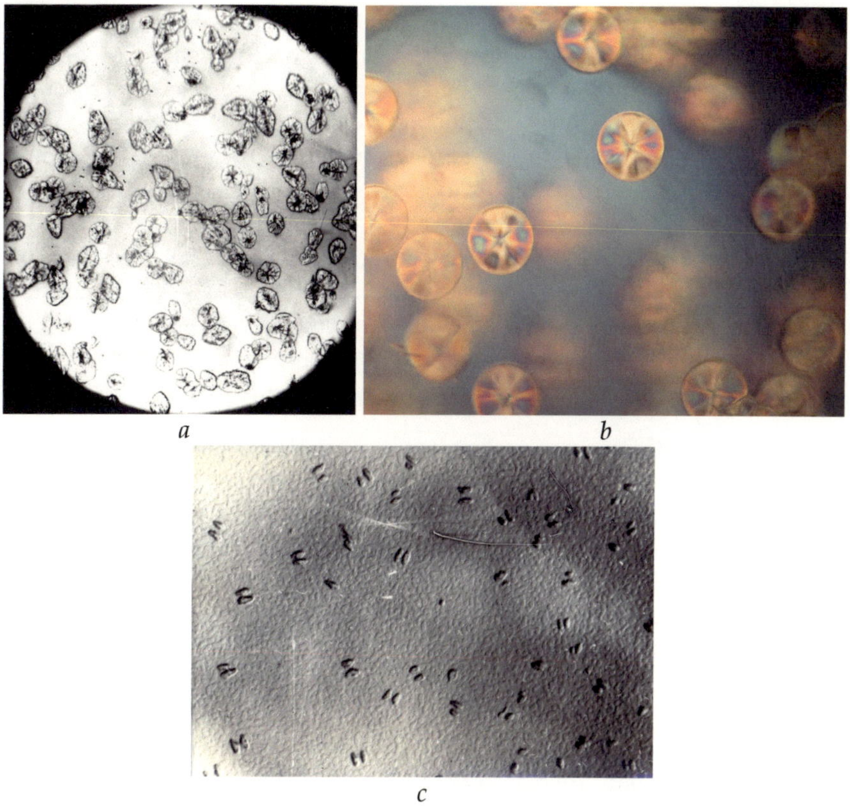

Fig. 5. Shape of lithium disilicate crystals in: (*a*) glass no. 4 after preliminary heat treatment at the temperature 420 °C for 4 h; (*b*) glass no. 2 after preliminary heat treatment at the temperature 400 °C for 4 h; (*c*) glass no. 3 after heat treatment at the temperature 440 °C for 8 h. All samples were developed at the temperature 600 °C for 10 min. Image (*a*) was obtained in reflected light with the use of the Neophot 32 optical microscope at a magnification of .150. Image (*b*) was obtained in transmitted light with the use of the Jenaval optical microscope at a magnification of .200. Image (*c*) was obtained in reflected light with the use of an EM-125 transmission electron microscope at a magnification of .28000.

3.2 Determination of the crystal nucleation rate

The stationary nucleation rate I_{st} was determined from the experimental dependences of the number of crystals $n(t)$ as the slope of the stationary portion of these dependences on the holding time of glasses at each holding temperature. In this work, the rate of stationary homogeneous crystal nucleation was determined by the cross section method (in reflected light) and by direct counting of the number of crystals (in transmitted light). The glass

samples were subjected to preliminary heat treatment and rapidly cooled in air to room temperature. Then, the samples were repeatedly heat treated at the development temperature of 600 °C for 10 min. With the aim of examining the samples in reflected light, the surface layer was ground off and the prepared surface was polished. In order to increase the contrast of the boundaries between crystals and glass, the sample surface was etched in a 0.01 *M* HF solution for 10 s. The number and size of cross sections of crystals per unit surface area were determined from the micrographs with the use of the Neophot optical microscope or directly in the field of vision of the Jenaval microscope.
The number of crystals per unit volume n and the number of cross sections of crystals in the micro section N_S are related by the expression $n = N_S/(SD)$, where S is the micro section area and D is the diameter of a maximum particle in the micro section. The nonstationary nucleation time τ was evaluated using the induction period t_{ind}, which differs from the time τ by the temperature independent factor.
The induction period t_{ind} was determined as the intersection point between the continuation of the linear portion of the dependence $n(t)$ and the time axis. Then, we constructed the temperature dependences of all the aforementioned quantities.

4. Experimental results on crystal nucleation and their discussion

The nucleation of the lithium disilicate was observed in all glasses. The dependences of the number of developed crystals n in the nucleation temperature range 400–520 °C (development at 600 °C for 10 min) were used to determine the stationary nucleation rate and the nonstationary nucleation time at each specific temperature. The induction period t_{ind} decreases monotonically with an increase in the temperature. The temperature dependence of the induction period $t_{ind}(T)$ allows us to determine the activation energies for nucleation of lithium disilicate crystals E_τ from the formula $E_\tau = R d\ln t_{ind}/d(1/T)$. The corresponding activation energies are equal to 129, 128, 127 and 126 kcal/mol for glasses no. 1–no. 4, respectively. The temperature dependences of the stationary nucleation rate of lithium disilicate crystals $I_{st}(T)$ for the glass with the stoichiometric lithium disilicate composition and glasses with displaced compositions are shown in Fig. 6. It can be seen from this figure that the nucleation rate of lithium disilicate crystals $I_{st}(T_{max})$ at the maximum decreases from 160 to 60 $mm^{-3}\ min^{-1}$; in this case, the position of the maximum T_{max} is shifted toward higher temperatures by 12 °C. This relatively small change in the quantity $I_{st}(T_{max})$ can be explained by the fact that the glasses with displaced compositions correspond to the phase separation region in the Li_2O–SiO_2 system, so that the nucleation of the lithium disilicate occurs in phase separated inhomogeneities with the composition that is displaced toward the lithium disilicate composition and depends weakly on the temperature. The shift in the maximum T_{max} by 12°C can be explained by the increases in the viscosity and the glass transition temperature of the melt in the phase separated (matrix) region [106,107] enriched by SiO_2 as compared to the lithium disilicate composition, which provides an increase in the temperature T_{max} according to the relationship for the maximum of the nucleation rate

$$T_{\max} = \frac{T_{melt}}{3}\frac{1 + H_A/\Delta\Phi_K^*}{1 + H_A/3\Phi_K^*}, \quad H_A = \Phi_A + TS_A \quad S_A = -\frac{\partial \Phi_A}{\partial T} \tag{6}$$

where T_{melt} is the melting temperature and the enthalpy H_A, activation entropy S_A, and the nucleation barrier correspond to $T = T_{max}$. It can be seen from Fig. 6 that the temperature T_{max} of the maximum nucleation rate of lithium disilicate crystals for all the compositions under investigation is close to the glass transition temperature T_g. The closeness of the temperatures T_{max} and T_g for the glass of the stoichiometric compositions and the glasses with displaced compositions can be explained as follows. The relative displacements of structural units that are necessary for crystal nucleation and occur through the breaking and switching of chemical bonds become sufficiently fast with an increase in the temperature beginning from the glass transition point T_g. The presence of the viscosity determined by the free activation energy $\Phi_\eta = \Phi_A$ leads to a shift in the maximum of the stationary nucleation rate $I_{st}(T)$ according to relationship (6) toward higher temperatures ($T_{max} = T_g = 2/3T_{melt}$) as compared to its position $T_{max} = 1/3T_{melt}$ that would be observed at $H = 0$.

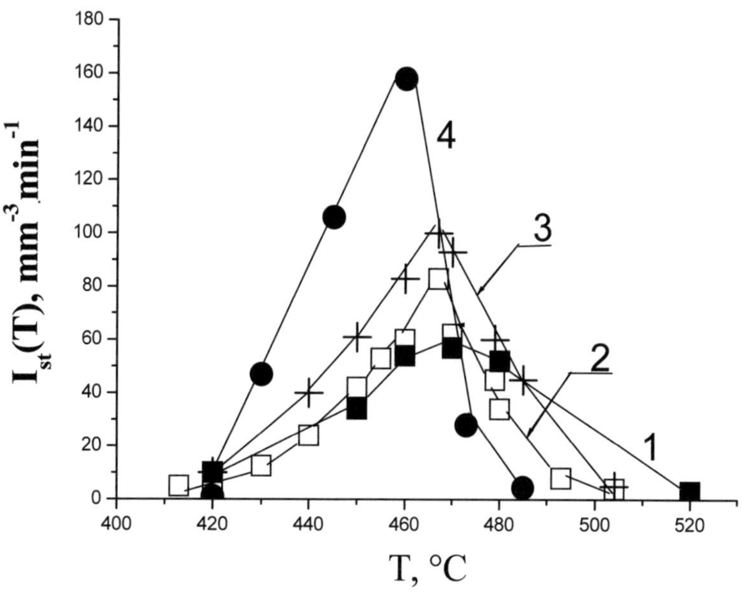

Fig. 6. Temperature dependences of the stationary nucleation rate of lithium disilicate crystals $I_{st}(T)$ for the glasses (*1*) no. 1, (*2*) no. 2 and (*3*) no. 3 with displaced compositions of the glasses and (*4*) with the stoichiometric lithium disilicate composition.

5. Crystal growth rate

Before proceeding to the discussion of the results of the experiment on the determination of the growth rate of lithium disilicate crystals in glasses no. 1–no. 3, let us consider the main notions used when determining the crystal growth. The crystal growth occurs as a result of accidental attachments and detachments of structural units from the surface of a supercritical nucleus (the supercritical nucleus is a nucleus with the size exceeding a critical size). The growth of the super-critical nucleus is thermodynamically favorable because this is accompanied by a decrease in the free energy of the system ΔΦ. The linear growth rate of the nucleus is determined by the equation

$$(U = \frac{dr}{dt} = l(\beta_+ - \beta_-), \tag{7}$$

where l is the thickness of the monomolecular layer growing on the crystal for some time interval, equal to the linear size of the structural unit), and β_+ and β_- are the probabilities of attachment and detachment of one structural unit from the crystal surface per unit time:

$$\beta_+ = \frac{1}{\tau_0}\exp\left(-\frac{\Phi_A}{kT}\right),\ \beta_- = \beta_+\exp\left[\frac{\left(\frac{\partial\Delta\Phi}{\partial a}\right)}{kT}\right] \tag{8}$$

Here, τ_0 is the time of the order of the period of vibrations of atoms in a chemical bond, whose switching provides the transition of the structural unit to the ordered state of the crystal structure of the nucleus; ΔΦ is the change in the free energy of the system; and a is the linear size of the structural unit. Substitution of these expressions into formula (7) gives

$$U = \frac{1}{\tau_0}\exp\left(-\frac{\Phi_A}{kT}\right)\left\{1 - \exp\left[-\frac{\Delta\varphi}{kT}\left(1 - \frac{r^*}{r}\right)\right]\right\}, \tag{9}$$

where $\Delta\varphi$ is the difference between the free energies of the glass and nucleus per unit volume of the nucleus and r^* is the radius of the critical nucleus. It follows from this relationship that the growth rate of the critical nucleus is equal to zero. Since the critical nucleus is in an unstable equilibrium state, both the increase and decrease in its size is accompanied by a decrease in the difference $\Delta\varphi$ and is equally probable. With an increase in the size of the supercritical nucleus, the growth rate increases and becomes constant when the nucleus radius is larger than the critical radius ($r >> r^*$).

Now, we consider the temperature dependence of the stationary growth rate. Under the assumption that $\Delta\varphi \approx q(1 - T/T_{melt})$, where q is the specific heat capacity, we have

$$U(T) \approx \exp\left(-\frac{\Phi_A}{kT}\right)\left\{1 - \exp\left[-\frac{q(1 - T/T_{melt})}{kT}\right]\right\} \tag{10}$$

According to expression (10), the growth rate is equal to zero at $T \to 0$ and $T \to T_{melt}$. At some temperature ($0 << T_{melt}$), the growth rate is maximum. This temperature can be evaluated. We assume that $\exp[-q(1 - T/T_{melt})/kT] \approx 1 - q(1 - T/T_{melt})/kT$ and substitute into expression (10). After the differentiation with respect to the temperature, we obtain

$$\frac{dU}{dT} \approx \frac{1}{T^2}\exp\left(-\frac{\Phi_A}{kT}\right)\left[\frac{H_A}{k}\left(\frac{T_{melt}}{T} - 1\right) - T_{melt}\right] \tag{11}$$

From the extremum condition

$$\left[\frac{H_A}{k}\left(\frac{T_{melt}}{T}-1\right)-T_{melt}\right]\Big|_{T=T_{U_m}}=0,$$

we find that

$$T_{U_M}=\frac{T_{melt}}{1+kT_{melt}/H_A} \tag{12}$$

The quantity kT_{melt}/H_A in formula (12) usually does not exceed 1/10. Therefore, the temperature of the maximum growth rate is close to the melting temperature, and it is very difficult to experimentally determine this temperature. This inference is very important for experimenters.

6. Experimental results on the crystal growth and their discussion

6.1 Crystal growth rate: Dependence on the temperature

There is a large number of works on the rate of crystal nucleation (see Table 1) in lithium silicate glasses, whereas investigations of the concentration dependences of the growth rate are absent. There exist works in which the results are presented only for specific compositions of glasses in this system [108, 99]. The photographs of characteristic lithium disilicate crystals grown at a development temperature of 600 °C in glasses no. 1–no. 3 are displayed in Fig. 5. In order to determine the growth rate *U*, the dependences $R_{max}(t)$ were constructed for a series of samples held for different times *t* at a specified temperature. It was revealed that these dependences exhibit a linear behavior, which allows us to determine the crystal growth rate from their slope. In the case of spherical particles, their growth rate *U* is determined from the expression $U = dR_{max}/dt'$, where t' is the time of holding of the glass at the development temperature. In the case of ellipsoids of revolution, the growth rate *U* is determined from the change in their average radius defined as the half sum of the major and minor radii of the ellipsoid. The crystal growth rates were studied at temperatures of 570, 590, 600, 640, 685, 718, 720, and 730 °C. The growth rates *U(T)* of lithium disilicate crystals in glasses no. 1-no. 4 as a function of the temperature are given in Table 4. Figure 7 shows the temperature dependences of the growth rate *U* for glasses no. 1–no. 3. The data presented in Fig. 7c allow us to compare the growth rates *U* determined in our work with those obtained in [109]. As can be seen from Fig. 7c, Burgner et al. [109] performed investigations at higher temperatures. We more thoroughly studied the growth rate at low temperatures. In the temperature range 575–600 °C, the results of both studies are in good agreement. It can be seen from Fig. 7 that the growth rate *U(T)* increases with an increase in the temperature. In the temperature range 640–718 °C, the growth rate increases linearly. It is difficult to judge the accurate position of the maximum of the growth rate *U(T)*, because the error in the measurement of the size at temperatures above 720 °C can increase in an uncontrollable manner due to the fact that the time of heat treatment of the sample (1–2 min) is comparable to the time of its heating to a specified development temperature and we cannot state with assurance that crystals grew accurately at this temperature. Therefore, although the temperature dependence of the growth rate should have a maximum, we cannot argue that its true position is recorded. It is quite probable that the maximum of the growth rate is located at higher temperatures. The maxima of the nucleation rates are located at lower temperatures than the maxima of the growth rates. Theoretically, they can either be located at a large distance from each other or overlap. Since the accurate position of the maximum of the growth rate is not known, we can only to make the inference that, for the compositions under investigation, the positions of the temperature maxima of the nucleation and growth rates are spaced along the temperature scale by an uncertain value no less than 260 °C (720–460 °C).

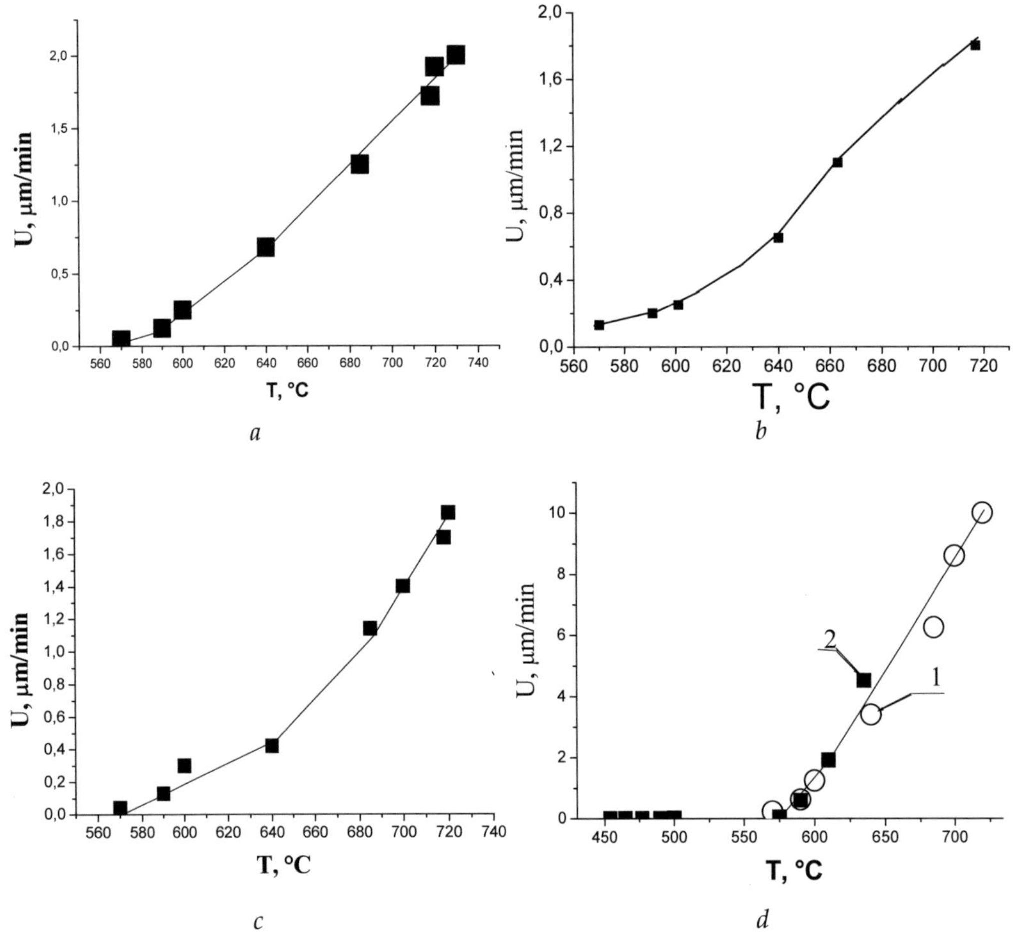

Fig. 7. Temperature dependences of the growth rate *U* for glasses (*a*) no. 1, (*b*) no. 2, (*c*) no. 3, and (*d*) no. 4 according to the data (*1*) obtained in this work and (2) taken from [109].

6.2 Crystal growth rate: Dependence on the time of holding of the glasses at the temperature of 450 °C

The growth rates of lithium disilicate crystals as a function of the holding time glasses at a constant temperature were investigated at the temperature of 450 °C. The results of the measurements are presented in Fig. 8. It can be seen from this figure that the dependence of the growth rate of lithium disilicate crystals in glasses no. 1, no. 2 and no. 3 on the time of low-temperature heat treatment exhibit an oscillatory behavior with a gradual decay of the oscillation amplitude and a retardation of the growth process. In the framework of the existing theories of nucleation and growth, it was demonstrated for metal alloys [110] that the retardation of the growth process is caused simultaneously by nucleation, growth, and dissolution of particles.

Glass no.	Growth temperature, °C								
	570	590	600	640	685	700	718	720	730
	Growth rate $U(T)$, μm/min								
1	0.045	0.125	0.25	0.68	1.25	-	1.72	1.92	2.0
2	0.036	0.125	0.31	0.44	1.12	-	1.66	1.58	-
3	0.042	0.130	0.40	0.42	1.14	1.40	1.70	1.85	-
4	0.225	0.625	1.25	3.4	6.25	8.6	-	10.0	-

Table 4. Growth rates $U(T)$ (μm/min) of lithium disilicate crystals in glasses no. 1–no. 4 as a function of the temperature

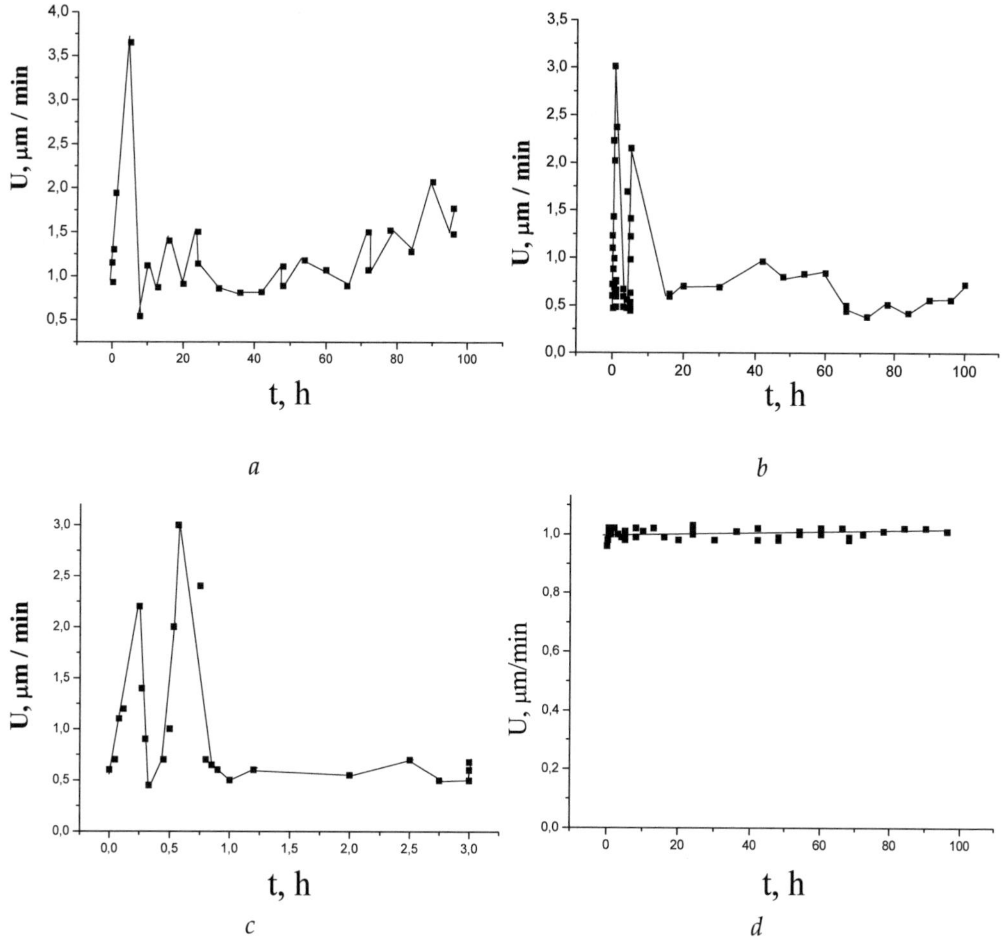

Fig. 8. Dependences of the growth rate U on the time of heat treatment at 450 °C for glasses (*a*) no. 1, (*b*) no. 2, (*c*) no. 3 and (*d*) no. 4. Samples were developed at the temperature of 600 °C for 10 min.

For glass no. 4 (with the stoichiometric lithium disilicate composition), the crystal growth rate does not depend on the time of low-temperature heat treatment (Fig. 8*c*).
The gradual decrease in the crystal growth rate with an increase in the time of holding of the glass at a temperature of 450 °C for glasses no. 1, no. 2 and no. 3 (Figs. 8*a*, 8*b*, 8*c*) can be explained by the fact that the crystals grow in the glass with the composition lying in the metastable phase separation region. A part of the material is spent for forming the crystal layer, and the growing crystal appears to be surrounded by the region depleted in the building material (the so-called diffusion zone) (Fig. 9).

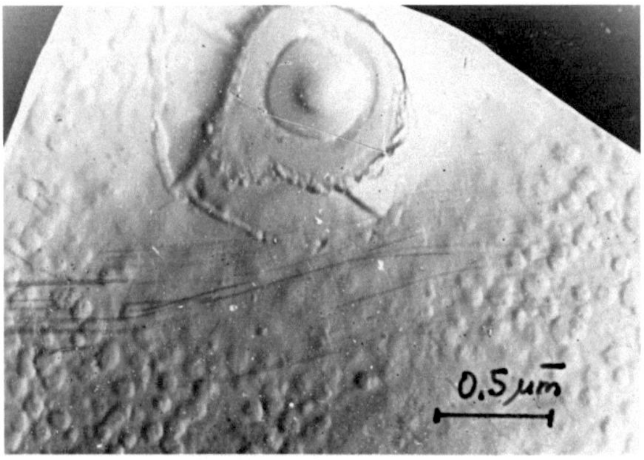

a

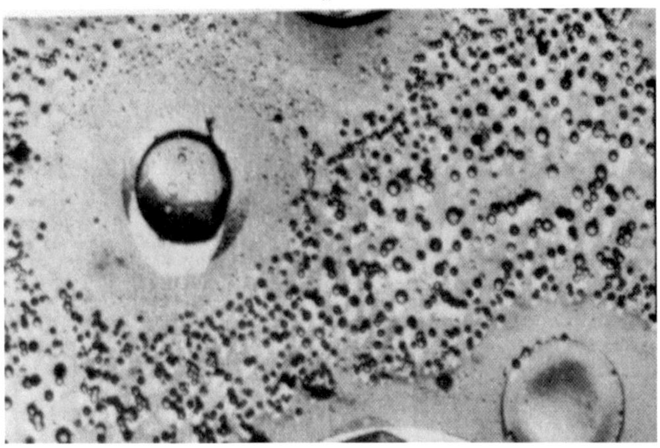

b

Fig. 9. Electron microscopic images of glasses (*a*) no. 1 and (*b*) no. 2 preliminarily heat treated at the temperature of 440 °C for 2 h and developed at 600 °C for 10 min.

The formation of the next crystal layer requires the time it takes for phase separated droplets to come to the crystal growth region. In this case, the phase separated droplets consist predominantly of SiO_2. When the droplets come closer to the growing lithium disilicate crystal, they will block its growth, which manifests itself in the dependences (Figs. 8*a*, 8*b*, 8*c*) as a decrease in the crystal growth rate.

7. Conclusions

Thus, we have investigated the kinetics of phase separation in three lithium silicate glasses containing 23.4, 26.0 and 29.1 mol % Li_2O with the compositions lying in the phase separation region in the Li_2O-SiO_2 system. It has been found that the temperature dependences of the ratio of the number of traces to the maximum number of traces $N_S/N_{S\max}$ and the ratio of the radius of the particle trace to the maximum radius R/R_{max} exhibit maxima. The crystal nucleation rate has been studied in glasses with the compositions displaced from the stoichiometric lithium disilicate composition toward an increase in the SiO_2 content. The following features have been revealed.

i. The absolute values of the crystal nucleation rate in glasses with the lower content Li_2O than that in the glass with the stoichiometric lithium disilicate composition vary insignificantly.
ii. The position of the temperature maximum of the nucleation rate for the glass with the lower lithium oxide content is shifted on the temperature scale by 12 °C. This relatively small change in the quantities $I_{st}(T_{max})$ and $T_{\max}$ is explained by the fact that the glasses with the displaced composition correspond to the phase separation region in the Li_2O-SiO_2 system, so that the lithium disilicate in glasses with displaced composition nucleates in phase separated inhomogeneities with compositions that are displaced toward the lithium disilicate composition and depend weakly on the temperature.

Variations in the crystal growth rate have been investigated as a function of the temperature and time of heat treatment. It has been shown that the temperature dependences of the crystal growth rate tend to an extremum. It has been found that the crystal growth rate depends substantially on the heat treatment time for phase separated glasses and that the growth rate is constant for the glass with the stoichiometric lithium disilicate composition. The dependence of the crystal growth rate on the heat treatment time in phase separated glasses exhibits a damped oscillatory behavior, which is explained by the depletion of the building material for crystals and the necessity of some time for its supply. Therefore, the performed investigations allow us to draw the following conclusion. Unlike the universally accepted opinion, the phase separation not only does not facilitate the conditions for the formation and growth of crystals but, in some cases (see, for example, the dependences of the crystal growth rate on the time of low-temperature heat treatment), even retards the crystal growth process. The data obtained on the nucleation and growth of crystals help to find optimum compositions and temperature-time conditions for the preparation of many glass-ceramic nanomaterials in lithium silicate system, including photostructured materials.

8. References

[1] *Hinz W., Kunth P.* Vitrokeramprodukte mit niedrigem Warmeausdehnungskoeffizienten // Silikattechnik. 1960. V. 11. N 12. P. 506–511.

[2] *Vogel W., Gerth K.* Catalyzed Crystallization in Glass, *Proceedings of the Symposium on Nucleation and Crystallization in Glasses and Melts, Columbus, Ohio, United States, 1962,* Reser, M.K., Ed., Columbus, 1962, pp. 11–22.

[3] *Filipovich V.N.* Initial Stages of Crystallization of Glasses and Formation of Glass Ceramics, in *Stekloobraznoe sostoyanie,* Vyp. 1: *Katalizirovannaya kristallizatsiya stekla* (Vitreous State, vol. 1: Catalyzed Crystallization in Glass), Moscow: Nauka, 1963, pp. 9–24 (In Russian).

[4] *Maurer R.* Crystal Nucleation in a Glass Containing Titanium // J. Appl. Phys. 1962. V. 33. N 6. P. 2132–2139.
[5] *Ohlberg S.M., Golob H.R., Strichler D.* Crystal Nucleation by Glass in Glass Separation, *Proceedings of the Symposium on Nucleation and Crystallization in Glasses and Melts, Columbus, Ohio, United States, 1962,* Reser, M.K., Ed., Columbus, 1962, pp. 55–62.
[6] *Stookey C.D.* Catalyzed Crystallization of Glass in Theory and Practice, *Int. Glasskongress,* 1959, vol. 32K, no. 5, p. 1.
[7] *Nakagawa K., Izumitani T.* Relationship between Phase Separation and Crystallization in $Li_2O \cdot 2.5SiO_2$ Glass and a Lithium Silicate Containing a Large Amount of Titanium Oxide // Phys. Chem. Glasses. 1969. V. 10. N 5. P. 179–184.
[8] *Phillips S.V., McMillan P.W.* Phase Separation and Crystallization in Li_2O–SiO_2 Glass and Li_2O–SiO_2–P_2O_5 Glasses // Glass Technol. 1965. V. 6. N 2. P. 46–51.
[9] *Tomozawa M.* Liquid Phase Separation and Crystal Growth in Li_2O–SiO_2 Glass // Phys. Chem. Glasses. 1973. V. 14. N 6. P. 112–113.
[10] *Milyukov E.M.* On the Phase Separation and Micro crystallization of Glasses in Li_2O–SiO_2 and Li_2O–Al_2O_3–SiO_2 Systems // Opt._Mekh. Prom_st. 1965. N 3. P. 31–36. (In Russian).
[11] *Harper H., James P.F., McMillan P.W.* Crystal Nucleation in Lithium Silicate Glasses // Discuss. Faraday Soc. 1970. N 50. P. 200–205.
[12] *Liebau F.* Untersuchungen an Schichtsilicaten des Formeltypes $A_m(Si_2O_5)_n$. I. Die Kristallstructur der Zimmertemperaturform des $Li_2Si_2O_5$ // Acta Cryst. 1961. Bd.14. S.389-395.
[13] *Jaccodine R.J.* Study of devitrification of lithium glass // J. Amer. Ceram. Soc. 1961. V. 44. N 11. P. 8-11.
[14] Kalinina A.M., Filipovich V.N., Kolesova V.A., Bondar I.A. On the products of crystallization of lithium silicate glasses // in: The Vitreous State. Vyp.1. Katalizirovannaya kristallizatsiya Stekla. M.-L.: Izv. Ak*ad. Nauk SSSR.*1963. P. 53-66. (In Russian).
[15] *Glasser F.P.* Crystallisation of lithium disilicate from Li_2O-SiO_2 glasses // Phys. Chem. Glasses. 1967. V.8. N6. P. 224-232.
[16] *Ito M., Sacaino T., Moriya T.* Study on the process of crystallization of the Li_2O2SiO_2 glass. I. Rates of crystal growth and nucleation // Bulletin of the Tokyo Institute of Technology. 1968. N 88. P. 127-149.
[17] *Ito M., Sacaino T., Moriya T.* Study on the process of crystallization of the Li_2O2SiO_2 glass I Variation of electric properties with crystallization // Bulletin of the Tokyo Institute of Technology. 1968. N 88. P. 151-171.
[18] *Filipovich V.N., Kalinina A.M.* On the nature of the influence of heat treatment on the crystallization kinetics of lithium silicate glasses // Izv. Akad. Nauk SSSR, Neorg. Mater. 1968. V. 4. N 9. P. 1532–1538. (In Russian).
[19] *Kinser D.L., Hench L.L.* Effect of a metastable precipitate on the electrical properties of an Li_2O - SiO_2 glass // J. Am. Ceram. Soc. 1968. V. 51. N 8. P. 445-448.
[20] *Ogura T., Hayami R., Kadota M.* Kinetics and Mechanism of crystallization of lithium silicate glasses // J.Ceram. Assoc. Japan. 1968. V. 76. N 8. P. 277-284.
[21] *Freiman S.W., Hench L.L.* Kinetics of crystallization in Li_2O - SiO_2 glasses // J. Am. Ceram. Soc. 1968. V. 51. N 7. P. 382-387.

[22] *Kashchiev D.* Solution of the Non-Steady State Problem in Nucleation Kinetics // Surf. Sci. 1969. V. 14. P. 209-220.

[23] *Filipovich V.N., Kalinina A.M.* Crystallization kinetics of lithium disilicate from simple and complex glasses // *Izv. Akad. Nauk SSSR, Neorg. Mater.* 1970. V. 6. N. 2. P. 351-356. (In Russian).

[24] *Kalinina A.M.* On the Polymorphism of Lithium Disilicate // *Izv. Akad. Nauk SSSR, Neorg. Mater.* 1970. V. 6. N. 5. P. 907-913. (In Russian).

[25] *Hench L.L., Frieman S.W., Kinser D.L.* The early stages of crystallization in Li_2O - $2SiO_2$ glass // Phys. Chem. Glasses. 1971. V. 12. N 2. P.58-63.

[26] *Sycheva G.A* Temperature dependence of the nucleation rate of crystals in some silicate glasses / Brief abstracts of the conference of young specialists. Leningrad (Sankt-Petersburg), Institute of Silicate Chemistry behalf I.V. Grebenshchikova (24-25 October 1972.). P. 15-16. (In Russian).

[27] *Matusita K., Tashiro M.* Rate of homogeneous nucleation in alkali disilicate glasses // J. Non-Crystalline Solids. 1973. V. 11. P. 471-484.

[28] *James P.F., Keown S.R.* Study of crystallization in lithium silicate glasses using high-voltage microscopy // Philosophical Magazine. 1974. N 30. P. 789-802.

[29] *ames P.F.* Kinetics of crystal nucleation in lithium silicate glasses // Phys. Chem. Glasses. 1974. V. 15. N 4. P. 95-105.

[30] *Kalinina A.M., Fokin V.M., Filipovich V.N.* On the method of determining the parameters characterizing the crystal nucleation in glasses // Fizika i Khimiya Stekla. 1976. V. 2. N 4. P. 298-305. (In Russian).

[31] *Kalinina A.M., Fokin V.M., Filipovich V.N.* Induction period of crystal nucleation in $Li_2O{\cdot}2SiO_2$ glass and its temperature dependence // Fizika i Khimiya Stekla. 1977. V. 3. N 2. P. 122-129. (In Russian).

[32] *Fokin V.M., Filipovich V.N., Kalinina A.M.* The Effect of Preliminary Heat Treatment on Crystal Nucleation in Li_2O-$2SiO_2$ Glass // Fizika i Khimiya Stekla. 1977. V. 3. N 2. P. 129-36. (In Russian).

[33] *ames P.F., Scott B., Armstrong P.* Kinetics of crystal nucleation in lithium disilicate glass. A comparison between melts prepared in platinum and silica crucibles and between melts prepared from ordinary and high purity starting materials // Phys. Chem. Glasses. 1978. V. 19. N 2. P. 24-27.

[34] *Rowlands E.G., James P.F.* Analysis of steady state crystal nucleation rates in glasses. Part 1. Methods of analysis and application to lithium disilicate glass // Phys. Chem. Glasses. 1979. V. 20. N 1. P. 1-8.

[35] *Rowlands E.G., James P.F.* Analysis of steady state crystal nucleation rates in glasses. Part 2. Further comparison between theory and experiment for lithium disilicate glass // Phys. Chem. Glasses. 1979. V. 20. N 1. P. 9-14.

[36] *Matusita K., Sakka S.* Kinetic study of the crystallisation of glass by differential scanning calorimetry // Phys. Chem. Glasses. 1979. V. 20. N 4. P. 81-84.

[37] *Ghoneim N.A., El Batal H.A., Ahmed A.A., Khalifa F.A.* Crystallization of lithium trisilicate glasses // Transactions and Journal of the British Ceram. Soc. 1979. V. 78. N 1. P. 15-22.

[38] *Gonzalez-Oliver C.J.R., Johnson P.S., James P.F.* Influence of water content on the rates of crystal nucleation and growth in lithia-silicate and soda-lime-silica glasses // J. Mater. Sci. 1979. V. 14. P. 1159-1169.

[39] *Fokin V.M., Kalinina A.M., Filipovich V.N.* Nucleation in silicate glasses and effect of preliminary heat treatment on it // J. Cryst. Growth. 1981. V. 52. P. 115-121.
[40] *Marotta A., Buri A., Branda F., Saiello S.* Nucleation and Crystallization of $Li_2O \cdot 2SiO_2$—a DTA Study. Advances in Ceramics, Nucleation and Crystallization in Glasses, 4, eds., J. H.Simmons, D. R.Uhlmann, and G. H.Beall. American Ceramic Society, Columbus, 1982. P. 146–152.
[41] *Kalinina A.M., Sycheva G.A., Filipovich V.N.* Homogeneous and heterogeneous crystal nucleation in some silicate glasses // Glastechn, Ber. 1983. 56K. V. 2. P. 816-821.
[42] *31Huang Z.J., Yoko T., Kamiya K., Sakka S.* Effect of water on the crystallization of Li_2O - SiO_2 glasses and gels // Yogyo-Kyokai-Shi. 1983. V.81. N5. P. 215-221. (In Japanese).
[43] *Zanotto E.D., James P.F.* Experimental test of the classical nucleation theory for glasses // J. Non-Crystalline Solids. 1985. V. 74. P. 373-394.
[44] *Sycheva G.A., Kalinina A.M., Filipovich V.N.* Change of nucleation kinetic of lithium di - and metasilicate crystals in photosensitive glasses at transition from heterogeneous to homogeneous nucleation // Fizika i Khimiya Stekla (Russian). 1986. T. 11. N 1. C.123-125. (In Russian).
[45] *Kalinina A.M., Sycheva G.A., Filipovich V.N. et. al.* Auto-catalytic Nucleation of Lithium Silicate glasses // Fizika i Khimiya Stekla (Russian). 1986. T. 12. N 2. C.260-264. (In Russian).
[46] *Kalinina A.M., Fokin V.M., Sycheva G.A., Filipovich V.N.* Three Types of Catalysis of Lithium Disilicate Crystal Nucleation in Lithium Silicate Glasses, *Proceedings of the XIV International Congress on Glass, New Delhi, India, 1986,* New Delhi, 1986, vol. 1, pp. 366–373.
[47] *Kalinina A.M., Sycheva G.A., Filipovich V.N.* Catalyzed and Spontaneous Nucleation of Lithium Disilicate Crystals in Nonphotosensitive and Photosensitive Glasses, in *Katalizirovannaya kristallizatsiya Stekla. Trudy Gosudarstvennogo Instituta Stekla* (Catalyzed Crystallization of a glass: Transactions of the State Institute of Glass), Moscow: Gos. Inst. Stecla, 1986, pp. 56-59. (In Russian).
[48] *Hishinuma A., Uhlman D.R.* Nucleation kinetics in some silicate glass-forming melts // J. Non-Crystalline Solids. 1987. V. 95-96. P. 449-456.
[49] *Braetsh V., Frischat G.H.* Crystallization and phase separation in the system Li_2O-SiO_2 // J. Non-Crystalline Solids. 1987. V. 95-96. P. 457-464.
[50] *Barket M.F., Wang T.H., James P.F.* Nucleation and growth kinetics of lithium disilicate and lithium metasilicate in lithia-silica glasses // Phys. Chem. Glasses. 1988. V. 29. N 6. P. 240-248.
[51] *Filipovich V.N., Kalinina A.M., and Sycheva G.A.* Glass Formation and Catalyzed Crystal Nucleation, in *Trudy VIII Vsesoyuznogo soveshchaniya po stekloobraznomu sostoyaniyu* (Proc. VIII All-Union Conf. on the Vitreous State), Leningrad: Nauka, 1988, pp. 87-96. (In Russian).
[52] *Schlesinger, M.E. and Lynch, D.C.* Effect of VB and VIB oxides on nucleation parameters in lithium disilicate glass // Journal of Non-Crystalline Solids. 1989. V. 108. P. 237–248.
[53] *Sycheva G.A., Kalinina A.M., Filipovich V.N.* Transition kinetic from the Heterogeneous to the Homogeneous Nucleation in Some Silicate Glasses, in: Glass and fine ceramics, Varna, Bulgaria, 18-20 Sep. 1987: *Proceedings of the* IX nat. Sci. and techn. conf. with intern. participation: Sophia 1989. P. 239-245.

[54] *Ray C. S., Day D. E.* Determining the Nucleation Rate Curve for Lithium Disilicate Glass by Differential Thermal Analysis // J. Am. Ceram. Soc. 1990. V. 73. P. 439-442.

[55] *Boiko G.G., Sycheva G.A., Valjuk L.G.* Influence of Synthesis Conditions of Photosensitive Glass on its Crystallization Kinetics, in: Glass and fine ceramics, Varna, Bulgaria, 18-20 Oct. 1990: X nat. Sci. and techn. conf. with intern. Participartion: summaries - S.1.: S. n., [1990]. - Vol.1. - P. 282.

[56] *Kalinina A.M., Filipovich V.N., Sycheva G.A.* The Mutual Influence of Crystal Nucleation in Complex Glasses, in *Amorfno-kristallicheskie materially: sintez, struktura, svoistva, primenenie. Sbornik trudov MKhTI im. D.I.Mendeleeva* (Amorphous-Crystalline Materials: Synthesis, Structure, Properties, and Applications; A Collection of Articles of the Mendeleev University of Chemical Technology), Moscow, 1991, pp. 77-85. (In Russian).

[57] *Deubener J., Bruckner R., Sterenitzke M.* Induction time analysis of nucleation and crystal growth in di- and metasilicate glasses // J. Non-Crystalline Solids. 1993. V. 163. P. 1-12.

[58] *Boiko G.G., Sycheva G.A., Valjuk L.G.* Influence of Synthesis Conditions on the Kinetics of Crystallization of Photosensitive Lithium Silicate Glasses // Glass Physics and Chemistry. Original Russian Text Copyright 1995 by Fizika i Khimiya Stekla. 1995. T. 21. N 1. P. 45-52. (In Russian).

[59] *Davis M.J.* Influence of water and thermal history on viscosity and nucleation kinetics of lithium disilicate melt, p. 159. 1996. Geology and Geophysics, Yale University, Connecticut.

[60] *Sycheva G.A.* Transition from the Heterogeneous to the Homogeneous Nucleation in Some Silicate and Phosphate Glasses, *Proceedings of the International Symposium on Glass Problems, Istanbul, Turkey, September 4–6, 1996*, Istanbul, 1996, vol. 2, pp. 373–379.

[61] *Mishima N., Ota R., Wakasugi T., Fukunaga J.* Analysis of crystallization behavior in Li_2O $2SiO_2$ glass by DTA method based on a liquid model // Journal of Non-Crystalline Solids. 1996. V. 197. P. 19–24.

[62] *Narayan, K.L., Kelton, K.F., and Ray, C.S.* Effect of Pt doping on nucleation and crystallization in Li_2O $2SiO_2$ glass: experimental measurements and computer modeling. Journal of Non-Crystalline Solids. 1996. V. 195. P. 148–157.

[63] *Zanotto E.D., Gomes Leite M.L.* The nucleation mechanism of lithium disilicate glass revisited // Journal of Non-Crystalline Solids. 1996. V. 202. P. 145-151.

[64] *Davis, M.J., Ihinger, P.D., and Lasaga, A.C.* Influence of water on nucleation kinetics in silicate melt. Journal of Non-Crystalline Solids. 1997. V. 219. P. 62–69.

[65] *Kalinina A.M., Filipovich V.N., Sycheva G.A.* Heterogeneous nucleation in photosensitive glasses // Journal of Non-Crystalline Solids. 1997. V. 219. P. 80-83.

[66] *Holland D., Iqbal Y., James P., Lee B.* Early stages of crystallization of lithium disilicate glasses containing P_2O_5 -An NMR study // Journal of Non-Crystalline Solids. 1998. V. 232-234. P. 140-146.

[67] *Sycheva G.A.* Surface Energy at the Crystal Nucleus-Glass Interface in Alkali Silicate Glasses // Glass Physics and Chemistry. Original Russian Text Copyright 1998 by Fizika i Khimiya Stekla. Proceedings of the X conference on the vitreous state (St. Petersburg, Russia, October 21-23, 1997). 1998. T. 24. N 4. P. 342-347.

[68] *Iqbal Y., Lee B., Holland D., James P.* Metastable phase formation in the early stage crystallization of the lithium disilicate glass // Journal of Non-Crystalline Solids. 1998. V. 224. P. 1-16.

[69] *Hasdemir I., Bruckner R., Deubener J.* Crystallization of lithium di- and metasilicate solid solutions from Li_2O-SiO_2 glasses // Phys. Chem. Glasses. 1998. V. 39. N 5. P. 253-257.

[70] *Davis, M.J., Ihinger, P.D.* Heterogeneous crystal nucleation on bubbles in silicate melt // American Mineralogist. 1998. V. 83. P. 1008–1015.

[71] *Davis M.J., Ihinger P.D.* Influence of hydroxyl on glass transformation kinetics in lithium disilicate melt and a re-evaluation of structural relaxation in NBS 710 and 711 // Journal of Non-Crystalline Solids. 1999. V. 244. P. 1-15.

[72] *Burgner L.L., Weinberg M.C., Lucas P., Soares P.C. Jr., Zanotto E.D.* XRD investigation of metastable phase formation in Li_2O-SiO_2 glasses // Journal of Non-Crystalline Solids. 1999. V. 255. P. 264-268.

[73] *Burgner L. I., Lucas P., Weinberg M. C., Soares P. C. Jr., and Zanotto E. D.* Metastable Phase Formation in the Early Stage Crystallisation of Lithium Disilicate Glass // Journal of Non-Crystalline Solids. 2000. V. 274. P. 188–194.

[74] *Kitamua N., Fukumi K., Mizoguchi H., Makihara M., Higuchi A., Ohno N., Fukunaga T.* High pressure densification of lithium silicate glasses // Journal of Non-Crystalline Solids. 2000. V. 274. P. 244-248.

[75] *Kanert O., Kuchler R., Suter D., Shannon G.N., Jain H.* Effect of devitrification on the ionic diffusion of Li-disilicate // Journal of Non-Crystalline Solids. 2000. V. 274. P. 202-207.

[76] *Burgner L.L., Weinberg M.C.* Crystal growth mechanisms in inorganic glasses // Phys. Chem. Glasses. 2001. V. 42. N 3. P. 184-190.

[77] *Burgner L.L., Weinberg M.C.* An assessment of crystal growth behavior in lithium disilicate glass // Journal of Non-Crystalline Solids. 2001. V.279. P. 28-43.

[78] *Soares P.C. Jr., Zanotto E.D., Fokin V.M., Jain H.J., Weinberg M.C.* TEM and XRD Srudy of early crystallization of lithium disilicate glasses // Journal of Non-Crystalline Solids. 2003. V. 331. P. 217-227.

[79] *Filipovich V.N., Kalinina A.M.* On the Relation between the Maximum of the Crystal Nucleation Rate in Glasses and the Glass Transition Temperature, *Izv. Akad. Nauk SSSR, Neorg. Mater.*, 1971, vol. 7, no. 10, pp. 1844–1848.

[80] *Gonzalez-Oliver C.J.R.* Crystal nucleation and growth in soda-lime-silica glasses. Ph.D. Thesis. The University of Sheffield, 1979. 175 p.

[81] *Gonzalez-Oliver C.J.R., James P.F.* Crystal nucleation and growth in a $Na_2O{\cdot}2CaO{\cdot}3SiO_2$ glass // J. Non-Crystalline Solids. 1980. V. 38&39. P. 699-704.

[82] *Kalinina A.M., Filipovich V.N., Fokin V.M.* Stationary and non-stationary crystal nucleation rates in the $2Na_2O{\cdot}CaO{\cdot}3SiO_2$ glass of stoichiometric composition // Journal of Non-Crystalline Solids. 1980. V. 38&39. P. 723-728.

[83] *ames P.F.* Nucleation in glass-forming systems - a review // Advances in Ceramics. Vol. 4. Nucleation and crystallization in glasses. American Ceramic Society, Columbus, Ohio, 1982. P. 1-48.

[84] Ramsden A.H., James P.F. The effect of amorphous phase separation on crystal nucleation kinetics in BaO-SiO2 glasses. Part 1. General survey // J. Mater. Sci. 1984. V. 19. N 5. P. 1406-1419.

[85] Ramsden A.H., James P.F. The effect of amorphous phase separation on crystal nucleation kinetics in BaO-SiO2 glasses. Part 2. Isothermal heat treatments at 700 °C // J. Mater. Sci. 1984. V. 19. N 9. P. 2894-2908.
[86] James P.F. Experimental studies of crystal nucleation in glasses // Ceramic Transactions. Vol. 30. Nucleation and cristallization in liquids and glasses. American Ceramic Society, Westerville, Ohio. 1993. P. 3-12.
[87] Gránásy L., Wang T., James P.F. Kinetics of wollastonite nucleation in CaO·SiO2 glass // J. Chem. Phys. 1998. V. 108. N 17. P. 7317-7326.
[88] Filipovich V.N., Kalinina A.M., Sycheva G.A. Temperature Dependence of the Rate of Nucleation of Acicular Crystals in Sodium Silicate Glass // Izv. Akad. Nauk SSSR, Neorg. Mater. 1975. V. 11. P. 1305-1308. (In Russian).
[89] Fokin V.M, Yuritsyn N.S. The Nucleation and Growth Rates of Sodium Metasilicate Crystals in Sodium Silicate Glass 44Na2O-56SiO2 // Fizika i Khimiya Stekla. 1997. V. 23. P. 236-239. (In Russian).
[90] Burgner L.L., Weinberg M.C. Crystal nucleation rates in a Na2O·SiO2 glass // Journal of Non-Crystalline Solids. 2000. V. 261. P. 163-168.
[91] Sycheva G.A. Evalution of the Surface Energy at the Crystal-Glass Interface in Sodium Silicate Glass 4Na2O · 54SiO2 // Glass Physics and Chemistry. Original Russian Text Copyright 1998 by Fizika i Khimiya Stekla. 1998. V. 24. N 1. P. 47-53.
[92] Burnett D.G., Douglas R.W. Nucleation and crystallization in the soda-baria-silica system // Phys. Chem. Glasses. 1971. V. 12. N 5. P. 117-124.
[93] Strnad Z., Douglas R.W. Nucleation and crystallization in the soda-lime-silica system // Phys. Chem. Glasses. 1973. V. 14. N 2. P. 33-36.
[94] Lakshmi Narayan K., Kelton K.F. First measurements of time-dependent nucleation as a function of composition in Na2O·2CaO·3SiO2 glasses // Journal of Non-Crystalline Solids. 1997. V. 220. P. 222-230.
[95] Potapov O.V., Fokin V.M., Suslova L.Ya., Filipovich V.N., Ugolkov V.L. Influence of Na2O content on the nucleation kinetics in glasses of composition close to the Na2O·2CaO·3SiO2 stoichiometry // Glass Physics and Chemistry. Original Russian Text Copyright 1998 by Fizika i Khimiya Stekla. 2000 V. 26. N 1. P. 39-47.
[96] Sycheva G.A. Nucleation Kinetics of Lithium Metasilicate in Photosensitive Lithium Aluminosilicate Glass // Glass Physics and Chemistry. Original Russian Text Copyright 1999 by Fizika i Khimiya Stekla. 1999. T.25. N 6. C. 501-511.
[97] Filipovich V.N., Kalinina A.M., Sycheva G.A. Catalyzed Crystallization of pyroxene glasses // Fizika i Khimiya Stekla. 1981. V. 7. N 1. P. 55-60. (In Russian).
[98] Sycheva G.A., Nasedkin V.V. Some physica-chemical parameters of volcanic glass formation. in: V.V.Nasedkin. Perlite genesis. Moscow. "Nauka" 1992. 188p.p. (P. 129-125. In Russian).
[99] Sycheva G.A. Influence of Tungsten and Molybdenum Oxides on the Crystallization Kinetics of Tin (II) Pyrophosphate in the SnO-SnO2-ZnO-P2O5 System // Glass Physics and Chemistry. Original Russian Text Copyright 1997 by Fizika i Khimiya Stekla. 1997. T. 23. N 5. P. 368-375.
[100] Sycheva G.A. Nucleation of Sodium-Zinc Phosphate Crystals in Sodium-Zinc Phosphate glass // Glass Physics and Chemistry. Original Russian Text Copyright 1998 by Fizika i Khimiya Stekla. 1998. V. 24. N 4. P. 342-347.

[101] Sycheva G.A, Sigaev V.N. Crystallization kinetics in Ca, Sr and Ba borophosphate systems Proceedings of Second Intern. Conf. on borate glasses, crystals and melts. Edited by A.C.Wright, S.H.Feller and A.C.Hannon. The Society of Glass Technology, Sheffild, 1997. 554s. P. 254-260.

[102] Sycheva G.A. Formation of the Bubble Structure in the 26Li2O · 74SiO2 Glass // Fiz. Khim. Stekla. 2009. V. 35. N 3. P. 342–351 [Glass Phys. Chem. (Engl. transl.), 2009, vol. 35, no. 3, pp. 267–273].

[103] Sycheva G.A. Influence of the Presence of Bubbles on the Parameters of Crystal Nucleation in the 26Li2O · 74SiO2 Glass // Fiz. Khim. Stekla. 2009. V. 35. N 6. P. 798–811 [Glass Phys. Chem. (Engl. transl.), 2009, vol. 35, no. 6, pp. 602–612].

[104] Mel'nichenko T.D., Rizak V.M., Mel'nichenko T.N., Fedelesh V.I. Parameters of the Fluctuation Free Volume Theory for Glasses in the Ge–As–Se System // Fiz. Khim. Stekla. 2004. V. 30. N 5. P. 553–564 [Glass Phys. Chem. (Engl. transl.), 2004, vol. 30, no. 5, pp. 406–414].

[105] Holler W.S., Blackburn D.H., Simmons J.H. Miscibility Gaps in Alkali-Silicate Binaries—Data and Thermodynamic Interpretation // Am. Ceram. Soc. Bull. 1974. V. 57. N 3. P. 120–126.

[106] Andreev N.S., Mazurin O.V., Porai_Koshits E.A., Roskova G.P., Filipovich V.N. Yavleniya likvatsii v steklakh, Leningrad: Nauka, 1974. Translated under the title Phase Separation in Glasses, Amsterdam: North-Holland, 1984.

[107] Mazurin O.V., Raskova G.P., Aver'yanov V.I., Antropova T.V. Dvukhfaznye stekla: Struktura, svoistva, primenenie (Two_Phases Glasses: Structure, Properties, and Applications), Varshal, B.G., Ed., Leningrad: Nauka, 1991.

[108] Braetsch V., Frischta G.H. Crystallization and Phase Separation in the System Li2O–SiO2 // Journal of Non-Crystalline Solids. 1987. V. 95–96. P. 457–462.

[109] Burgner L.L., Weinberg M.C., Symmons J.H. Early Stage Crystallization Kinetics of Lithium Disilicate and Sodium Silicate Glasses, Final Report of US Department of Energy Grant DE_FG03ER45500, Tucson, Arizona, United States: The University of Arizona, p. 85721;

[110] Ardell A.S. The Effect of Volume Fraction on Particle Coarsening: The Theoretical Consideration // Acta Metall. 1972. V. 20. N 1. P. 61–71.

3

Elaboration of a Specific Class of Metamaterial: Glass in Single Crystal

Bertrand Poumellec, Matthieu Lancry, Santhi Ani-Joseph,
Guy Dhalenne and Romuald Saint Martin
*Institut de Chimie Moléculaire et des Matériaux d'Orsay,
Université de Paris Sud 11, Orsay,
France*

1. Introduction

Metamaterials are composite materials, which can exhibit some interesting optical properties, which is usually not found in natural materials. Some examples of the properties that can be obtained and controlled in these materials are negative refraction and artificial magnetism [1, 2]. The wave vector k of a wave propagating through such a left-handed substance is anti-parallel to its Poynting-vector S. This remarkable property has far-reaching consequences. Usually these materials have some periodic or quasi-periodic structure. Such photonic metacrystals are composed of periodic assembly of dielectric or metallo-dielectric structures that are designed to affect the propagation of electromagnetic waves. In particular, they should have refraction indices significantly different. The assembly can be unidimensional as in periodic multilayered materials, or two-dimensional with periodically spaced rods or fibers aligned in arrays, or 3D like silica spheres in Opal.

When the characteristic length scale of constituting elements is much smaller than the radiation wavelength, they are called photonic crystals (PC). This is because these structures avoid diffraction. The electromagnetic wave passing through the PC will feel only the effective parameters of the material such as the effective magnetic permeability, effective permittivity, etc. They therefore can in good approximation be considered as continuous media with some effective electrodynamic properties. For these reasons, metamaterials do not necessarily possess a photonic band gap (PBG), which is a striking difference compared to PCs. Nevertheless it is interesting to look for the possibilities of opening up a PBG, for extensive applications and also for better insight into the optical properties of the material [3, 4].

There are various methods to fabricate ordered composites like metamaterials. They are collected, for instance, in the book Nanophotonics by Prasad [5] or in Chemistry of Nanomaterials by Rao et al.[1]. They are classified into two categories, those, which are carried out by engineering, and those, which are obtained by self-organization. In the former category, there are various techniques including nanolithography, the deposit of the multi-layer films (CVD, Langmuir-Blodget) etc. The most commonly practiced method for preparing these materials are largely based on lithography, which uses a pre-fabricated photo mask as a master from which the final pattern is derived. Self-organization is an unconventional way of manufacturing photonic crystals by purely chemical method and it would open up new fields of application avoiding the problems related to engraving. Especially, in self organization, no one has described a way of synthesis starting from high temperatures, although it has some advantages like low cost, large volume and weak OH content (absorbing in the near-IR range). This last advantage can be made profitable to obtain materials compatible with the existing optical fiber networks.

In the past, researchers [6-12] showed the possibility of obtaining such ordered structures starting from eutectic compositions. Eutectics correspond to particular points in the phase diagram between the liquid phase and a solid phase, the composition of the liquid is fixed but it decomposes into two solid phases of different compositions, on cooling. A microstructure appears, but very often it is disordered. Nevertheless for some sets of elaboration parameters, ordered microstructures settle. Those can be defined by the kinetics of the growing environment, speed of cooling and also by some intrinsic parameters like diffusion, chemical kinetics of phase formation etc.

In this paper, we report the use of this method for synthesizing a new category of composites: equidistant glass fibers in a single crystal matrix. This is carried out at high temperature, by a technique called 'floating zone'/'melting zone' associated with an image furnace [7, 10]. The apparatus is described in Figure 1. A floating zone is established between a single crystal (the seed) and a polycrystalline rod (the feed rod) by using the radiative heat provided by two halogen lamps in an image furnace. These ways of development are essentially adapted to the development of large volume of single crystal (see Fig. 1).

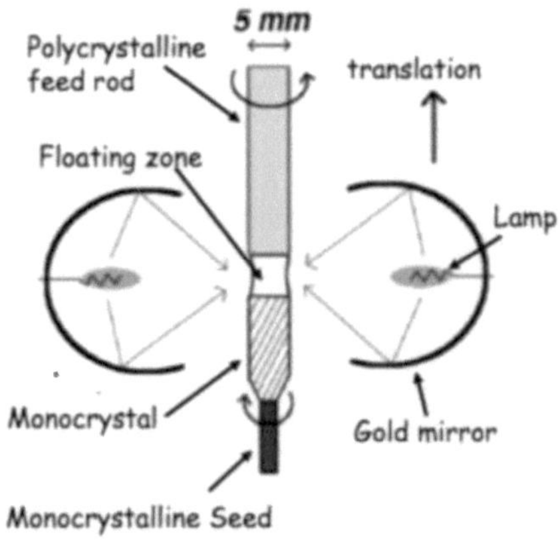

Fig. 1. Scheme of the floating zone apparatus using a double image furnace.

2. Synthesis route

The system that we investigated is $CuO\text{-}SiO_2\text{-}GeO_2$ around the composition $CuGe_xSi_yO_{1+2(x+y)}$ with $1 \leq x+y \leq 1.4$ and $0 \leq y/x \leq 0.4$. $CuGeO_3$ is an inorganic crystal that was studied in our laboratory for other properties [6, 10]. This crystal is orthorhombic with a space group Pbmm and the crystalline parameters: a = 4.79Å, b = 8.49 Å, c = 2.94 Å. It can dissolve Ni/Mn on its network of octahedral Cu and Si on its tetrahedral Ge network. Even though it can dissolve Si into its atomic tetrahedral chain, there is a limit (typ.: a fraction of a few %) in the atomic fraction of Si which can thus replace Ge, beyond which it will not produce a single phase on solidification. We discovered this while looking for methods to obtain a self-organized composite. Moreover, as we increase the atomic fraction of Si in the starting composition, the nature and quality of the crystal varies. In fact, we could not confirm if we really have a eutectic or not in our case. This is because, a detailed chemical analysis by the electron microprobe method [6], both on the crystalline matrix and on the fibers, confirmed the presence of Si at these two locations. If it were a eutectic, more chances would be that Si would not replace Ge in the crystalline matrix, instead stand alone as silica fibers.

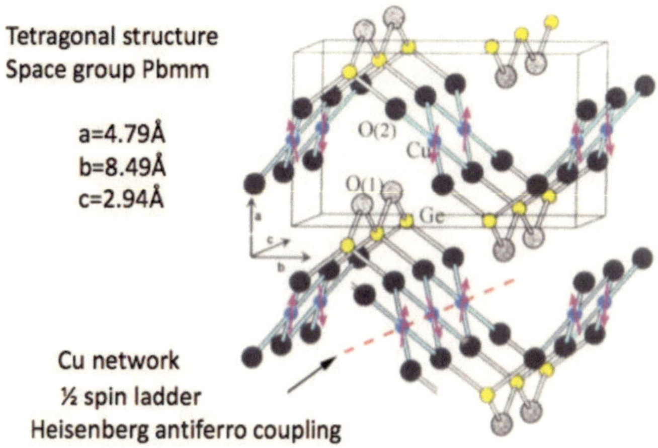

Fig. 2. $CuGeO_3$ crystallographic stacking.

Phase diagram from Figure 3(a) shows that the cooling of $CuGeO_3$ from liquid phase is congruent. So, there is no phase separation in the pure compound. With silicon dioxide substitution a priori to GeO_2, the pseudo ternary diagram Figure 3(b) leads to expect that this property is conserved until 50% but our observations contradict this result. On the other hand, the phase diagram in Fig. 3 shows the existence of a multiphase domain limited by the single phase line $CuGe_{1-x}Si_xO_3$. The idea is thus to exploit this domain for inducing the second phase precipitation more strongly. The compositions of the starting material were $CuGe_xSi_yO_{1+2(x+y)}$ with x between 0.89 and 1.2 and for y between 0 and 0.3 and $1 \leq x+y \leq 1.4$, $0 \leq y/x \leq 0.33$ (see Figure 4). They break up from the liquid state into a monocrystalline matrix, incorporating some precipitates.

Crystal growth technique: The technique of crystalline growth known as 'floating zone method' (see Figure 1), that we used here, makes it possible to obtain monocrystals of a few cm long without stacking faults. By the way, the advantage of this technique is the possibility to grow high-purity single crystals since the melt has no contact with a crucible during the experiment. Homogeneous and dense feed rods were prepared using >99.99% pure powders of CuO, GeO_2 and SiO_2 as described in reference [7]. Required amounts of these powders were accurately weighted and mixed thoroughly. This mixture was then pressed into cylindrical rods of about 4mm in diameter and 8 cm in length under isostatic pressure of about 2500 bar. The feed rods, were then sintered in air at 1050°C for 24h. All composites (varying in the composition x and y) were grown from a pure $CuGeO_3$ monocrystalline seed aligned along its 'a' axis that defined the growth axis (see Figure 5). Growth was carried out with the 'a' axis of $CuGeO_3$ along the growth direction. The essential idea of this crystal growth technique is to establish a liquid zone between the single crystal seed and the polycrystalline feed rod. The vertical high-temperature gradients required to stabilize the molten zone are ensured by focusing the image of two halogen lamps with two ellipsoidal mirrors. The molten zone temperature is precisely controlled by adjusting the lamp electric current magnitude. The crystal growth was carried out in an enclosed quartz tube where 2-atm oxygen pressure was established in order to prevent vaporization of CuO from the molten zone as much as possible. The growth rate (0.6 to 9 mm/h) is the rate at which the ensemble moves down from the furnace and this has a major influence on the characteristics of the grown crystal. The slow relative displacement of the heat source with respect to the liquid zone makes it possible to force the solidification of the fluid on the correctly directed monocrystal and thus the solidified material presents a continuous crystallographic character. This method yields single crystals of $Cu(Ge_{1-x}Si_x)O_3$ as well as two phase composite in the vicinity of the monocrystalline domain.

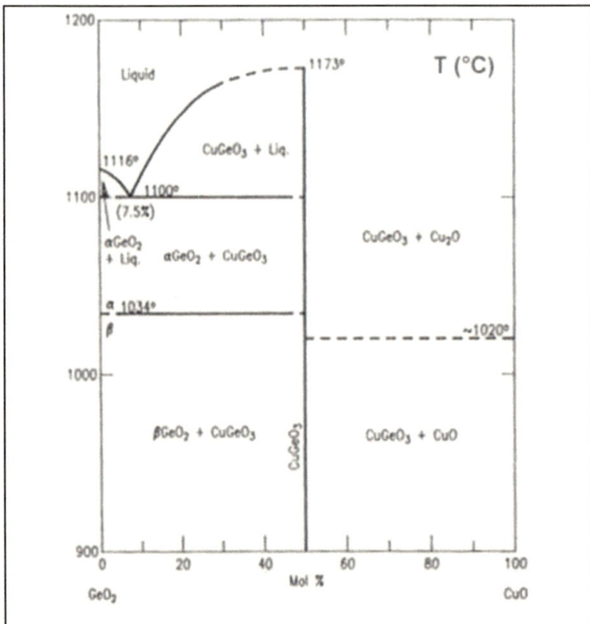

(a): CuO-GeO_2 binary phase diagram [12].

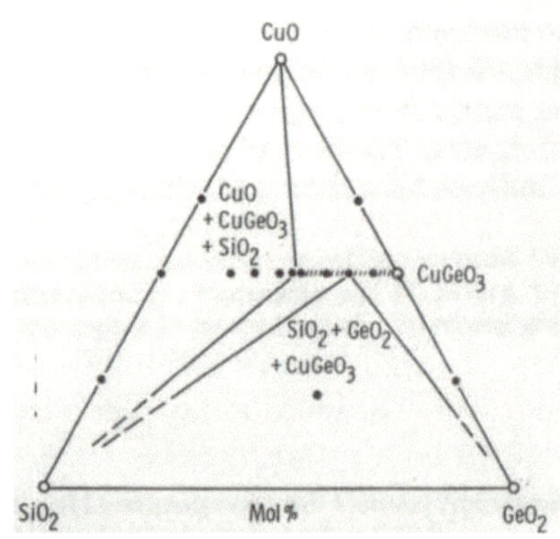

(b): Pseudo ternary diagram of CuO-SiO_2-GeO_2 at 600°C and 1kb. Lines are only indicative. [13]

Fig. 3. Phase diagrams for $CuGeO_3$.

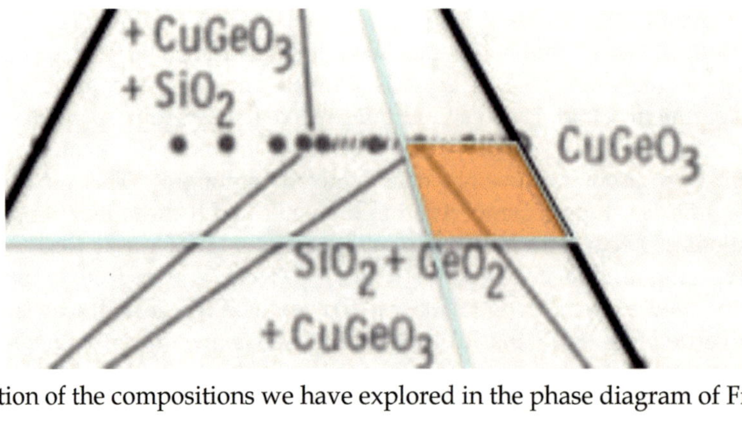

Fig. 4. Location of the compositions we have explored in the phase diagram of Figure 3

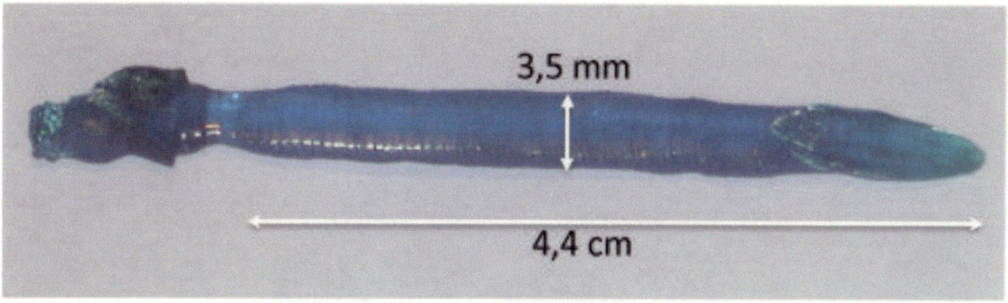

Fig. 5. Enlarged photograph showing the typical size of a crystal having the composition $CuGe_{0.915}Si_{0.085}O_3$. Growth rate 0.75-1.3 mm/h. 'a' axis is the growth axis. On the right hand, we can see the seed that defined the growth axis.

The growth rate has an important influence on the final structure of the crystal that evolves from the system. By carrying out the experiments at different rates, we conclude that, at this particular composition it is not possible to obtain ordered fibers if the rate exceeds 4 mm/h. However, there is a trade-off between the periodicity and the size and spacing of these fibers, as we increase the growth rate. Even though the size and spacing becomes smaller with increasing rate, it becomes more disordered. Therefore, we fixed this value at 1.3 mm/h and it was possible to obtain single crystals of about 14-cm long. The optimum growth rate varies with the composition, stoichiometry, etc. In this case also, an increase in the Si content beyond a limit (10%) results in more chaotic distribution of the fibers even if it is advantageous in obtaining smaller fiber diameter. It also restricts us in the range of feasible growth rates. Therefore, it is the compromise between the Si content and the growth rate that acts as the limitation to the minimum fiber size and spacing obtainable by this method. By working at slightly over stoichiometric composition, we are able to reduce this size up to 0.6 μm with spacing as small as 1.5 μm. What we observed was that, beyond a limit, the increase in Si will inhibit the ordered formation of silica fibers in the matrix.

3. Characterization

The crystals are deep blue in color (see Figure 6), translucent, and grow with an elliptical cross section, with the b and c axes of the structure being the minor (0.30 cm) and major (0.75 cm) axes of the ellipse, respectively. They can be easily cleaved along the (100) plane. Characterization of the samples was done by optical microscopy, SEM (secondary electrons and backscattered electrons), X-ray diffraction (XRD), EBSD (electrons backscattered diffraction), electron microprobe analysis, Raman micro-spectroscopy.

3.1 Optical microscopy and SEM

The obtained composites were cleaved and thin slices were observed by optical microscopy using polarized light. Precipitates are easily seen when they are larger than the micron. This indicates that the refraction index difference is large enough. Upon rotation of the sample about the 'a' axis, the brightness of the matrix varies whereas that of the rods does not change, indicating that the matrix is not isotropic but the rods are. The optical microscopy always showed two phases and not 3 for some composition as it is shown in Fig. 3b. It shows also the regularity of the precipitates, the heterogeneity of the distribution in some regions.

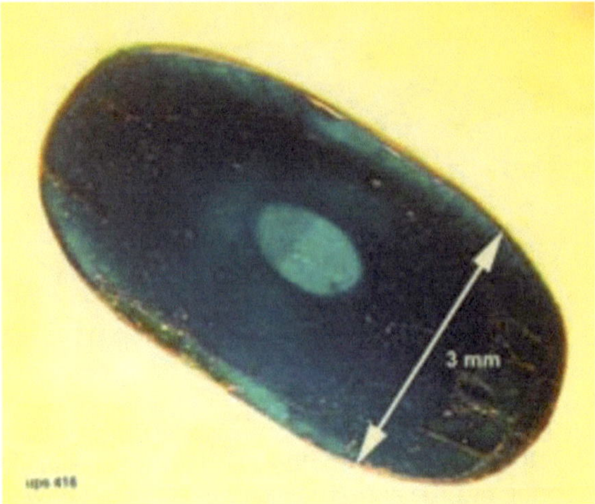

Fig. 6. Global view of a sample cleaved perpendicularly to its growth axis, the « a » axis. The sample composition was $CuGe_{0.915}Si_{0.085}O_3$ and it was grown at the rate 0.75-1.3 mm/h.

As we can see in Figure 6, the samples show concentric heterogeneities. There is a central region (brighter) containing precipitates. There is an external region clear from precipitates. There are regions with ordered precipitates of the same size e.g. the central region (appearing brighter in Fig. 6), and region rather disordered with various sizes and distances e.g. intermediate region. The outside part near the surface of the crystal never shows precipitates. Close to the borderline between intermediate region and external one, precipitates are aligned along equally spaced lines. The lines are more distant than in the central region.

Figure 7 shows precipitates in the different region of the sample. In the central region, there is only a short range order. When moving to the outside, the precipitates` distribution varies. The distance and size increases. When reaching the surface, they disappear. In this region fibers are no more perpendicular but they are all tilted at the same angle. This is probably due to the curvature of the solid-liquid interface characteristic of the floating zone technique. Then, further from the center, the arrangement is more chaotic as we can see in Figure 7b. Figure 8 show the regularity of the precipitates in the central zone that is about 1 mm in diameter, in four of the numerous crystals that we elaborated. The smallest size and distance that we have obtained in the phase domain investigated is 0.6 and 1.4 μm respectively (see Figure 8d). Further observations by means of optical microscope in transmission and by successive cleaving perpendicular to 'a' axis showed that, in the central zone, these precipitates were long, columnar and perpendicular to the cleavage surface thus 'a' axis is along the crystal rods. They are therefore analogous to fibers.

One of the best ordered samples showing nearly periodically distributed fibers, was selected for the observation along a longitudinal section containing 'a' axis, to analyze whether these fibers run uniformly along the entire length of crystal without crossing over. A small-cleaved piece of sample was cut into two pieces. It was then fixed in epoxy resin and polished using SiC discs of grain sizes 600 and 1200 mesh for 10 minutes each and then using felt discs and diamond powder of granule sizes 6 μm, 3 μm, 1 μm and 0.25 μm respectively for 10 minutes each. This was then observed using the optical microscope and the result is given in Figure 9. The photographs show that the fibers are quite uniform and grow along a great length of the crystal. It was observed that the length of the fiber as revealed from the photograph is almost 750 μm. The diameter ϕ of the fibers in this photograph from the sample with initial composition $CuGe_{0.915}Si_{0.085}O_3$ is around 4 μm and is more or less uniform throughout the length.

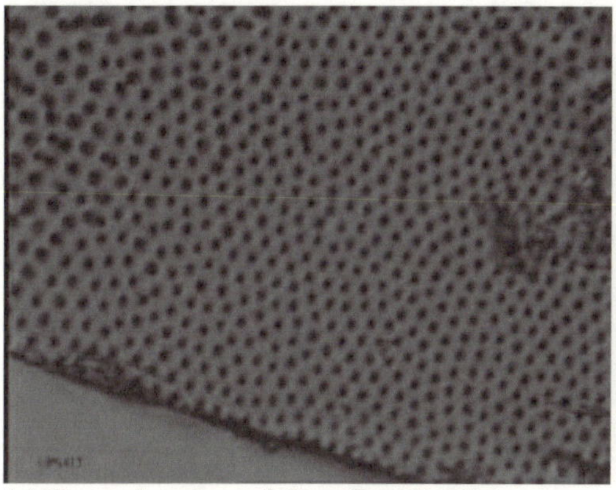

(a) Central region of $Cu_{1.0}Ge_{0.9}Si_{0.1}O_{3.0}$, 1mm/h (UPS417b) Precipitate distance is 4.1 microns, their size is 2.5 microns

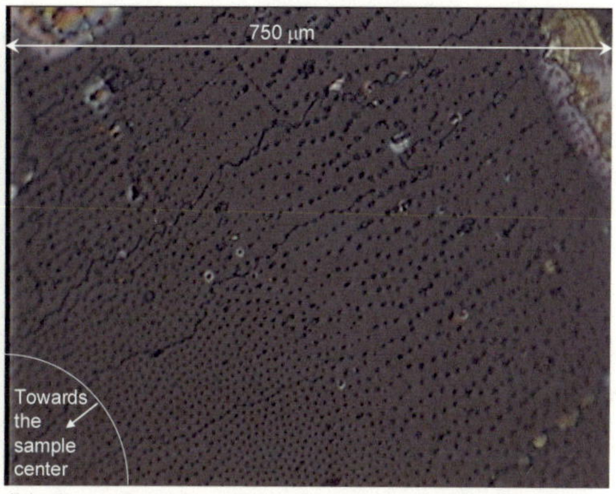

(b) $Cu_{1.0}Ge_{0.9}Si_{0.1}O_{3.0}$, 1mm/h (UPS417a2)

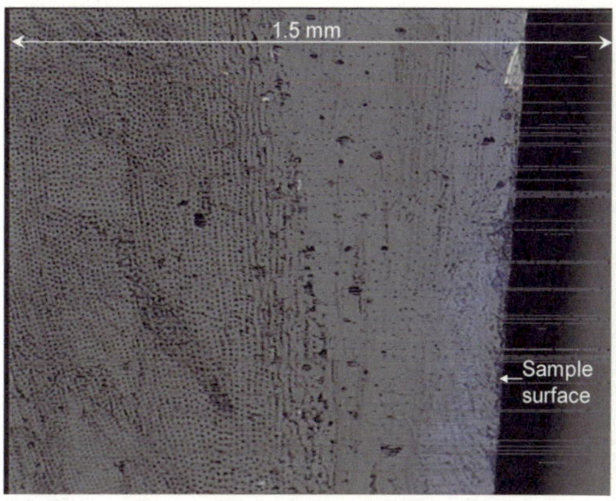

(c) $Cu_{1.000}Ge_{0.915}Si_{0.085}O_{3.000}$ (UPS416)

(d) Intermediate region of $Cu_{1.000}Ge_{0.915}Si_{0.085}O_{3.000}$ (UPS416) 0.75-1.3 mm/h, size 2.5 microns, distance 11.5 microns

Fig. 7. Example of the structure of the samples: (a) typical image of the central region of the samples. Fibers are ordered on a short distance. (b) When moving out of the center, precipitate distance and size increase (c) when we are going closer to the surface the precipitates disappear. (d) In the intermediate region corresponding to c, the fibers appear tilted.

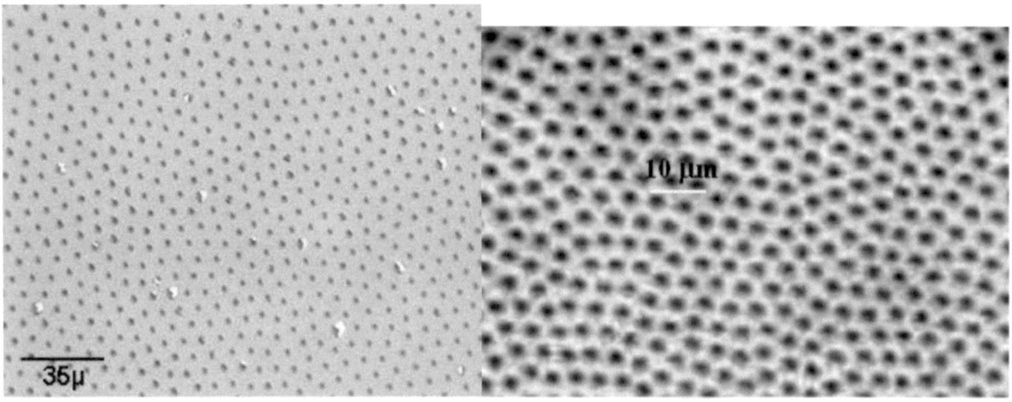

(a) Sample $Cu_{1.000}Ge_{0.915}Si_{0.085}O_{3.000}$ grown at 1 mm/h. The average distance between precipitates is 10 µm and the precipitate size is 4 µm.

(b) Sample with average distance between precipitates is 5 µm and the precipitate size is 3 µm.

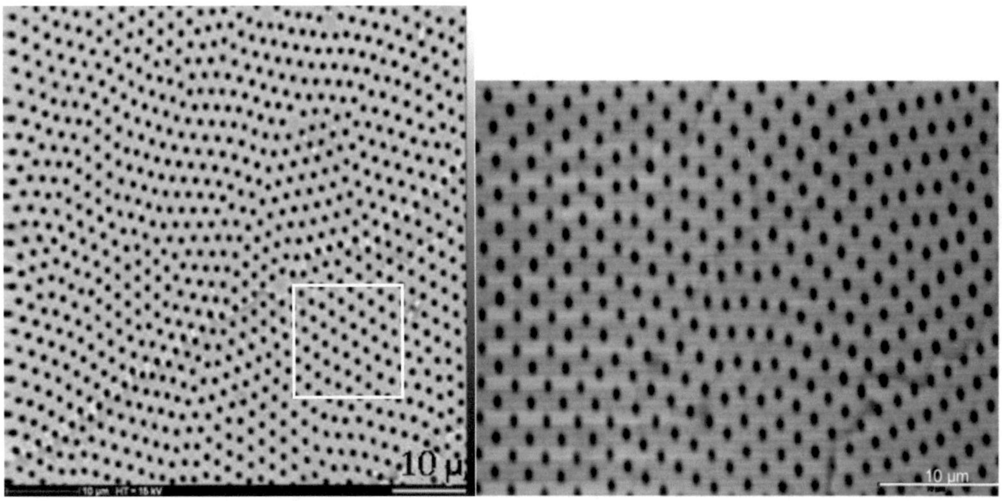

(c) Sample $CuGe_{0.9}Si_{0.2}O_{3.2}$ grown at 2 mm/h (UPS430), fiber distance is 2.5±0,3 microns and diameter is 1.2±0,13 micron (filling factor 0.18).

(d) Sample $CuGe_{1,1}Si_{0.3}O_{3.8}$ grown at 2 mm/h (UPS436), fiber distance is 1.4±0,3 microns and diameter is 0.65±0,13 micron (filling factor 0.17).

Fig. 8. Variation of the central region of the samples when elaboration parameters are varied (SEM-backscattered electrons).

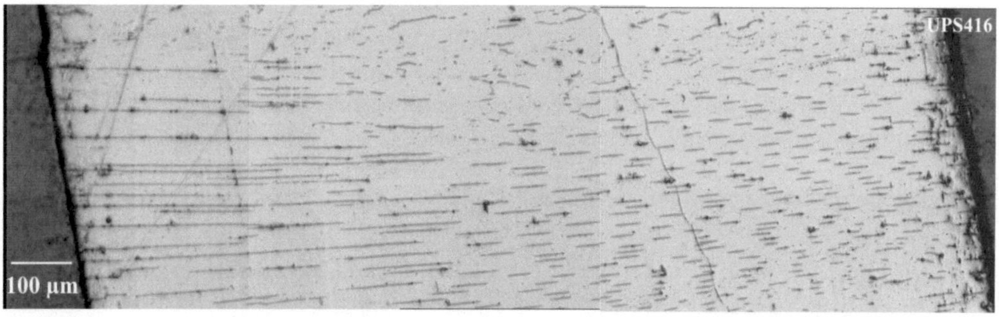

Fig. 9. Three images of the longitudinal section of the crystal were taken and carefully joined together to see the extension of the fibers running without distortion. The maximum length is around 750 microns and the fiber diameter is around 4 microns. Of course, as it is not possible to be exactly parallel to fiber axis, they are often cut during polishing. Sample $Cu_{0.20}Ge_{0.183}Si_{0.017}O_{0.60}$ (UPS416) growth rate 0.75-1.3 mm/h, fiber size 2.5µm, fiber distance 11.5 microns has been used here.

3.2 SEM-SE on tilted samples

Secondary electrons in SEM have a large field depth that is convenient for imaging on a large scale the surface of the sample that shows some topographies. However, it is necessary to tilt the sample sharply. Some images are displayed in Figure 10.

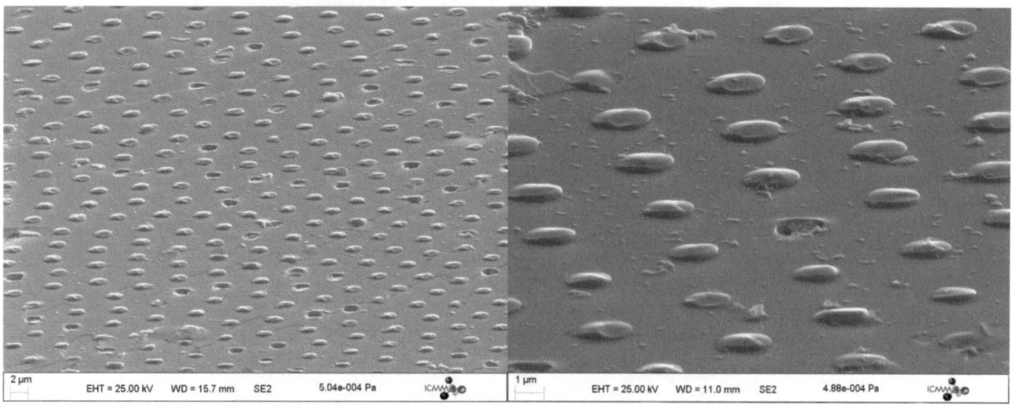

Fig. 10. Secondary electron image of the surface of a cleaved sample tilted under 70° upward. The fibers appearing here elliptic due to the tilt are actually circular.

As we can see, the fibers seem to protrude the surface of the sample. We checked using a phase shift interferometer whether it is actually the case. This observation can be made on both faces of the cleavage. Sometimes, fibers are broken below the surface level but commonly, they protrude. This means that a stress relaxation occurs and that they were in compression before cleaving. This can be due to smaller thermal expansion coefficient for the fiber than for the matrix.

In Fig. 11, the surface of the sample has not been perfectly cleaved and allows us to view the layer structure of the single crystal along an edge. It also reveals the channels where the fibers were but in this case they disappeared during the cleaving. However, in the second image, a fiber remains showing that the crystallization around the fiber is not disturbed and the interface is very well defined. A transition region is not appearing. However, in Figure 12, we see better that a ring is appearing around the fiber compatible with a compression effect.

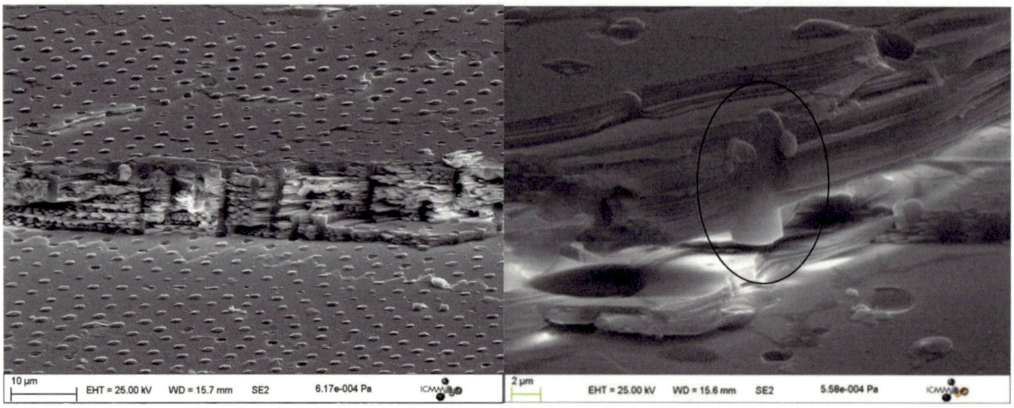

Fig. 11. Secondary electron image of surface of a sample not perfectly cleaved and tilted 70° upward. The ellipse in the photograph on the left shows a fiber end.

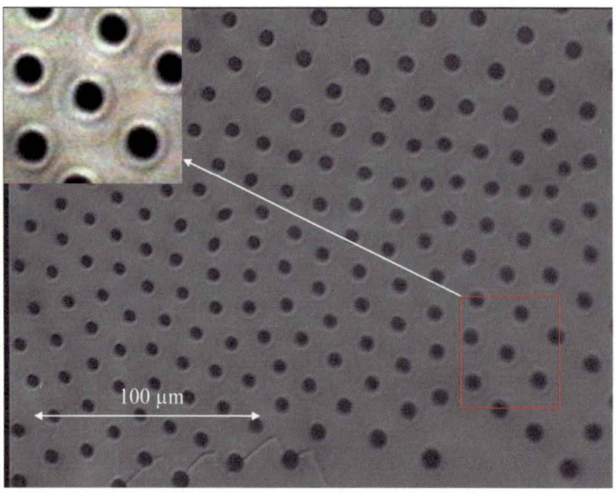

Fig. 12. This picture shows that around the fiber, a contrast is detected that seems to show a compression effect in agreement with the topography seen in SEM (sample UPS416).

3.3 X-ray diffraction

We have performed X-ray diffraction experiment (see fig.13) in order to identify additional lines among those of the well known cuprogermanate. Surprisingly, we detected the presence of only one crystalline phase, in contrast with the two phases observed by optical microscopy or SEM-EBS. This leads to think that the fibers are not crystallized.

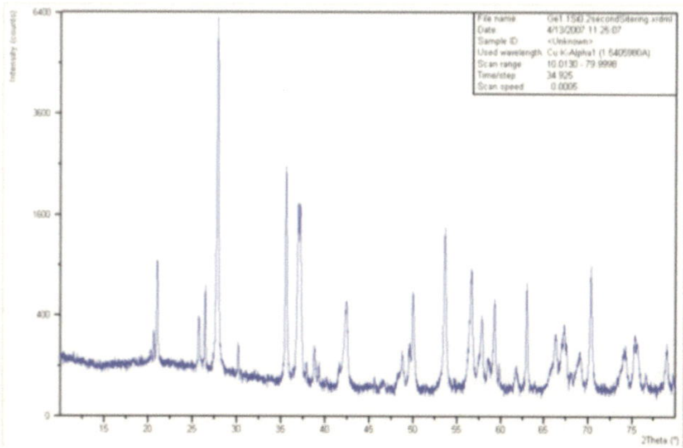

Fig. 13. X-ray diffraction pattern of a typical sample of $CuGe_xSi_yO_{1+2(x+y)}$. All the peaks in this spectrum can be labeled from $CuGeO_3$ space group.

3.4 SEM-EBSD (complementary observations, glass fibers are vitreous)

EBSD has been used to show that all fibers do not diffract the electron beam. The electron beam is incident on the sample under at an angle 70°. Electrons diffracted by crystal planes form Kossel cones. They are intercepted on a plane where they organized into lines called Kikuchi lines. Their indexation can lead to finding of crystal orientation if the space group and unit cell parameters are known. In our case, just the existence of diffraction is searched for. The resolution of the SEM-FEG used is good enough to analyze only one fiber. The accelerating voltage was 5kV and the working distance 10 mm. As we can see in Fig. 14 precipitates appear black. They do not diffract the backscattered electrons. We can thus conclude that the precipitates are vitreous.

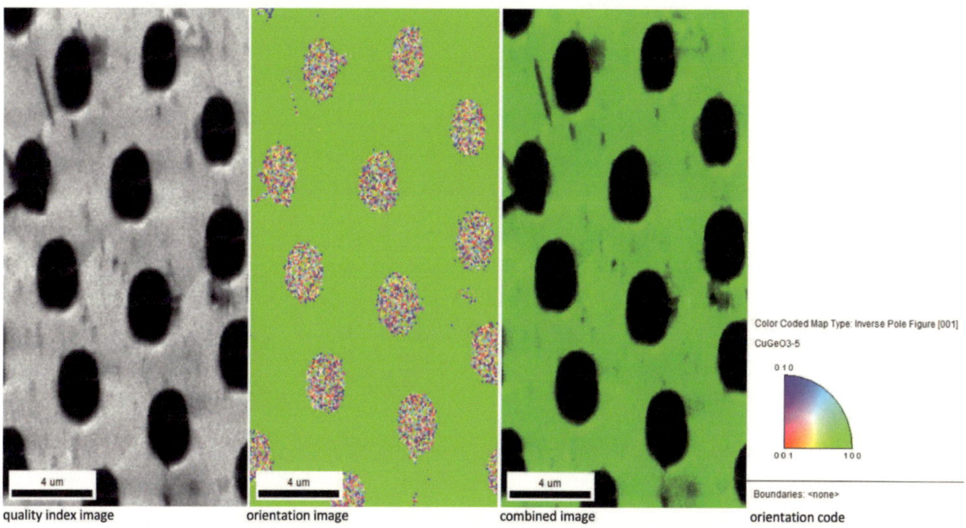

Fig. 14. Electron Backscattered Diffraction images. (a) EBSD image rebuilt from quality index of the Kikuchi diagrams (black = no diffraction). Precipitates appear actually black. The matrix has uniform brightness. (b) Orientation image, the matrix with uniform green color is a single crystal actually. (c) Combination of the two previous images.

3.5 Raman micro-spectroscopy analysis

Raman micro-spectroscopy can be used to identify a compound by its spectroscopic signature. It is based on the inelastic scattering of usually visible electromagnetic waves (energy exchange between the optical wave and the material vibration waves. Raman spectroscopy was performed in confocal mode with a pinhole of 300 microns and an objective of magnification 100x. The source of excitation was the 514 nm emission from an Ar^+ laser. To record the Raman spectra, the beam was focused on the surface of the material. Under these conditions, the lateral resolution is of the order of the micron whereas the depth resolution is about 30 microns. The acquisition time for the spectrum was fixed at 60s. By taking measurements between 200 and 4000 cm^{-1}, it was observed that, for our materials, Raman features were located only between 200 and 1500 cm^{-1}. The Raman spectra were recorded on the reference sample $CuGeO_3$ and on the sample doped with 10 % Si/(Si+Ge). Figure 15 shows the spectra collected between fibers and on a fiber. The spectrum of the

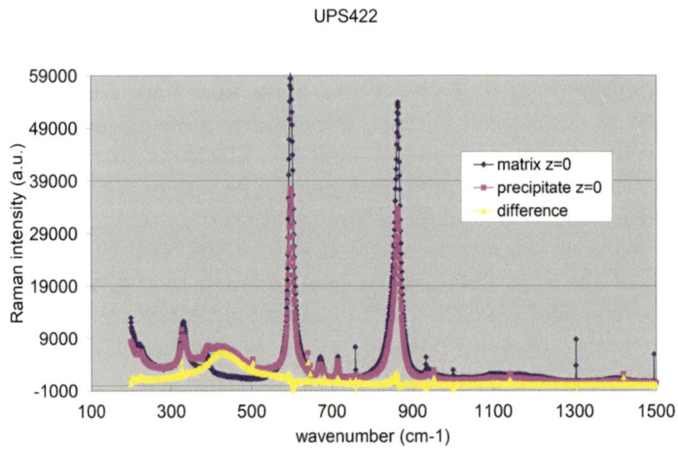

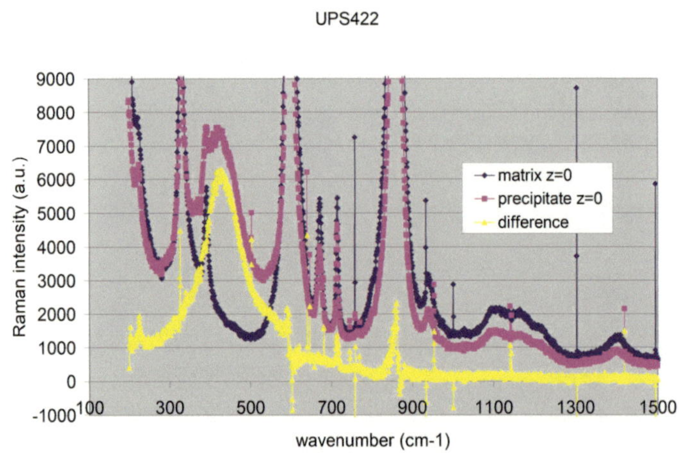

Fig. 15. Raman scattering spectra recorded at the surface of a cleaved sample with 10 % Si/(Si+Ge). One of the spectra has been recorded in the matrix surrounding the precipitate and the other one on the fiber itself. The third spectrum has been computed by weighted difference of the two others in order to eliminate the matrix spectrum as much as possible (UPS422). The second picture is an enhancement of the first one.

matrix, which is perfectly crystalline, is made of rather fine structures corresponding to the solid state vibration modes allowed in Raman scattering. Concerning the spectrum collected from the fiber region, even though the lateral resolution is larger than the fiber diameter, it can be seen however that it is identical to the spectrum of the matrix, except for an additional band around 430 cm^{-1}. To clearly express the difference between the two spectra, we performed a weighted difference between the spectrum of the precipitate and that of the matrix so as to eliminate, as much as possible, the lines from the matrix. One can thus clearly see the additional band in Figure 15, which has its maximum at 433 cm^{-1}. Its width is 107 cm^{-1}, which is significantly larger than the width of the crystal lines. It is thus possible to conclude that, in the fibers, there is no long-range order, in contrast to the crystals. Taking into account the results of the chemical analysis of the samples, it can be concluded that the fibers have a glass structure with an average composition intermediate between GeO_2 and SiO_2. We find the Raman spectra for this type of glass in the literature recently [14, 15], Raman spectra for SiO_2 and GeO_2 glasses are also well known. They present both a wide and intense band, respectively at 440 cm^{-1} and 416 cm^{-1}. Their respective widths are 234 cm^{-1} and 94 cm^{-1} [16]. The values that we measured are thus intermediate between these values in agreement with [15] and we can thus conclude from the Raman measurements that the probed precipitate is indeed vitreous.

3.6 Chemical composition by EPMA (it reveals that compensation occurs on the metal lattice)

In order to understand the phase separation, we performed chemical analyses. They were done by electron microprobe analysis (EPMA). Cu, Ge, Si and O, were analyzed independently. In this way, one could exclude measurements presenting a sum deviating from 100 wt %. A composition profiling was carried out in the zones containing precipitates. Monocrystal of $CuGeO_3$ and a silica blade were the references of measurement. Correct measurements were then converted into atomic percentage. The error bars were obtained by statistical average. For a global view of the content variations, we have recorded chemical profiles in dynamic mode. We can see large variations in Si and Cu profiles. It means that the profile is crossing a fiber. However, as we have no chance to cross exactly a fiber in the center, contrarily to static mode, the variation is never extremum. Of course, precise positioning has been achieved in the static mode. We made profiles on different samples with SiO_2 content from 8.5 to 30 mol %, only three measurements are shown below for demonstration.

3.6.1 In UPS 416, the nominal composition is $Cu(Ge_{0.915}, Si_{0.085})O_3$ $=Cu_{0.20}Ge_{0.183}Si_{0.017}O_{0.60}$

The external area, without precipitate, was measured separately in static mode, it showed the composition $Cu_{0.203}Ge_{0.191}Si_{0.014}O_{0.593}$. The ratio of the number of oxygen to metal (Cu, Ge, and Si) is a useful quantity for analyzing the differences in chemical composition at various regions. It is around 1.43 in this area instead of 1.5 in the perfect crystal. The external matter is thus slightly under-stoichiometric. One can also see, by comparing the chemical formula at this place with the nominal one, that it presents a loss in Si.

At the central area, the matter surrounding the fibers has the chemical composition: $Cu_{0.205}Ge_{0.197}Si_{0.015}O_{0.583}$ and here the ratio is 1.4. Thus this matter is also under-stoichiometric and again Si depressed (0.015 instead of 0.017). Under-stoichiometry undoubtedly corresponds to a natural reduction of Cu at high temperature during the elaboration process. The single crystal therefore should include oxygen vacancies.

The fibers show an average composition of $Cu_{0.05}Ge_{0.20}Si_{0.09}O_{0.66}$ with a ratio of oxygen to metal of 1.94. This is clearly a different chemical formula than the one for the matrix. This leads us to consider a formal composition of type MeO_2; like SiO_2 or GeO_2. In addition, we can also observe a clear enrichment of the net content in Si and a reduction in the quantity of Cu on comparison with the nominal composition. On the other hand, the sum of the oxygen quantities for saturating the metal content should yield 1.91 instead of 1.94. It can be thus considered that the compound is over-stoichiometric in oxygen. It should also be noted that the quantity of Si in the precipitates is higher than it can be obtained by a simple expulsion of CuO from the starting composition as can be seen by comparing the nominal composition rewritten in the form of $CuO+(Ge_{0.915}Si_{0.085})O_2$ with that of the precipitates $(Cu_{0.15}Ge_{0.61}Si_{0.27})O_2$. The chemical profile of Cu, Ge, Si and O made across the center (containing precipitates) and the surrounding area (clear from precipitate) is displayed in Fig. 16. It shows that O and Ge are almost constant along the profile. In contrast Si and Cu vary when a precipitate is crossed. We suspected a balance between only Si and Cu variation between the phases. The curve Cu+2Si actually confirms this observation. This curve does not show large variation anymore and suggests a valence charge balance between Cu and Si atoms.

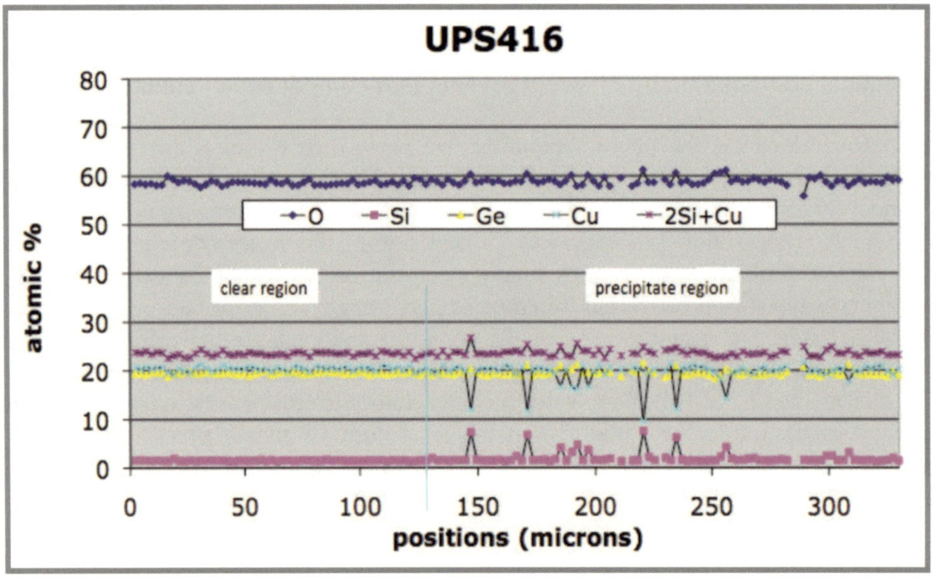

Fig. 16. Chemical profile of typical sample with the composition $Cu_{0.20}Ge_{0.183}Si_{0.017}O_{0.60}$. The region with precipitates is on the right hand side, the one without is on the left. Note the constancy of Ge atomic content from matrix and fibers. Note also the Cu variation, which is roughly the double of the Si one (Cu+2.Si curve).

3.6.2 In UPS 417, the nominal composition is $Cu(Ge_{0.900}, Si_{0.100})O_3$ $=Cu_{0.20}Ge_{0.180}Si_{0.020}O_{0.60}$

In this case also, several regions are analyzed (Fig. 17). The outermost region is without any precipitate and going towards the center; there is a region with disordered precipitates; a clear region and a small central region with ordered precipitates. The clear area exhibits a Si rich content with 2.5 to 3 at% and another area with a content poor in Si close to the biphased area. The average composition is $Cu_{0.189\pm0.007}Ge_{0.187\pm0.002}Si_{0.027\pm0.004}O_{0.597\pm0.004}$ with oxygen to metal ration of 1.48. The other one shows the composition: $Cu_{0.206\pm0.005}Ge_{0.185\pm0.004}Si_{0.017\pm0.001}O_{0.592\pm0.006}$ with oxygen to metal ratio of 1.45. In the central region, the matrix composition is $Cu_{0.203}Ge_{0.187}Si_{0.018}O_{0.592}$. The precipitates have the following composition: $Cu_{0.055}Ge_{0.197}Si_{0.104}O_{0.644}$ with an oxygen-to-metal ratio of 1.91. We observe thus similar contrasts between matrix and precipitates similar to the previous sample.

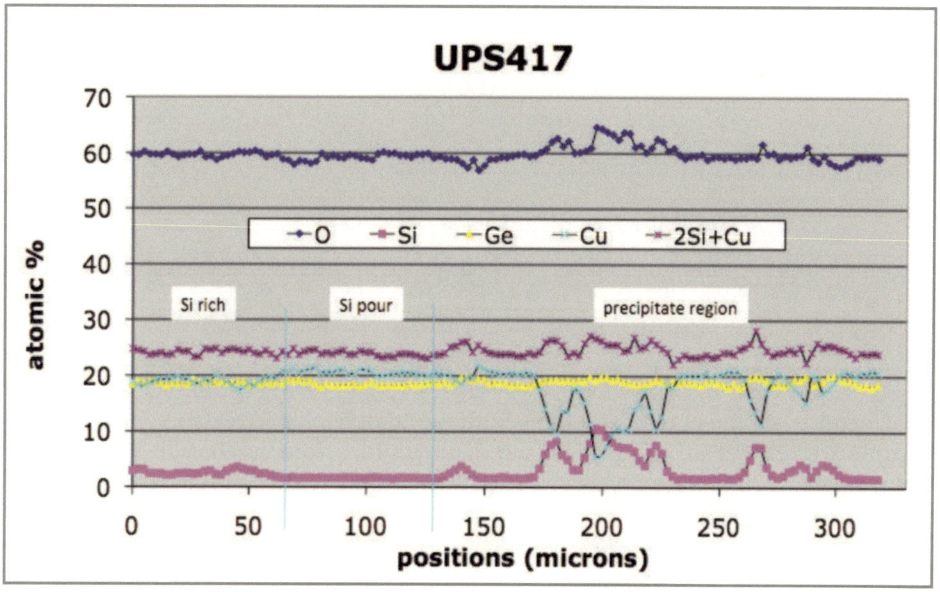

Fig. 17. Chemical profile of typical sample with the composition $Cu_{0.20}Ge_{0.180}Si_{0.020}O_{0.60}$. The region with precipitates is on the right hand side, the one without is on the left. Note the constancy of Ge atomic content from matrix and fibers. Note also the Cu variation, which is roughly the double of the Si one (Cu+2.Si curve).

3.6.3 In UPS 423, the nominal composition is $Cu(Ge_{0.800}, Si_{0.200})O_3$ $=Cu_{0.20}Ge_{0.160}Si_{0.040}O_{0.60}$. This sample contains precipitates but they are disordered

In the central region, the matrix appears with composition: $Cu_{0.210\pm0.005}Ge_{0.177\pm0.003}Si_{0.028\pm0.001}O_{0.580\pm0.004}$ with oxygen to metal ratio of 1.4 and with an depletion in Si compared to the nominal composition. The average composition of the precipitates is $Cu_{0.125\pm0.015}Ge_{0.160\pm0.004}Si_{0.11\pm0.01}O_{0.62\pm0.01}$ with oxygen to metal ratio of 1.57. Compared to the previous sample, here, the nominal composition contains more Si and therefore the precipitates contain more silicon too but the exchange is not in proportion. For starting compounds with 8.5 and 10% mol $SiO_2/(SiO_2+GeO_2)$, the fiber contains about 27% of Si/(Si+Ge+Cu), 59% of Ge/(Si+Ge+Cu) and 14% Cu/(Si+Ge+Cu). In this sample with 20% mol $SiO_2/(SiO_2+GeO_2)$ in the nominal compound, the fiber is made with 28% Si/(Si+Ge+Cu) and 41% Ge/(Si+Ge+Cu) but 31% Cu/(Si+Ge+Cu) i.e. more than twice the Cu content of the previous samples. Therefore, the oxygen to metal ratio does not reach 2 and even it is under-stoichiometric in oxygen. Nevertheless, the Ge content remains the same in all parts of the sample. However, the Cu/Si balance is now degraded (see Cu+2Si curve in Figure 18). What we observe also, is that the oxygen content does not show variation in proportion for balancing. We can suspect an electric charge of the fiber.

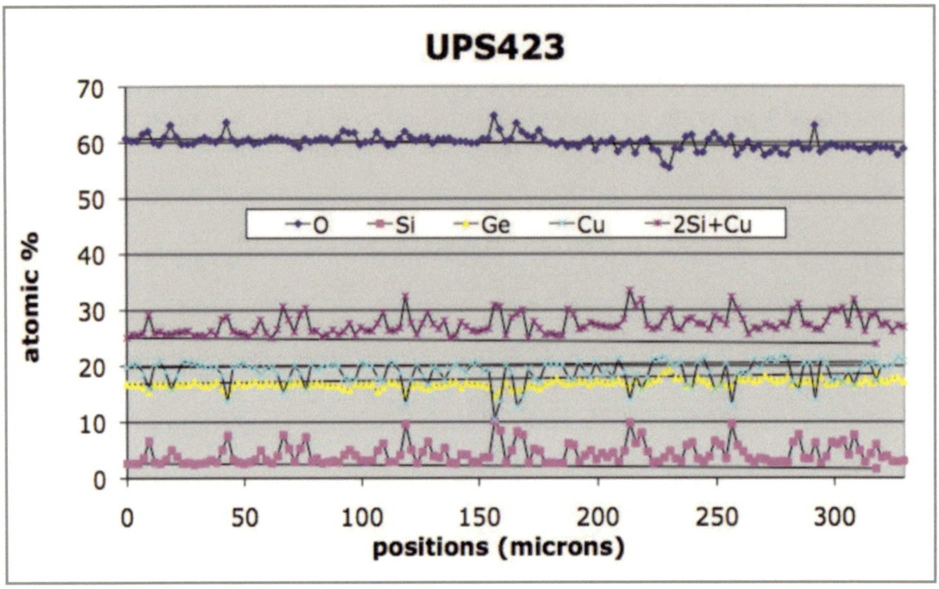

Fig. 18. Chemical profile of typical sample with the composition $Cu_{0.20}Ge_{0.160}Si_{0.040}O_{0.}60$. In this sample precipitates are everywhere across the profile. Note the constancy of Ge atomic content from matrix and fibers. Note also the Cu variation, which is roughly the double of the Si one (Cu+2Si curve).

3.6.4 Discussion and conclusions on the chemical observations

When precipitates are correctly ordered, we observed an over-stoichiometry in oxygen in fiber and an under-stoichiometry in the matrix. Although all the samples exhibit heterogeneities, the chemical composition of the cuprogermanate phase is rather homogeneous and close to the nominal one. Some small fluctuations in the Si content are nevertheless observed. They are compensated by Cu content variations.

For all the samples studied, which contain varying amounts of Si, we observed that the fibers are vitreous and have a $(Si+Ge)O_2$ type composition containing a small amount of CuO whereas the matrix has a composition close to the nominal one. We also observed that the fibers present a clear enrichment in Si in comparison with the surrounding matrix. More surprising is that this exchange does not disturb the Ge content. Quantitatively, the GeO_2 proportion is constant but the SiO_2 variation is 1/2 the one of CuO i.e. 2 CuO molecules in matrix structure is replaced by 1 molecules of SiO_2 in precipitate when they are ordered. We observe that the oxygen content increases consistently (we remind that oxygen content has been measured independently of the other constituents). The most external region is slightly silicon depressed when the central region is ordered. The constancy of Ge throughout the sample is remarkable. This is not the case for Si and Cu. Even more surprising is the balance between Cu and Si. There should be no constraint on oxygen diffusion in the liquid and so we expected no relation between Cu and Si. Therefore, the existence of such a relation means that oxygen migration is much lower than Cu and Si.

Now, one has to find an answer to three questions in order to master metamaterial elaboration: why is a second phase formed, why does it appear as micro rods (or fibers) and why are these rods regularly distributed?

For a ternary system like the one we used ($CuO-SiO_2-GeO_2$), when the starting composition falls into a 3-phase domain, the composition of the phases is defined by the intersection of a complex line (the connodal) with the domain boundaries. In this case, the molar proportion of each phase is defined by the position of the starting composition from each phase boundaries along the connodal, there is no free parameter. It is different with 2-phases case, the compositions of each phase depends on the starting composition. The phase molar proportion is defined by other constraint (not thermodynamically) e.g. the constancy of the Ge content whatever the phase.

In our case, we obtain a two phase composite with varying phase compositions when we change the starting composition. This is in principle, in agreement with the phase diagram of Breuer et al. [17]. However, the compositions of our phases are not the expected ones. We should obtain a combination of the $CuMeO_3$ phase and an almost pure GeO_2 phase. Instead, the fibers contain about 30% of Si, 40-60% of Ge and 15-30% Cu. We thus deduce that the formation of a vitreous phase allows germanosilicate glass to dissolve CuO and changes significantly the ternary $CuO-SiO_2-GeO_2$ phase diagram in which the two-phase domain is extended (see Figure 19).

The rod-type shape of the second phase and the regularity of the microstructure can then be explained as characteristic of a eutectic structure resulting from the decomposition, upon solidification, of a liquid into two solid phases. It is the right compromise between the thermodynamical energy and the surface energy. In such a eutectic transformation, taking place in a temperature gradient like the one available in the growth technique used in this study, the resulting aligned microstructures found are generally either lamellar or fibrous, depending on the volume fraction of the minor phase: when this fraction is smaller than about 30% the microstructure is rod-shaped, which is obviously the case here. Above 30% it is lamellar [18, 19]. The space period and also the size of the precipitates changes in the same samples and also the size but not continuously. There are region with precipitates of the same size and rather region rather disordered with various size and distance. When the size is constant, this means that the theory above is applicable.

The last question is why there is a spatially regular distribution? Some interpretation is given by Flemings [20] for eutectic solidification but can be applied more generally to our case. Decomposition of the liquid into two solid phases by eutectic reaction is limited by diffusion and reaction rate. This leads to a constrained control of the size and period of the minority phase distribution. In order to understand the relevant forces that lead to this situation, we tried a rationalization of our observations on a kinetics basis. We studied the differences in volume chemical content between the two solid phases (Table 1). Knowing the number of molecules per cm^{-3} of each phase ($CuMeO_3$ for the matter surrounding fibers and germania-silica glass for fibers themselves), we computed the changes per unit volume of each species in Table 1. The concentration of germanium and oxygen decreases only a little. This is apparently in contradiction with the atomic fraction (second column of Table 1) but change in atomic fraction does not mimic directly the change in the unit volume (third and above columns in Table 1) as we have to take into account the molecule volume. About the difference in Cu and Si, they appear complementary really in a ratio of 1 Si to 2 Cu. Therefore, we can conclude that during solidification of the liquid phase, there is an exchange of Cu and Si, on one hand, and a slight decrease in GeO_2 content in the fiber compared to the matrix.

Sample UPS417 $Cu_{0.20}Ge_{0.18}Si_{0.02}O_{0.60}$	Chemical formula (atomic %)	Number of atoms 10^{22} per cm^3				
		Cu	Ge	Si	O	Total
Matrix composition	$Cu_{0.203}Ge_{0.187}Si_{0.018}O_{0.592}$	1.70	1.56	0.15	4.94	8.35
Fiber composition	$Cu_{0.055}Ge_{0.197}Si_{0.104}O_{0.644}$	0.39	1.39	0.74	4.56	7.08
Difference (matrix to fiber)	$Cu_{-0.148}Ge_{0.010}Si_{0.086}O_{0.052}$	-1.31	- 0.15	0.594	- 0.33	-1.27

N.B.: There are 1.67 10^{22} molecules per cm^3 for $CuGeO_3$ crystal and 2.36 10^{22} molecules per cm^3 about for $1/3SiO_2+2/3GeO_2$ glass.

Table 1. Atomic content in each phase and their difference.

Sample UPS416 $Cu_{0.20}Ge_{0.183}Si_{0.017}O_{0.60}$	Chemical formula (atomic %)	Number of atoms 10^{22} per cm^3				
		Cu	Ge	Si	O	Total
Matrix composition	$Cu_{0.205}Ge_{0.197}Si_{0.015}O_{0.583}$	1.71	1.64	0.125	4.87	8.35
Fiber composition	$Cu_{0.05}Ge_{0.20}Si_{0.09}O_{0.66}$	0.354	1.43	0.637	4.67	7.08
Difference (matrix to fiber)	$Cu_{-0.148}Ge_{0.010}Si_{0.086}O_{0.052}$	-1.35	- 0.21	0.61	- 0.20	-1.27

N.B.: There are 1.67 1022 molecules per cm3 for CuGeO3 crystal and 2.36 1022 molecules per cm3 about for 1/3SiO2+2/3GeO2 glass.

Table 2. Atomic content in each phase and their difference.

We can now arise the question on the transformation of the system under cooling in the phase diagram. As a matter of fact, the observation of a first phase with composition $CuMeO_3$ where Me is a mixture between Ge, Si, and a second phase with composition $Me'O_2$ where Me' is also a metallic mixture, means that the starting liquid did not keep the composition of the rod in particular, $CuMeO_3$ in the case considered in the table. Unfortunately, it was not possible to know the composition of the liquid phase, but we can deduce it by calculation. In this direction, considering the sample with 8.5% at Si, from the optical and SEM observations, we know the filling factor in the central region in 2D is 16%. With this information, and considering that there is invariance along the growth axis, we can achieve the calculation by combining the chemical formula of one phase with the other one. This is done in the Table 3.

Table 3 shows for the central region of the sample that the Ge or Si content have not changed whereas Cu and O show a similar decrease. It is thus clear that the liquid has been losing CuO under heating. We can remind here that from chemical analysis, we found that the external region without precipitates is SiO_2 depressed. As the GeO_2 content remains almost constant, and because a flux of SiO_2 is compensated by a flux of CuO, it yields that CuO migrated from the center to the surface and SiO_2 migrated in the reverse direction. This is likely due to thermal gradient. Therefore, for this specific composition, this migration moved the system (corresponding to the ordered region) from the congruent line to the two phase domain as we show in Figure 18.

Sample UPS416 $Cu_{0.20}Ge_{0.183}Si_{0.017}O_{0.60}$	Chemical formula (atomic %)	Number of atoms 10^{22} per cm^3				
		Cu	Ge	Si	O	Total
Matrix filling factor=0.84		1.71	1.64	0.125	4.87	8.34
Fiber filling factor=0.16		0.354	1.43	0.637	4.67	7.08
Average composition in at.cm^{-3} i.e. estimate for the liquid phase		1.49	1.60	0.207	4.84	8.14
Nominal composition in at.cm^{-3} i.e. composition of the ceramics rod		1.67	1.53	0.14	5.01	8.35
Difference average minus nominal		-0.18	0.07	0.07	-0.17	-0.21

Table 3. Computation of the liquid phase composition.

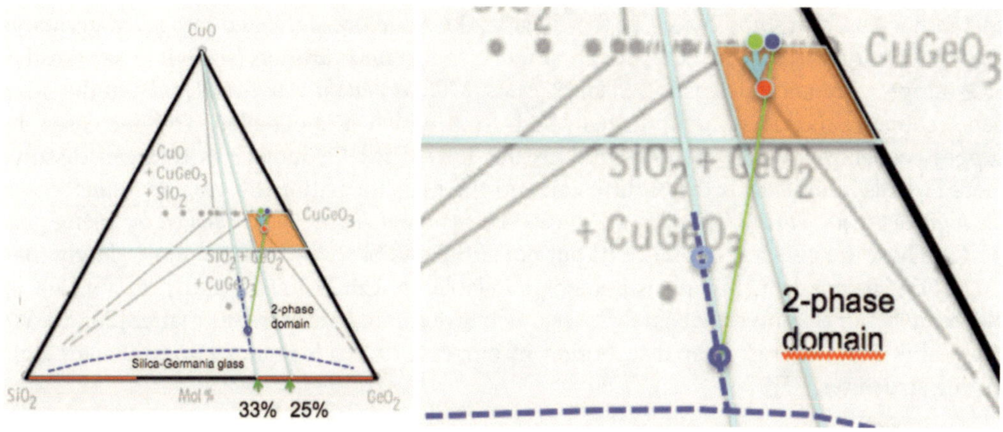

Fig. 19. Ternary diagram for CuO-SiO_2-GeO_2 from [13] (1kb, 600°C) on which we have placed the estimated composition for the liquid phase (red point), the matrix composition (open blue circle) and the fiber composition (solid blue circle). Dashed blue lines are re-estimate of the location of the phase boundaries. The green point is the starting composition for samples $Cu_{0.20}Ge_{0.18}Si_{0.02}O_{0.60}$

Figure 18 shows the location of the previous example in the phase diagram. As we can see, the rod composition in green moved to the red point on losing CuO, and then under cooling, the two phase composition is defined by the line linking the two blue points and the matter conservation. On one hand, the matrix composition surrounding the fiber is defined by the congruent line. On the other hand, since the fibers contain CuO, this means that a phase boundary (dashed blue line) is missing in the diagram. Below this line, there exists a silica glass containing CuO. In addition, we can say, that between the two blue points, it is not a three phases domain but a two phases one. Finally, as we had seen that the liquid phase corresponding to ordered area, moved to glass boundary, we made further experiments with increasing content in glass formers. This gave then results with the smallest size and distances between fibers (See Figure 8d).

In order to obtain spacing between fibers compatible with the second telecom window (1.5 µm), we studied several parameters of growth. For a fixed filling factor f: $d^2R=a$, $\pi\phi^2/4d^2=f$,

$\phi^2 R=b$ where d is the fiber distance, R is the growth rate, ϕ is the fiber diameter. f is dependent of the phase diagram. 'a' and 'b' are constants depending on several thermodynamic and kinetic quantities themselves depending on nominal chemical composition and temperature, R can be considered as a free parameter [20]. Experimentally, we proved that the distance between fibers could be reduced at least down to 1.4 μm and fiber diameter to about 0.6 μm keeping a rather good order. Good fiber ordering is obtained for filling factor lying between 0.126 and 0.140. Below this range, the order may be good enough but the fiber distance is quite large. Above that, the ordering degrades. Since the filling factor is partly defined by the nominal composition, the increase in silicon content leads to an increase in the filling factor but it may lead also to disordering.

There is so a series of conditions for obtaining good ordering. This ensemble of conditions suggests an instability like the Turing one [21]. As a matter of fact, we can remark that at the beginning there is too much MO_2 in the liquid to elaborate a single crystal (this is deduced from phase diagram). This induced Si content fluctuations achieved by diffusion. As Si content increased in some places, at these places the structure stabilized into silica-germania glass increasing the driving force for Si to increase its concentration (since it is known that SiO_2 adopts vitreous state easily). *In other words, MO_2 formation is self-catalyzed.* On the other hand, copper is hardly soluble in this phase from which it is expelled. This increases the copper concentration out of fibers and creates new $CuMeO_3$ molecules that can dissolve more Si. This inhibits thus the Si diffusion and the MO_2 formation. So, *we find in this scheme the ingredients for Turing instability: a species self-catalyzed (MO_2) and inhibited by another one (CuO)*. These are necessary conditions but not sufficient, Si should diffuse more slowly than Cu. There is no diffusion measurement available in $CuGeO_3$ (contrarily to $(Si,Ge)O_2$). Nevertheless, it is known that Cu diffuses much faster than Si in vitreous matter [22, 23]. We thus think that ordered fiber distribution in our case is a self-organization relevant from Turing structures.

4. Optical properties

Figure 20 is a comparison of Fourier transforms of the two samples shown in Figure 8. We observe a circle around the central point in Figure 19(a) with a non-zero thickness whereas there are just a few points in Figure 20(b). This corresponds to isotropy of equally distant point distribution in Figure 8(b) whereas (c) is well ordered at the first and subsequent neighbors. If sample in Figure 20(b) is recognized as a **metacrystal**, sample in Figure 20(a) exhibits properties of a glass with a rather well defined first neighbor coordination shell but an isotropy due to random long range order. We can thus call this structure **metaglass**.

From Fourier transforms (FT), we can extract several structural quantities:

- The size of the central point yields coherence length, it is 50 μm horizontally and 33 μm vertically for sample (a). This means 4.3 to 6.0 fiber periods, whereas for sample (b), it is 11.5 μm which corresponds to 5.5-7.7 fiber periods.
- The distance from the first point (Figure 20(b)) or circle (Figure 19(a)) to the center point yields the fiber distance by simple inversion. It is 7.7 μm vertically and 8.3 μm horizontally for sample (a) whereas it is 2.1 μm and 1.5 μm respectively for sample (b). For sample (a), the dispersion of the first neighbor distance is around 20%.
- In FT image of sample (a) image, we can see a larger intensity at the higher and lower bounds of the circle. This means that a texture begins to appear in this sample. We

have elaborated other samples with a texture more developed, intermediate between FT of sample (a) and sample (b). Diffraction directions appear in that case as spots on a circle.

The next point is to observe that only one circle (sample (a)) or one series of points located on an ellipse (sample (b)) is detected. This arises from the ratio of the fiber size to the period or the filling factor. Due to the large value of this ratio (0.43), the FT domain is strongly limited to the first points around the central spot. This can be good for application as the intensity is concentrated only in several directions [24].

A thin slice of the sample was cleaved along the (100) plane or perpendicular to the 'a' axis, for observation under optical microscope. This can be done both in reflection and transmission modes using either natural or polarized white light. The precipitates are highlighted easily with respect to their monocrystalline surroundings, because of sufficiently high contrast in refractive index (the difference is about 0.32, matrix index=1.77 (average value), fiber index 1.45). The samples are not homogeneous in the conditions we used as it can be seen from the difference in color in the center and on the outer regions of the sample surface (see Fig. 2). These regions are concentric, indicated by a lighter shade of blue at the center and a darker one outward. The use of polarized light indicates that the matrix is not isotropic. 'b' and 'c' are the neutral crystal axes perpendicular to 'a' axis (unfortunately we were not able to measure this birefringence). On the contrary, the precipitates do not provide any contrast, which means that they have an isotropic structure in the cleavage plane.

5. Towards Photonic Crystals (PC) working in the IR

Optical microscopy of the sample shows regular arrangement of a large number of silica fibers. Therefore, it was selected for further chemical treatment to develop a material with characteristics close to that of a PC. The essential idea is to etch out the fibers from the crystalline matrix, using HF acid, and thus obtain a higher refractive index difference between the two dielectric phases. To begin with, a piece of pure $CuGeO_3$ was HF treated for 30 min, in 40% HF acid, to make sure that the crystal is not etched away significantly. It is because the aim is to remove the fibers from the sample, without affecting its monocrystalline surroundings. There is no observable difference in the structure of the material before and after the procedure. A thin-cleaved slice of the sample (of thickness 0.5mm) was taken for the experiment. It was immersed in 20% HF acid for 30 min and was observed under the optical microscope (Figure 21a and 21b). There is significant difference in the structure of the sample after the procedure. Apparently, the surface of the material has been cleaned and the etched sample reveals lamellar structures around the fibers. Also the circular cross section of the fiber changes to a rhomboid shape after etching. This sample was again immersed in acid and the microscope images are given in Figure 22a and 22b. The purpose was to confirm, whether the fibers are completely etched off the sample. Figure 21a is the image after 1 h and Figure 22b is taken after keeping it in acid for a longer time. The image of the crystal in Figure 22b shows some white dots at the center of the periodic structures, which were the locations of fibers in the non-etched sample. This means that the fiber from that region has been etched off across the entire thickness of the sample. We further confirmed this by the surface topography from a phase shift interferometer and also by the AFM studies, which we do not include in this article since it carries no more information than those we already showed in the SEM and optical microscope images. However, to ensure that the fibers are indeed etched away completely, we did a layman experiment, placing it over a colored transparent slide to observe the image under optical microscope, in the transmission mode. The results were positive, but it is to be mentioned that the thickness of the sample was only less than 0.5mm. We measured the etching rate of the sample by comparing its thickness with a reference, after repeated etching procedure. It is about 12 μm/h.

SEM analysis: The SEM images of this sample using the backscattered (BSE) and secondary electron (SE) modes are given in Figure 23a and b, respectively. The BSE results give an indication of the chemical differences at various locations, whereas the SE measurements are more related to the topography of the material. One interesting feature that we observed is the change in the shape of the cross sectional area occupied by the fibers. The fibers themselves had circular cross section. However, after their removal a location assumes a rhomboid shape. One might be lead thinking that there is a preferential direction along which the etching took place faster, which is the direction of the minor axis of the crystal. Even if the reasons are not clear, we propose two hypotheses. (1) It is the surface tension of the fibers while in liquid form that enabled them to assume circular cross section. When the fibers are removed, the tension releases, and the surrounding matrix takes the shape decided by its crystalline properties. (2) On the border of the fibers the crystalline matrix is also etched away slightly, and the rhomboid shape is an indication of the crystallographic axis in the growth direction [25]. It can also be seen that the appearance of the white lines and the variations in the color of the surrounding matrix are perfectly repetitive. We further confirmed using phase shift interferometer technique and AFM that these correspond to variation in the depth profile.

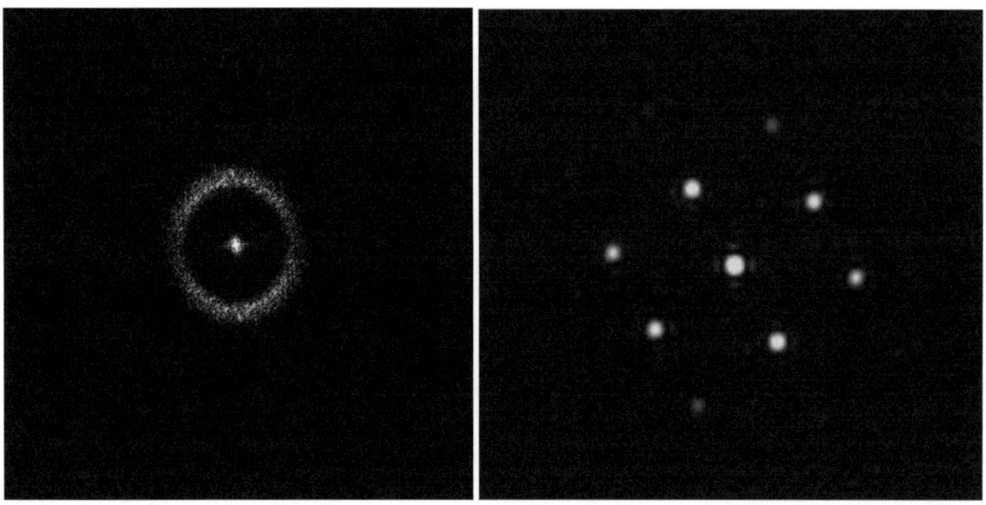

(a) Fourier transform associated to Figure 8b. The radius of the first circle is 0,125 μm^{-1} in average.

(b) Fourier transform associated to the region delimited by white square in Figure 8c. The distance of the neighbouring central point is 0.58 μm^{-1} in average.

Fig. 20. Fourier Transforms from pictures in Figure 8.

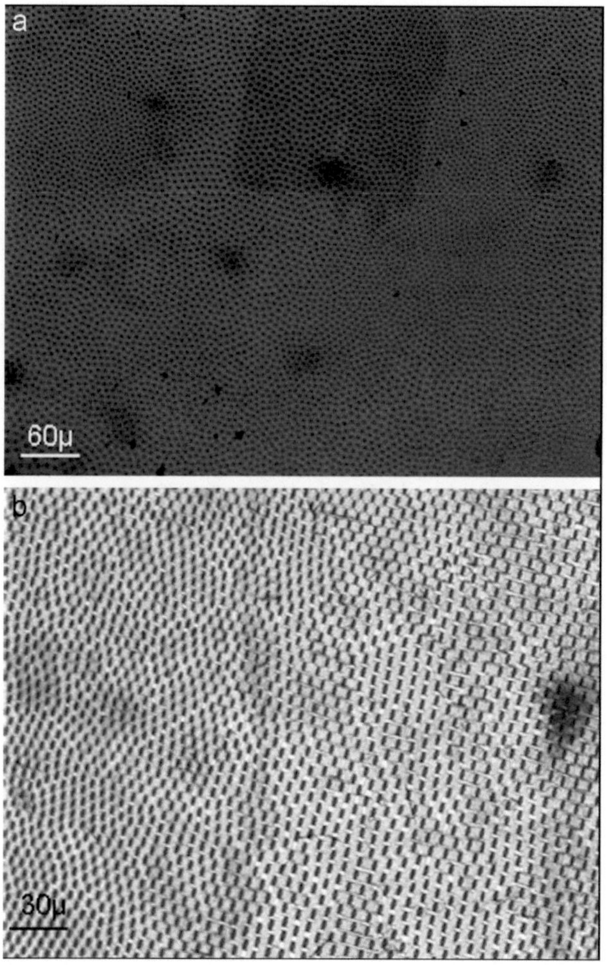

Fig. 21. Photograph of the sample before (a) and after (b) HF treatment for 30 min in 20% acid. There are obvious differences in the structure of the crystalline matrix after the procedure.

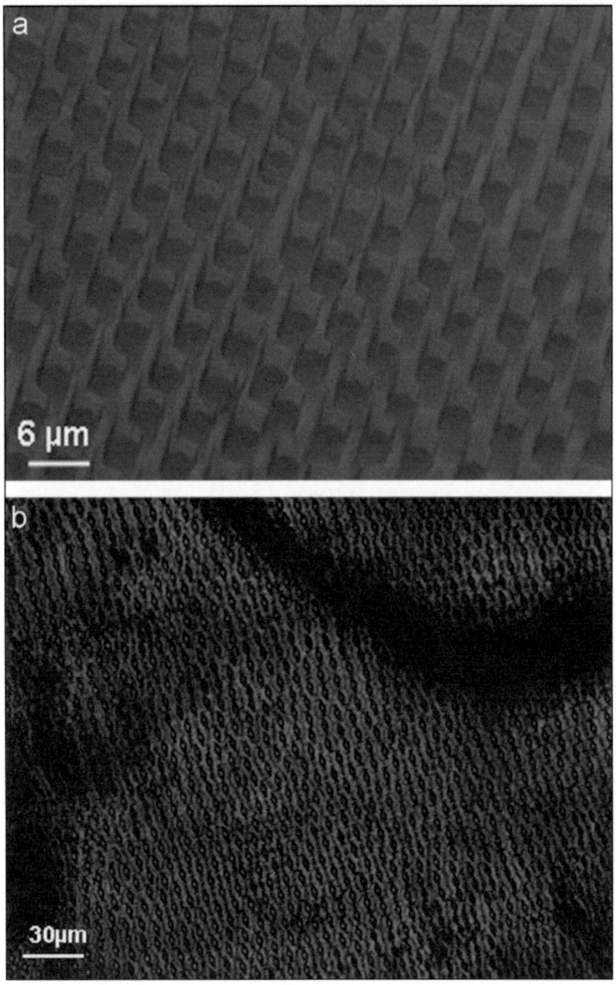

Fig. 22. Photograph of the sample after HF treatment for 3 h (a) at high magnification and (b) shows the picture after keeping the sample in acid for a longer time such that the fibers are etched off. This is indicated by the appearance of white dots.

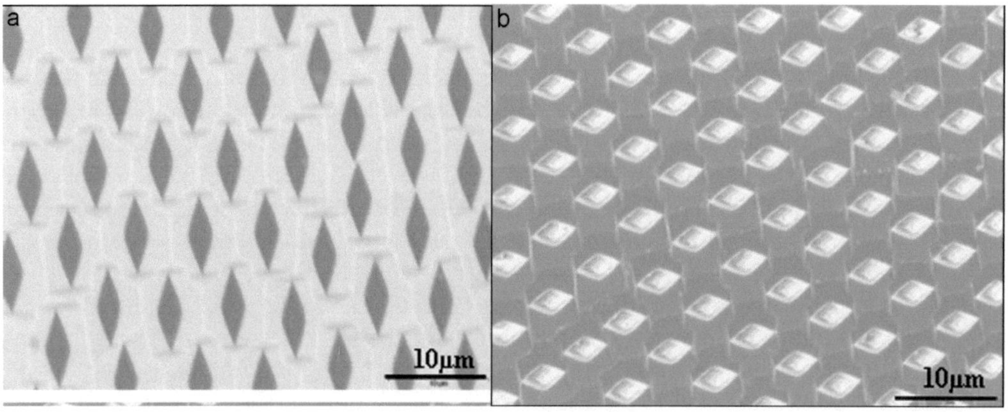

Fig. 23. A thin-cleaved slice of the sample (ofthickness0.5mm) was taken for the experiment. It was immersed in 20% HF acid for 30 min and was observed Under the Optical microscope. There is significant difference in the structure of the sample after the procedure. Apparently, the surface of the material has been cleaned and the etched sample reveals lamellar structures around the fibers. Also the circular cross section of the fiber changes to a rhomboid shape after etching.

6. Conclusion

We synthesized from the melt an aligned two-phase material whose minority phase is vitreous and the majority one, crystalline through self-organization. This is original, as far as we know. Another example of glass precipitation (Si_3N_4) in a crystalline matrix (Fe) has been published [26] but no order appears at all. The samples that we have synthesized are monocrystals of $CuMeO_3$ type (with Me = Ge or Si) with orthorhombic crystalline structure containing germania-silica glass fibers $(Ge_{0,7}Si_{0,3})O_2$. The fibers can have sub-micrometer size, regularly spaced, close to a compact hexagonal stacking with a period slightly above the micron. This regular microstructure was obtained only for certain ranges of Si content and growth rate.

Furthermore, the manipulation of the parameters controlling the microstructure of composite presented above (diameter and distribution of fibers) and the beginning of rationalization in the frame of Turing structure, opens up ways for finding other systems exhibiting similar structures. This will be valuable for a better control of the degree of ordering and for finding chemical systems with much lower light absorption in the telecom window. In the present composition, the absorption coefficient is around 2 cm^{-1}, quite flat for wavelength ranging from 1 to 4 μm.

These two dielectric structures react differently to HF acid, which we exploit to remove the fibers from the matrix without affecting the crystal. So far this has been successful until a length limit of 0.5mm. There are research results indicating that photonic band gaps can also be observed in quasi-periodic dielectric structures [16, 27]. Some researchers have already reported that PCs show complete band gaps, if the dielectric contrast is sufficiently high [28, 29]. We used the finite difference time domain method for photonic band gap (PBG) calculation assuming a hexagonal lattice. Even if the results are not conclusive, what we can say at this stage is that, before etching there was no indication of a band gap; however, after

improving the refractive index difference from 0.32 to 0.77, we observed that a PBG opened up in the TE mode (but not in the TM mode).
The next question, here, is the use of such a composite. We have to firstly underline that we have fabricated real 2D ordered composite as the translation symmetry along the thickness is much larger than the material period in the perpendicular plane. Thus, vertically, the intensity remains concentrated without any diffraction. One application is wavelength extraction: with sample with optimal order, it is possible to diffract light at 1.5 μm under an angle of 50° and to collect 1 nm wide channel for injection in a 10 μm wide waveguide 3cm away from the center of the crystal. With a crystal with short coherence length, it is not necessary to turn the crystal for fulfilling a Bragg condition and thus a star demultiplexing is achievable. But this is quite modest application with the system we used here. We think possible that our metamaterials could be useful for bio-optical applications. We are optimistic to reach metamaterial applications in the visible and maybe photonic crystal ones by investigating other composites achievable by the same method.

7. Acknowledgements

This work has been performed with a postdoctoral support of the University of Paris Sud 11. The authors are deeply grateful to Prof. Revcolevschi for his great scientific help on this subject.

8. References

[1] C. N. R. Rao, A. Müller, and A. K. Cheetham, *The chemistry of Nanomaterials, Synthesis, properties and applications.* (Wiley-VCH, 2004), Vol. 1-2.

[2] C. M. Soukoulis, S. Linden, and M. Wegener, "Metamaterials are designed to have structures that provide optical properties not found in nature. If their capacity can be extended, new kinds of devices for imaging and control of light will be possible.," Science 315., 47-49 (2007).

[3] J. B. Pendry, "Negative Refraction Makes a Perfect Lens," Physical Review Letters 85, 3966-3969 (2000).

[4] A. Starr, P. Rye, D. Smith, and S. Nemat-Nasser, "Fabrication and characterization of a negative-refractive-index composite metamaterial," Physical Review B 70, 113102 (2004).

[5] P. N. Prasad, *Nanophotonics* (John Wiley and sons, 2004).

[6] M. Bassi, P. Camagni, R. Rolli, G. Samoggia, F. Parmigiani, G. Dhalenne, and A. Revcolevschi, "Optical absorption of CuGeO3," Physical Review B 54, 11030-11033 (1996).

[7] G. Dhalenne, A. Revcolevschi, J. C. Rouchaud, and M. Federoff, "Floating zone crystal growth of pure and Si or Zn substituted copper germanate $CuGeO_3$.," Material Research Bulletin 32, 939-946 (1997).

[8] J. P. Pouget, L. P. Regnault, M. Ain, B. Hennion, J. P. Renard, P. Veillet, G. Dhalenne, and A. Revcolevschi, "Structural evidence for a spin peierls gound state in the quasi-one dimensional compound $CuGeO_3$.," Physical Review B 72, 4037-4040 (1994).

[9] A. Revcolevschi, "Nickel oxide – based aligned eutectics," in *Tailoring Multiphase and Composite Ceramics.*, R. E. M. Tressler, G L; Pantano, C G; Newnham, R E, ed. (Plenum publishing corporetion, Pennsylvania, USA, 1986).

[10] A. Revcolevschi and G. Dhalenne, "Engineering Oxide-Oxide and Metal-Oxide microstructures in directionally solidified eutectics.," Advanced Materials 5, 657-662 (1993).
[11] A. Revcolevschi, G. Dhalenne, and D. Michel, "Interfaces in directionally solidified oxide-oxide eutectics.," Material Science Forum 29, 173-198 (1988).
[12] E. I. Speranskaya, Inorganic Materials 3, 1271-1277 (1967).
[13] U. Müller and E. Conradi, "Fehlordnung bei Verbindungen MX_3 mit Schichtenstruktur.," Zeitschrift für Kristallographie 176, 233-261 (1986).
[14] G. S. Henderson, D. R. Neuville, B. Cochain, and L. Cormier, "The structure of GeO2-SiO2 glasses and melts: A Raman spectroscopy study," Journal of Non-Crystalline Solids 355, 468-474 (2009).
[15] O. Majérus, L. Cormier, J. P. Itié, L. Galoisy, D. Neuville, and G. Calas, "Pressure-induced Ge coordination change and polyamorphism in SiO2-GeO2 glasses," Journal of Non-Crystalline Solids 345, 34-38 (2004).
[16] F. L. Galeener, "The Raman spectra of defects in neutron bombarded and Ge-rich vitreous GeO_2.," J. of Non-Crystalline Solids 40, 527-533 (1980).
[17] K. H. Breuer, W. Eysel, and M. Behruzi, "Copper (II) silicates and germanates with chain structures II. Crystal chemistry*," Zeitschrift für Kristallographie 176, 219-232 (1986).
[18] K. Jackson and J. Hunt, "Lamellar and rod eutectic growth," AIME MET SOC TRANS 236, 1129-1142 (1966).
[19] W. Kurz and D. J. Fisher, "Fundamentals of solidification," Trans Tech Publications Ltd, Trans Tech House, 4711, Aedermannsdorf, Switzerland, 1986. 244 (1986).
[20] M. C. Flemings, *Flemings (1974) Solidification processing* (McGraw-Hill (New York), 1974).
[21] A. M. Turing, "The chemical basis of morphogenesis," Philosophical Transactions of the Royal Society of London. Series B, Biological Sciences 237, 37 (1952).
[22] J. D. McBrayer, R. Swanson, and T. Sigmon, "Diffusion of metals in silicon dioxide," Journal of the Electrochemical Society 133, 1242 (1986).
[23] H. A. Schaeffer, "Diffusion-controlled processes in glass forming melts.," J. of Non-Crystalline Solids. 67, 19-33 (1984).
[24] I. L. Gheorma, S. Haas, and A. Levi, "Aperiodic nanophotonic design," Journal of applied physics 95, 1420 (2004).
[25] D. A. Pawlak, K. Kolodziejak, S. Turczynski, J. Kisielewski, K. Rozniatowski, R. Diduszko, M. Kaczkan, and M. Malinowski, "Self-organized, rodlike, micrometer-scale microstructure of Tb3Sc2Al3O12-TbScO3: Pr eutectic," Chemistry of materials 18, 2450-2457 (2006).
[26] E. Mittemeijer, M. Biglari, A. Boettger, N. Pers, W. Sloof, and F. Tichelaar, "Amorphous precipitates in a crystalline matrix; Precipitation of amorphous Si {sub 3} N {sub 4} in {alpha}-Fe," Scripta materialia 41(1999).
[27] H. Gersen, T. Karle, R. Engelen, W. Bogaerts, J. Korterik, N. Van Hulst, T. Krauss, and L. Kuipers, "Direct observation of Bloch harmonics and negative phase velocity in photonic crystal waveguides," Physical review letters 94, 123901 (2005).
[28] S. G. Johnson and J. D. Joannopoulos, "Three-dimensionally periodic dielectric layered structure with omnidirectional photonic band gap," Applied Physics Letters 77, 3490-3492 (2000).

[29] J. H. Park, W. S. Choi, H. Y. Koo, J. C. Hong, and D. Y. Kim, "Doped colloidal photonic crystal structure with refractive index chirping to the 111 crystallographic axis," Langmuir 22, 94-100 (2006).

Part 2

Inorganic/Organic Materials for High Technology

4

Effect of Amino Acid Additives on Crystal Growth Parameters and Properties of Ammonium Dihydrogen Phosphate Crystals

P.V. Dhanaraj and N.P. Rajesh
Centre for Crystal Growth, SSN College of Engineering, Kalavakkam
India

1. Introduction

The isomorphous ammonium dihydrogen phosphate (ADP) and potassium dihydrogen phosphate (KDP) are technologically important crystals grown in large size for various applications. ADP crystal is of more appeal due to its piezo-electric property (Tukubo et al., 1989). Studies on ADP crystals attract interest because of their unique nonlinear optical, dielectric and antiferroelectric properties (Gunning et al., 2001). ADP crystals are widely used as the second, third and fourth harmonic generators for Nd: YAG, Nd: YLF lasers and for electro-optical applications such as Q-switches for Ti: Sapphire, Alexandrite lasers, as well as for acousto-optical applications. ADP crystal has found applications in NLO, electro-optics, transducer devices and as monochromators for X-ray fluorescence analysis.

The room temperature structure of ADP determined by X-ray diffraction analysis was reported by Ueda (1948). Tenzer et al (1958) and Hewat (1973) examined the structure by neutron diffraction analysis. The projection of the structure onto the b, c plane is shown in Figure 1.

ADP differs from KDP by having extra N-H-O hydrogen bonds which connect PO_4 tetrahedra with neighbouring NH_4 group. Each oxygen atom is connected with another oxygen atom in the neighboring PO_4 ion and with a nitrogen atom in a neighbouring NH_4 ion by two Kinds of bonds: (O-H-O) and (N-H-O). According to the positional refinements of each atom in ADP by X-ray diffraction study (Srinivasan, 1997), both above and below the phase transition point, each NH_4 ion at the potassium position in KDP structure is shifted to the off-center position by forming two shorter and two longer bonds with four PO_4 tetrahedra at low temperature phase. When an oxygen is connected with the shorter N-H-O bond, it tends to keep the other proton off in the O-H-O bond and when with the longer N-H-O bond it tends to take the acid proton nearby. Thus the extra hydrogen bonds produce a distorted NH_4 ion lattice at low temperature and co-operate with the acid protons in causing proton configurations different from those found at low temperature in KDP (Matsushita et al., 1987). As a representative hydrogen bonded material, ADP has attracted extensive attention in the investigation of hydrogen bonding behaviors in crystal and the relationship between crystal structures and their properties.

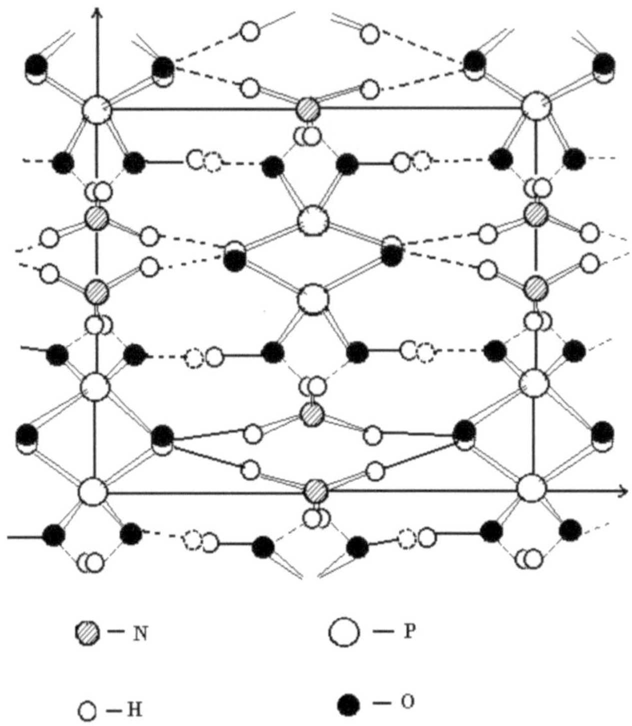

Fig. 1. The (100) projection of ADP structure.

Several researchers have carried out a lot of studies in pure and doped ADP crystals (Zaitseva et al., 2001; Ren et al 2008). In ADP and KDP crystal growth, the metallic cations present in the solutions, especially materials with high valency were considered to strongly affect the growth habit and optical properties of the crystals. The most dangerous impurities which affect the growth habit are trivalent metals Cr^{3+}, Fe^{3+} and Al^{3+} (Alexandru et al., 2003). Even after repeated recrystallization, the presence of small amount of those kinds of impurities in the solution suppresses the crystal quality and growth rate. Here comes the importance of beneficial effects of additives in the crystal growth. An additive can suppress, enhance or stop the growth of crystal completely and its effects depend on the additive concentration, supersaturation, temperature and pH of the solution. Some dopants are added to suppress the effect of metal ion impurities on ADP and KDP crystals. For example, EDTA and KCl reduces the effect of metal ion impurities and enhance the metastable zone width and increases the growth rate of the crystals (Rajesh et al., 2000; Podder 2002; Meenakshisundaram et al., 2009). The addition of such kind of dopants does not remove the impurities present in the solution; it just reacts with the metal ions and is making complexes. By making complex, the ions become bigger in size and it is not possible to enter into the growing crystal (Li et al., 2005; Asakuma et al., 2007). Studies have also been made about the effect of additives on growth, habit modification and structure of ADP (Davey et al., 1974; Boukhris et al., 1998). The adsorption of impurities at different sites can cause growth inhibitions, even block the growing surface and in consequence stop the growth process. However, the adsorbed impurities may simultaneously lead to a reduction in the edge free energy, which results in an increase in crystal growth rate (Rak et al., 2005). Several dopants help in the growth of ADP crystals at with higher growth rate and enhancement in the various properties of the crystals. The growth promoting effect is observed in the presence of organic additives (Kern et al., 1992; Bhagavannarayana et al., 2006) as well as inorganic additives (Shantha et al., 1997; Podder et al, 2001).

Amino acid family crystals exhibit excellent nonlinear optical and electro-optical properties. Reports are available in literature on the doping of amino acids in technologically important crystals and the enhancement of the material properties like nonlinear optical and ferroelectric properties. For example, enhancement of Second Harmonic Generation (SHG) efficiency has been reported in L-arginine doped KDP crystals (Parikh et al., 2007). Kumaresan et al (2008) reported the doping of amino acids (L-glutamic acid, L-histidine, L-valine) with KDP and studied its properties. The effects on various properties of L-theronine, DL-theronine and L-methionine admixtured Triglycine Sulfate (TGS) crystals were studied and the authors reported that the admixtured TGS crystal has different properties compared to pure TGS crystal (Meera et al., 2004). Batra et al (2005) investigated the growth kinetics of KDP and TGS crystals doped with L-arginine phosphate monohydrate. The addition of L-arginine decreases the value of dielectric constant of KDP crystals (Meena et al., 2008).

In the light of research work being done on ADP crystals, to improve their growth and other characteristics, it was thought interesting and worthwhile to investigate the effects of amino acid materials L-arginine monohydrochloride ($C_6H_{15}N_4O_2Cl$) and L-alanine ($C_3H_7NO_2$) on nucleation studies, growth and properties of ADP crystals for both academic and industrial uses. The reason for choosing the dopants is that L-arginine monohydrochloride and L-alanine are efficient NLO materials under the amino acid category. Monaco et al (1987) discovered NLO material L-arginine monohydrochloride, which belongs to space group $P2_1$ of monoclinic system with two molecules in the asymmetric unit. L-alanine crystallizes in orthorhombic system with noncentrosymmetric space group $P2_12_12_1$ (Razzetti et al., 2002).

2. Experimental studies

2.1 Determination of solubility and metastable zone width

Metastable zone width is an essential parameter for the growth of large size crystals from a solution, since it is the direct measure of the stability of the solution in its supersaturated region. Metastable zone width is an experimentally measurable quantity which depends on number of factors, such as stirring rate, cooling rate of the solution and presence of additional impurities (Nyvlt et al., 1970; Sangwal 1989; Zaitseva et al., 1995). Ammonium dihydrogen phosphate, L-arginine monohydrochloride (LAHCl) and L-alanine of GR grade from Merck and Millipore water of resistivity 18.2 MΩcm were used for all studies. No further purification was done. The solubility was determined gravimetrically for pure ADP and ADP doped with small amount (5 mol%) of LAHCl and L-alanine separately as additives. Polythermal method (Nyvlt et al 1970) was adopted for the metastable zonewidth studies of pure and doped ADP solutions. The ADP solution (600 ml) saturated at 30 °C was prepared according to the solubility diagram with continuous stirring using a magnetic stirrer and the solutions were filtered. Three similar beakers with 200 ml solution each were used; the first beaker contained pure ADP solution whereas the second and the third beakers contained 5 mol% LAHCl and 5 mol% L-alanine doped ADP solutions respectively. Then pure and doped ADP solutions were kept in a Constant Temperature Bath (CTB) with cooling facility. After stirring for 6 h, the solution was slowly cooled at a desired cooling rate of 4 °C/h, until the first crystal appeared. The experiments were repeated for different saturation temperatures (30–50) °C with the interval of 5 °C and the corresponding metastable zonewidths were measured. Several nucleation runs (7–9 times) were carried out under controlled conditions and reproducible results with the accuracy of ±0.25% were obtained. The metastability limit of LAHCl added solution and L-alanine added solution is shown in Figure 2 in comparison with the pure system.

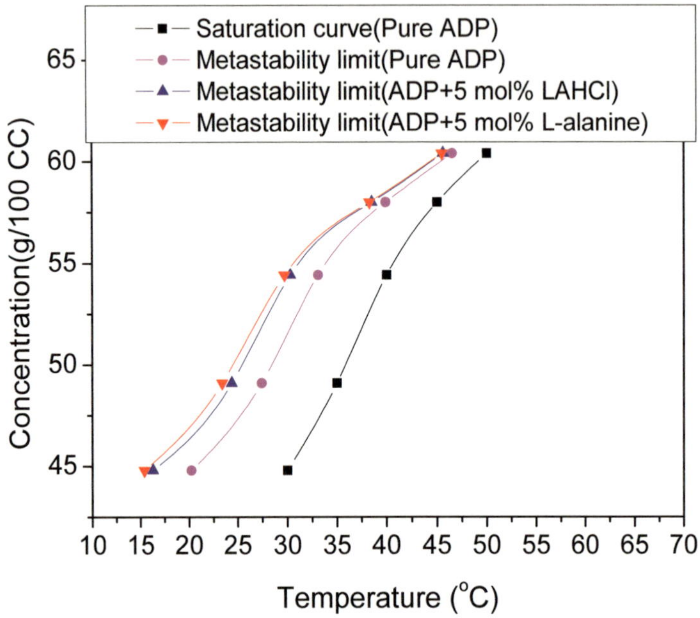

Fig. 2. Saturation and metastability limit curves of pure, LAHCl and L-alanine added ADP solutions.

It is obvious from the figure that the zone widths for all the solutions decrease as the temperature increases. At the same time, the addition of dopants enhances the metastable zonewidth of ADP solutions for all the temperatures studied in this work, and makes the ADP solution more stable. During the experiment, the number of tiny crystals formed by spontaneous nucleation was appreciably reduced in the case of the doped solutions compared with the pure one. Among the two additives, the additive L-alanine enhances the metastable zonewidth of ADP than the other additive LAHCl especially at lower temperatures. The addition of these amino acid additives can make ADP solution more stable.

2.2 Determination of induction period

The induction period, a measure of the nucleation rate was determined experimentally for ADP solutions with and without the presence of additives at different supersaturations by means of isothermal method (Zaitseva et al., 1995). The "direct vision method" by naked eye is used for measuring induction period. Aqueous solutions of various supersaturated concentrations were prepared. Supersaturation was obtained by natural cooling. Supersaturated solutions of equal volume (100 ml) were taken in the cells at a higher temperature. As the temperature of the cell reached the experimental temperature (35 °C), the time was noted. Once the nucleation occurred, it grew quickly and a bright sparkling particle was seen. The time of observation of the sparkling particle in the cell from the time at which the solution reaches the experimental temperature of nucleation (temperature of beginning of nucleation) gives the induction period of nucleation. The effect of heterogeneous nucleation due to scratching on the inner wall of the nucleation cell was reduced by choosing a glass beaker without scratches. Experiments were performed at selected degrees of supersaturation (c/c_o), viz., 1.25, 1.275, 1.3, c being the mole fraction of solute in supersaturated solution and c_o is the equilibrium concentration. Several nucleation runs (5-7 times) were carried out under controlled conditions and reproducible results with the accuracy of ±0.25% were obtained. The experimental results of induction period for pure and doped ADP solutions are presented in Figure 3.

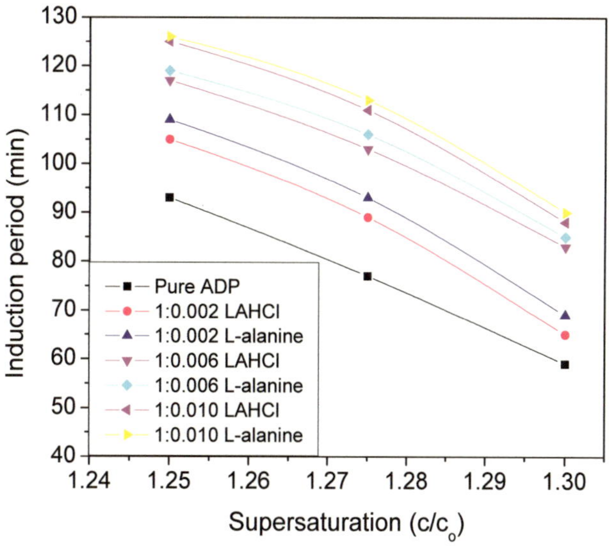

Fig. 3. Values of induction period against supersaturation.

Considering the principles of homogeneous and heterogeneous nucleation theories, the free energy of formation of a nucleus under heterogeneous nucleation is less than that of a homogeneous condition (Sangwal 1996; Srinivasan et al 1999). Considering the additive added system it can be noticed that the induction period of doped ADP is higher than that of pure and it increases with the increase in the additive concentration. Among the additives, L-alanine has a longer higher induction period than LAHCl at every concentration. The presence of additives in the system affects the nucleation behavior very considerably. This may be due to the suppression of chemical activity of the metal ions present in the ADP solution (Mullin 1993).

2.3 Growth rate measurements

The growth rate of a crystal is known to be changed by traces of certain impurities or additives. In this work, the influence of the additives on the growth rate of ADP crystals is determined by the weighing method put forward by Kubota et al (1995). By this method, the growth rate of a crystal is defined as

$$G_g = (m - m_o)/m_o\Delta\theta \quad (1)$$

where m_o is the initial mass of the crystal (kg), m is the final mass of the crystal (kg), and $\Delta\theta$ is the growth time. Here growth time is taken as 1 h. A single crystal with a size of 5-10 mm was used as the seed for the experiment. The seed crystal (of mass m_o) was suspended in the solution in 500 ml glass vessel (working volume: 400 ml) for 1 h ($\Delta\theta$) for growth. The solution was continuously stirred throughout the process. Supercooling was varied from 2 to 10 °C by changing the growth temperature. The same procedure is done for pure and doped ADP crystals. Figure 4 shows the growth rates of pure and doped ADP crystals.

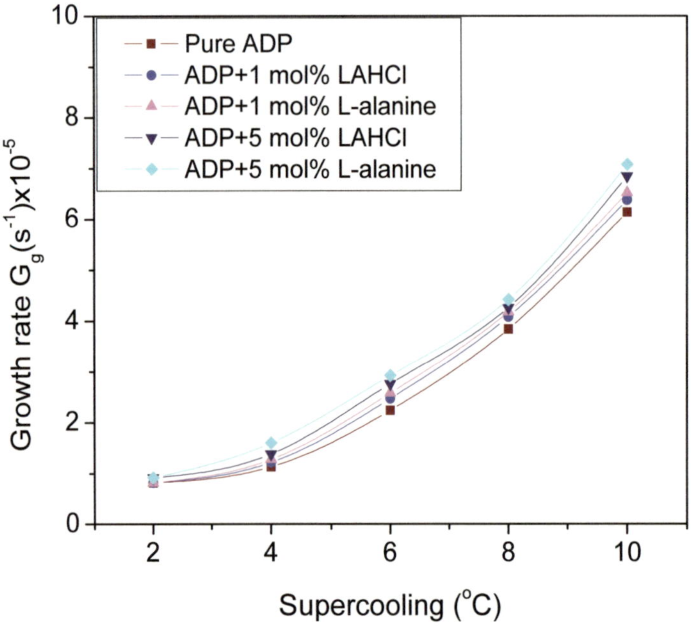

Fig. 4. Variation of mass growth rates for pure and doped ADP crystals.

The presence of additives is found to have an influence on the growth rate. The role of additives during the growth process could be visualized as follows. The additives LAHCl and L-alanine have higher solubility than that of the crystallizing substance (ADP). The additives might have changed the thermodynamic parameters, i.e. the surface concentration of the growth species and the surface energy. An increase in solubility by the addition of LAHCl and L-alanine may lead to decrease in the surface energy, which consequently decreases the rates of layer displacement that cause an increase in the growth rate (Sangwal 1996).

2.4 Crystal growth

In the present work, ADP crystals doped with 5 mol% LAHCl and L-alanine separately were grown from aqueous solution with a simple apparatus that can be applied in certain forced convection configurations to maintain a higher homogeneity of the solution. This apparatus consists of seed rotation controller coupled with a stepper motor, which is controlled by using a microcontroller based drive. This controller rotates the seed holder in the crystallizer. The seed crystal is mounted on the center of the platform made up of acrylic material and is fixed into the crystallizer. The seed mount platform stirs the solution very well and makes the solution more stable, which results in better crystal quality. The schematic diagram of the seed rotation controller designed for low temperature solution growth method is shown in Figure 5. The uniform rotation of the seed is required to avoid stagnant regions or re-circulating flows, otherwise inclusions in the crystals will be formed due to inhomogeneous supersaturation in the solution (Fu et al., 2000).

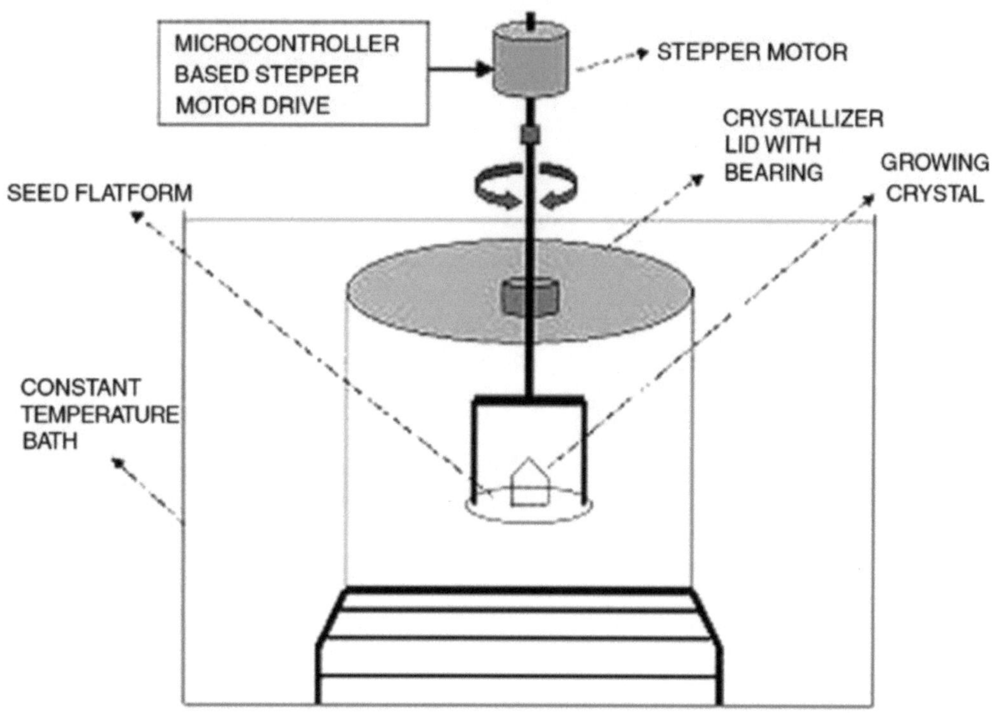

Fig. 5. Schematic diagram of the seed rotation controller.

The crystal growth was carried out in a 5000 ml standard crystallizer used for conventional crystal growth by using the method of temperature reduction. The temperature of solution in the crystallizer was controlled using a CTB and the temperature fluctuations are less than 0.01 °C. The saturation temperature was 50 °C. The solution was filtered by filtration pump and Whatman filter paper of pore size 11 μm under slight pressure in a closed system to remove extraneous solid and colloidal particles, which may act as the centers of spontaneous nucleation during growth. Then the solutions were overheated at 70 °C for 24 h. This duration of overheating was found to be effective to destroy the molecule clusters existing in the solution and to make the solution stable against spontaneous nucleation under a high supersaturation (Zaitseva et al., 1995; Nakatsuka et al., 1997). Then the temperature of the solution was reduced to 3-5 °C higher than saturation point (50 °C) at 1 °C/h. After that temperature was reduced to the saturation point at 1 °C/day and the seed crystal was mounted on the platform. The rotation rate of the platform was 40 rpm. From the saturation point, the temperature was decreased at 0.1 °C/day at the beginning of the growth. As the growth progressed, the temperature lowering rate was increased up to 1 °C/day. After the growth period of 30 days, crystals were harvested. The as-grown crystals are shown in Figure 6. In the figure, (a) is ADP doped with L-Alanine and (b) is ADP doped with LAHCl.

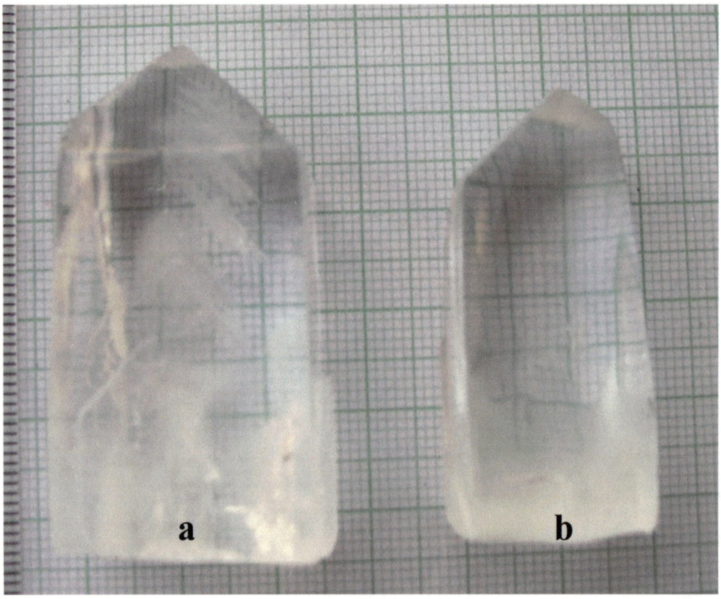

Fig. 6. Photograph of L-alanine doped ADP crystal and LAHCl doped ADP crystal

For various characterization techniques, pure and doped (in concentrations 1 and 5 mol %) ADP crystals were grown by slow cooling method under identical conditions.

3. Analysis of physicochemical studies

3.1 Powder XRD studies

The powder X-ray diffraction is useful for confirming the identity of a solid material and determining crystallinity and phase purity. Grown crystals were ground using an agate mortar and pestle in order to determine the crystal phases by X-ray diffraction. Powder X-ray diffraction study was carried out by employing SEIFERT, 2002 (DLX model) diffractometer with CuK_α (λ = 1.5405 Å) radiation using a tube voltage and current of 40 kV and 30 mA respectively. Figure 7 shows X-ray powder diffraction patterns of ADP doped with LAHCl (5 mol%) and ADP doped with L-alanine (5 mol%) compared with that of pure ADP crystal. X-ray powder diffraction patterns of pure ADP and doped ADP crystals are identical. As seen in the figure, no additional peaks are present in the XRD spectra of doped ADP crystals, showing the absence of any additional phases due to doping.

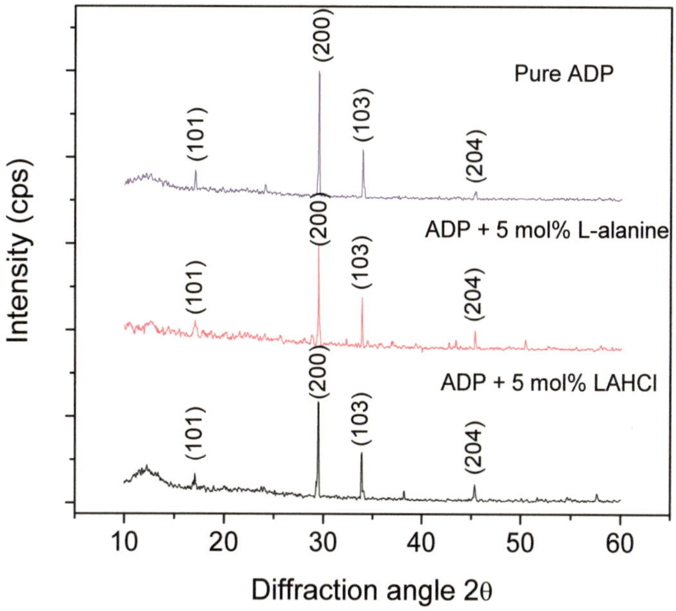

Fig. 7. X-ray powder diffraction patterns of ADP crystals.

3.2 High-resolution x-ray diffraction (HRXRD) analysis

The crystalline perfection of the grown single crystals was characterized by HRXRD analysis by employing a multicrystal X-ray diffractometer designed and developed at National Physical Laboratory (Lal et al., 1989). Figure 8 shows the schematic diagram of the multicrystal X-ray diffractometer. The divergence of the X-ray beam emerging from a fine focus X-ray tube (Philips X-ray Generator; 0.4 mm × 8 mm; 2kWMo) is first reduced by a long collimator fitted with a pair of fine slit assemblies. This collimated beam is diffracted twice by two Bonse-Hart (Bonse et al., 1965) type of monochromator crystals and the thus diffracted beam contains well resolved $MoK\alpha_1$ and $MoK\alpha_2$ components. The $MoK\alpha_1$ beam is isolated with the help of fine slit arrangement and allowed to further diffract from a third (111) Si monochromator crystal set in dispersive geometry (+, –, –). Due to dispersive configuration, though the lattice constant of the monochromator crystal and the specimen are different, the dispersion broadening in the diffraction curve of the specimen does not arise. Such an arrangement disperses the divergent part of the $MoK\alpha_1$ beam away from the Bragg diffraction peak and thereby gives a good collimated and monochromatic $MoK\alpha_1$ beam at the Bragg diffraction angle, which is used as incident or exploring beam for the specimen crystal. The dispersion phenomenon is well described by comparing the diffraction curves recorded in dispersive (+, –, –) and non-dispersive (+, –, +) configurations (Bhagavannarayana 1994). This arrangement improves the spectral purity ($\Delta\lambda/\lambda << 10^{-5}$) of the $MoK\alpha_1$ beam. The divergence of the exploring beam in the horizontal plane (plane of diffraction) was estimated to be << 3 arc s. The specimen occupies the fourth crystal stage in symmetrical Bragg geometry for diffraction in (+, –, –, +) configuration. The specimen can be rotated about a vertical axis, which is perpendicular to the plane of diffraction, with minimum angular interval of 0.4 arc s. The diffracted intensity is measured by using an in-house developed scintillation counter. To provide two-theta ($2\theta_B$) angular rotation to the detector (scintillation counter) corresponding to the Bragg diffraction angle (θ_B), it is coupled to the radial arm of the goniometer of the specimen stage. The rocking or diffraction curves were recorded by changing the glancing angle (angle between the incident X-ray beam and the surface of the specimen) around the Bragg diffraction peak position θ_B (taken as zero for the sake of convenience) starting from a suitable arbitrary glancing angle. The detector was kept at the same angular position $2\theta_B$ with wide opening for its slit, the so-called ω scan.

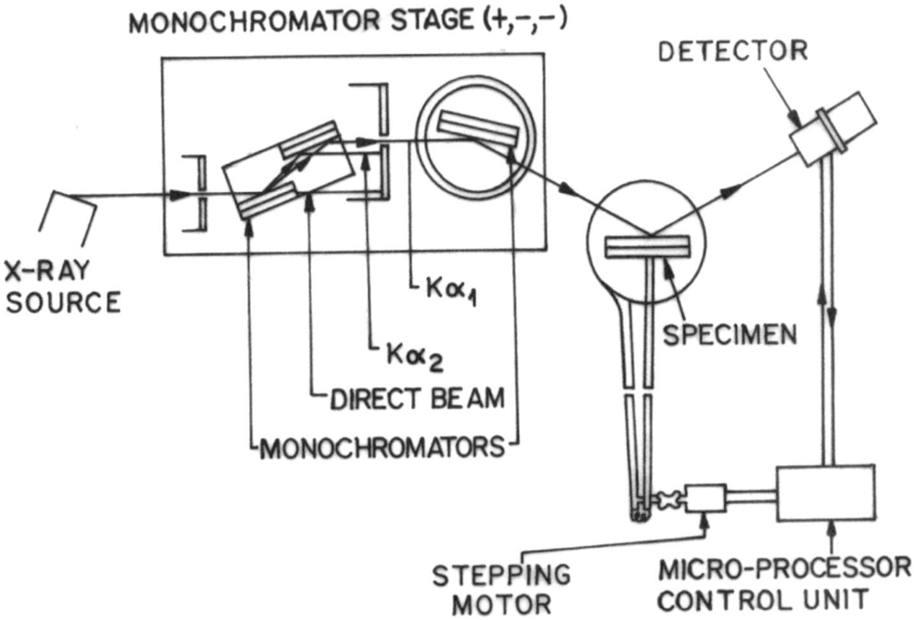

Fig. 8. Schematic line diagram of multicrystal X-ray diffractometer designed, developed and fabricated at National Physical Laboratory

Before recording the diffraction curve to remove the non-crystallized solute atoms remained on the surface of the crystal and also to ensure the surface planarity, the specimen was first lapped and chemically etched in a non preferential etchant of water and acetone mixture in 1:2 ratio.

Figure 9 shows the high-resolution diffraction curve (DC) recorded for LAHCl doped (5 mol%) ADP specimen and Figure 10 shows the DC recorded for L-alanine doped (5 mol%) ADP specimen using (200) diffracting planes in symmetrical Bragg geometry by employing the multicrystal X-ray diffractometer with $MoK\alpha_1$ radiation. The curves are very sharp having full width at half maximum (FWHM) of 8 arc s for LAHCl doped ADP and 5 arc s for L-alanine doped ADP crystals as expected for nearly perfect crystals from the plane wave dynamical theory of X-ray diffraction (Batterman et al., 1964). The absence of additional peaks and the very sharp DC shows that the crystalline perfection of the specimen crystals is extremely good without having any internal structural grain boundaries and mosaic nature. The high reflectivity (≈ 50% in LAHCl doped and ≈ 60% in L-alanine doped) and the very small value of FWHM indicate that even the unavoidable point defects like self interstitials and vacancy defects (Lal et al., 1989) are also extremely low. However, the quality of L-alanine doped ADP specimen is better than that of LAHCl doped specimen.

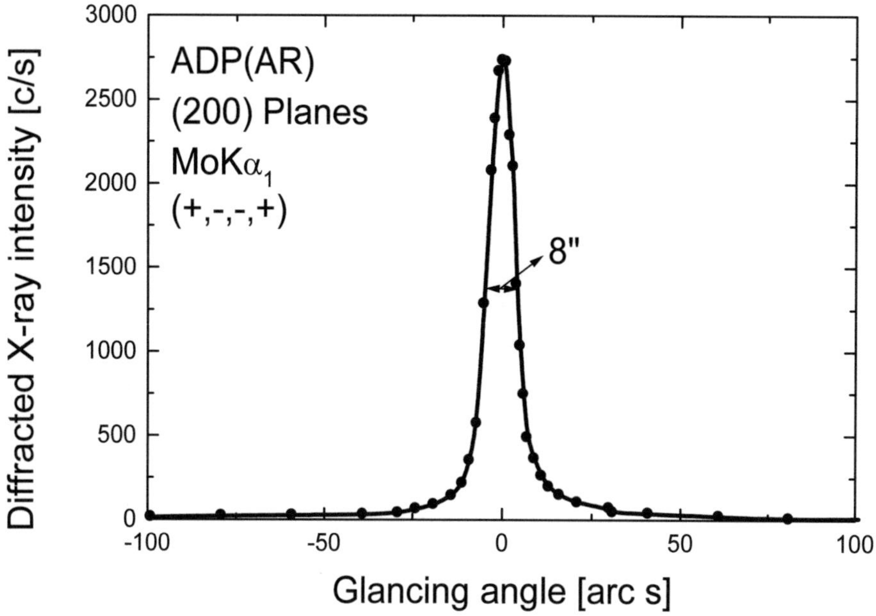

Fig. 9. Diffraction curve recorded for LAHCl doped ADP single crystal using (200) diffracting planes.

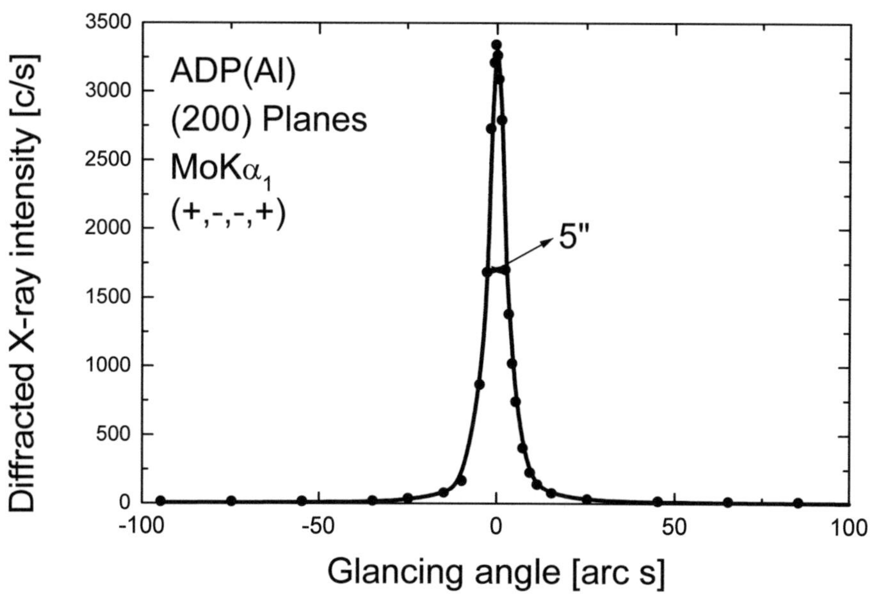

Fig. 10. Diffraction curve recorded for L-alanine doped ADP single crystal using (200) diffracting planes.

3.3 FTIR spectral analysis

The influence of additives used in this work on the vibration frequencies of functional groups of pure ADP crystal has been identified by FTIR spectroscopy. The FTIR spectra were recorded in the region 400–4000 cm^{-1} using a Perkin-Elmer FTIR Spectrum RXI spectrometer by KBr pellet technique. Figure 11 shows the FTIR spectra of the pure ADP, ADP doped with LAHCl (1 mol%) and ADP doped with L-alanine (1 mol%). The broad band in the high energy region is due to O-H vibrations of water, P-O-H group and N-H vibrations of ammonium (Rajesh et al., 2009). The broadness is due to the hydrogen bonding interaction with adjacent molecules (Rani et al., 2011). The peak at 2370 cm^{-1} is due to the combination band of vibrations occurring at 1293 and 1290 cm^{-1}. The bending vibrations of water give the peak at 1646 cm^{-1}. The peak at 1402 cm^{-1} is due to bending vibrations of ammonium (Rani et al., 2011). The P-O-H vibrations give the peaks at 1090 and 930 cm^{-1}. The PO_4 vibrations give their peaks at 544 and 470 cm^{-1}.

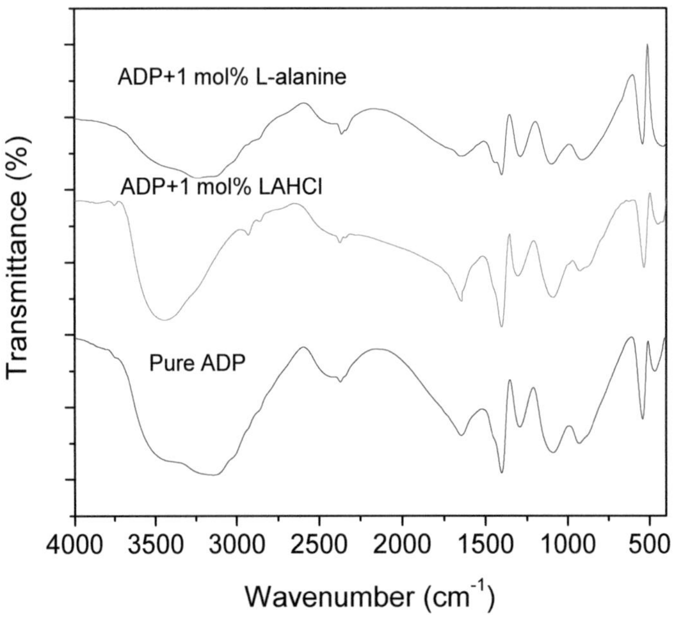

Fig. 11. FTIR spectra of pure ADP, ADP doped with LAHCl and ADP doped with L-alanine.

In the spectrum of ADP doped with LAHCl, the intense band appearing at 3442 cm^{-1} includes O-H vibrations and N-H vibrations of ammonium and amino acid. Although this spectrum carries similar features as that of ADP, there is a distinct evidence for the presence of LAHCl in the lattice of ADP. The peaks appearing at 2928 cm^{-1} and about 2890 cm^{-1} are due to CH_2 vibrations of LAHCl. In addition, shift in the peak positions of P-O-H and PO_4 vibrations compared to ADP established the presence of the additive in the lattice of ADP. In the spectrum of ADP with L-alanine also, there is a significant shift in the peak positions. For example, the PO_4 vibration of the parent is shifted from 470 to 419 cm^{-1}. Similarly the P-O-H vibrations at 1090 and 930 cm^{-1} of the parent are shifted to 1101 and 913 cm^{-1}. Such a shift establishes the presence of L-alanine in the lattice of ADP. As the vibrations of L-alanine are not clearly resolved from the spectrum of the parent, it might be in a trace amount below the deductibility limit. There is a slight evidence of CH_2

vibrations of L-alanine just below 3000 cm^{-1}. All these support the presence of L-alanine in the lattice of ADP.

3.4 Dielectric studies

The dielectric constant is one of the basic electrical properties of solids. Dielectric properties are correlated with the electro-optic property of the crystals (Aithal et al., 1997). The capacitance (C_{crys}) and dielectric loss (tan δ) of pure and doped ADP crystals were measured using the conventional parallel plate capacitor method for temperatures from 313 to 423 K with frequency (f) of 1 kHz. Good quality transparent crystals of size 7 × 7 × 2 mm^3 were used for the measurements. The dimensions of the samples were determined using a traveling microscope (LC = 0.001 cm). Samples were coated with good quality graphite in order to obtain a good ohmic contact. The measurements were done on a–b directions of the crystals. The samples were annealed up to 423 K to remove water molecules if present. The observations were made while cooling the sample and the air capacitance (C_{air}) was also measured. Several trials of experiments were conducted.

The dielectric constant of the crystal was calculated using the relation

$$\varepsilon_r = \frac{C_{crys}}{C_{air}} \tag{2}$$

As the crystal area was smaller than the plate area of the cell, parallel capacitance of the portion of the cell not filled with the crystal was taken into account and, consequently, the above equation becomes

$$\varepsilon_r = \left(\frac{C_{crys} - C_{air}\left(1 - \frac{A_{crys}}{A_{air}}\right)}{C_{air}} \right) \left(\frac{A_{air}}{A_{crys}} \right) \tag{3}$$

where A_{crys} is the area of the crystal touching the electrode and A_{air} is the area of the electrode.

Figure 12 shows the temperature dependence of dielectric constants of pure and LAHCl doped (1 and 5 mol%) ADP crystals. Temperature dependence of dielectric constants of pure and L-alanine doped (1 and 5 mol%) ADP crystals are depicted in Figure 13.

It is observed from the figures that the dielectric constant increases with increase in temperature. This is normal dielectric behaviour of an antiferroelectric ADP crystal. In the present study, it has been observed that the LAHCl and L-alanine doped ADP crystals have lower ε_r values compared to pure. Among these, 5 mol% doped crystals have lower ε_r values than 1 mol% doped ones. Suitable dopants added in suitable concentrations can reduce the ε_r value to a lower one as observed in the case of KDP single crystals added with urea (Goma et al., 2006). Thus, in effect, the present study indicates that LAHCl and L-alanine doped ADP crystals are not only potential NLO materials but also low ε_r value dielectric materials, which will be useful for microelectronic industries and electro-optic modulators.

The dielectric loss of the grown crystals for various temperatures at the frequency (1 kHz) is shown in Figures 14 and 15. It is observed that the dielectric loss increases with increase in temperature for the crystals. It reveals that doped crystals have lower dielectric loss values compared to pure crystals.

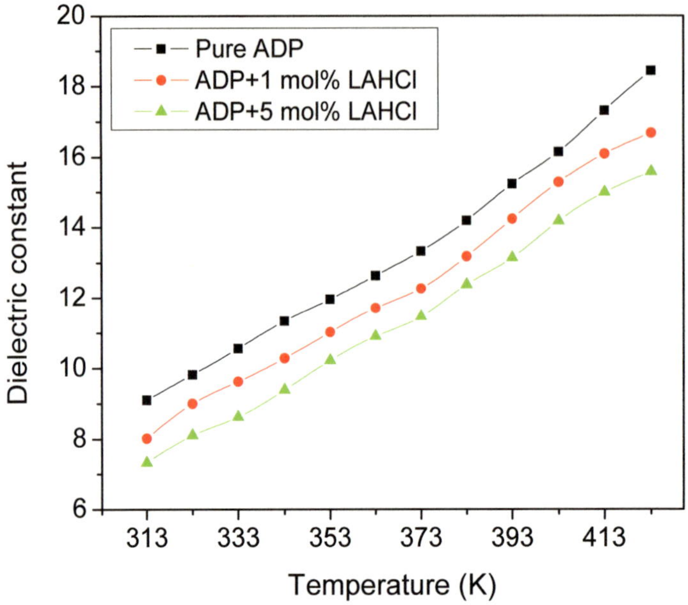

Fig. 12. Variation of dielectric constant with temperature for pure and LAHCl doped ADP crystals.

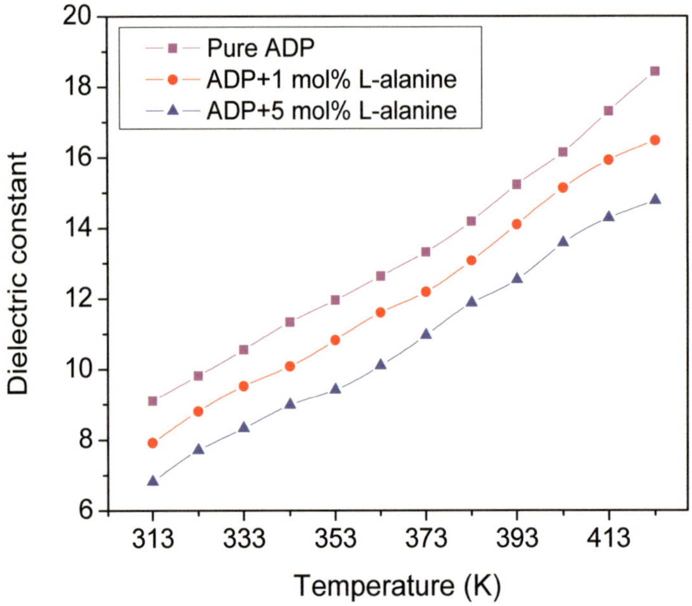

Fig. 13. Variation of dielectric constant with temperature for pure and L-alanine doped ADP crystals.

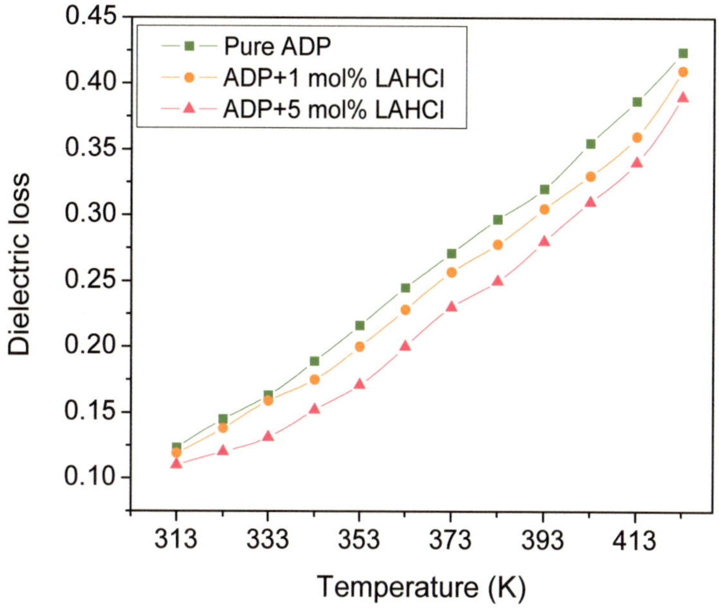

Fig. 14. Variation of dielectric loss with temperature for pure and LAHCl doped ADP crystals.

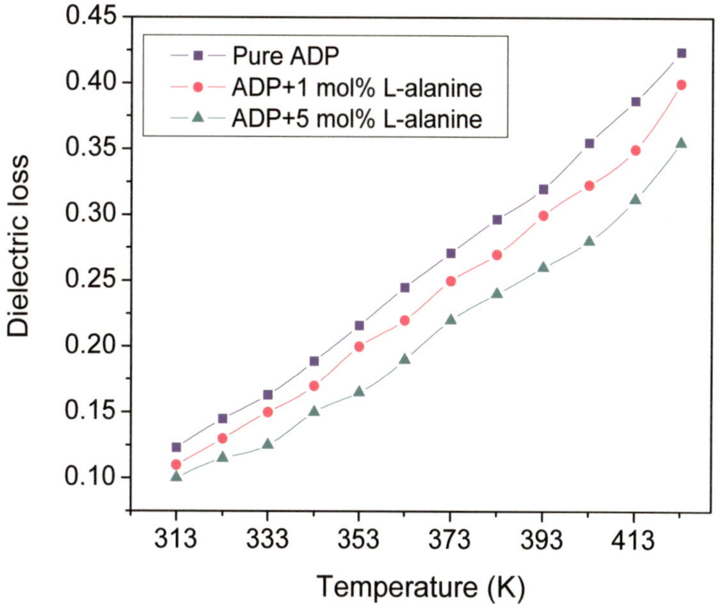

Fig. 15. Variation of dielectric loss with temperature for pure and L-alanine doped ADP crystals.

3.5 Optical transmission studies

Optical transmission spectra were recorded for the samples obtained from pure as well as from doped crystals grown by the slow cooling method. The spectra were recorded in the wavelength region 200-1100 nm using Lambda 35 spectrophotometer. Crystal plates with 2 mm thickness were used for the study. The reported value of the optical transparency for ADP is from 184 to 1500 nm (Dmitriev et al., 1991). The UV-vis-NIR spectra recorded for pure and doped ADP crystals are shown in Figure 16. It is clear from the figure that the crystals have sufficient transmission (pure ADP has 70% whereas LAHCl and L-alanine doped ADP have 78% and 82% respectively) in the entire visible and IR region. The optical transparency of the ADP crystal is increased by the addition of LAHCl and L-alanine. The addition of the amino acid dopants in the optimum conditions to the solution is found to suppress the inclusions and improve the quality of the crystal with higher transparency.

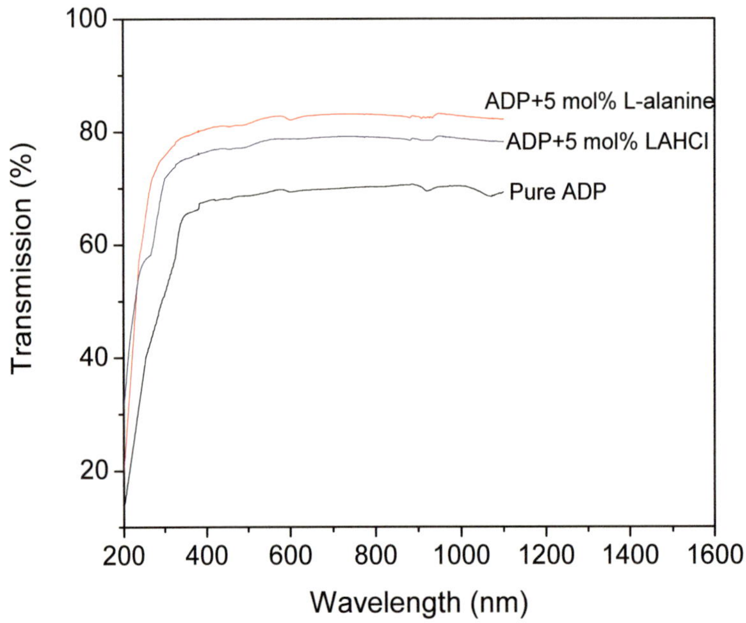

Fig. 16. Transmission spectra of pure and doped ADP crystals.

3.6 NLO property

Kurtz and Perry (1968) proposed a powder second harmonic generation method for comprehensive analysis of the second order nonlinearity. This is an important method for characterizing the materials before going through the long and tedious process of growing large optical quality crystals. To determine the SHG conversion efficiency of doped crystals, pure and doped crystals were ground into powder and densely filled into the cells. A Q-switched Nd: YAG laser (DCR11) was used as a light source. A laser beam of fundamental wavelength 1064 nm, 8 ns pulse width, with 10 Hz pulse rate was made to fall normally on the sample cell. The power of the incident beam was measured using a power meter. The transmitted fundamental wave was passed over a monochromator (Czemy Turner monochromator), which separates 532 nm (second harmonic signal) from 1064 nm, and absorbed by a $CuSO_4$ solution, which removes the 1064 nm light, and passed through BG34 filter to remove the residual 1064 nm light and an interference filter with bandwidth of 4 nm and central wavelength of 532 nm. The green light was collected by a photomultiplier tube (Hamamatsu). The input laser energy incident on the powdered sample was 1.35 mJ/pulse. An emission of green light was seen in all the samples. It was observed that the measured SHG efficiency of L-alanine doped (5 mol%) ADP was 1.75 and LAHCl doped (5 mol%) ADP was 1.5 that of pure ADP.

3.7 Piezoelectric measurements

Piezoelectricity is the ability of certain crystals to generate an electric charge when subjected to mechanical stress. The piezoelectric property is related to the polarity of the material (Ge et al 2008). The piezoelectric studies were made using piezometer system. A precision force generator applied a calibrated force (0.25 N) which generated a charge on the piezoelectric material under test. An oscilloscope gives the output in d_{33} coefficient in of pC/N units. Piezoelectric measurements were conducted for the grown crystals by without polishing of the crystal. Pure ADP crystal gives the piezoelectric coefficient (d_{33}) value of 0.37 pC/N. The obtained piezoelectric coefficient (d_{33}) values for LAHCl doped (5 mol%) ADP and L-alanine doped (5 mol%) ADP crystals are 0.68 and 0.8 pC/N. Thus, d_{33} value of 5 mol% LAHCl doped ADP crystal is 1.84, while and for 5 mol% L-alanine doped ADP crystal it is 2.16 times higher than that of pure ADP crystal. Greater crystalline perfection may be the reason for the increase in piezoelectric efficiency.

4. Conclusions

With the aim of improving the quality of ADP crystals with better optical properties for both academic and industrial uses, an attempt has been made in this present work to grow the ADP crystals by doping it with new additives L-arginine monohydrochloride and L-alanine. The addition of these amino acid materials enhances the metastable zone width and induction period of pure ADP solution. Also, during the experiment it was observed that the number of tiny crystals formed by spontaneous nucleation was appreciably reduced in the case of doped solution. It is believed that the addition of these amino acid materials suppresses the activities of the metal ion impurities present in the solution which enables larger metastable zone width and faster growth rate. HRXRD curves recorded for 5 mol% doped crystals have excellent crystalline perfection. The FTIR spectrum shows that amino acid additives have entered into the ADP crystals. The transmission spectrum reveals that the crystal has sufficient transmission in the entire visible and IR region. The SHG conversion efficiency and piezoelectric coefficient values of doped crystals are higher than that of pure. This study will help the growth of high quality large size ADP single crystals.

5. References

Aithal, P. S., Nagaraja, H. S., Mohan Rao. P., Avasti, D. K. & Sarma, A (1997). Effect of high energy ion irradiation on electrical and optical properties of organic nonlinear optical crystals, *Vacuum*, Vol. 48, (No. 12), pp. 991-994.

Alexandru, H. V. & Antohe (2003). Prismatic faces of KDP crystal, kinetic and mechanism of growth from solutions, *J. Cryst. Growth*, Vol. 258, (No. 1-2), pp. 149-157.

Asakuma, Y., Ukita, E., Maeda, K., Fukui, K., Iimura, K., Suzuki, M. & Hirota, M (2007). Surface topography of dyed potassium dihydrogen phosphate (KDP) Crystals, *Cryst. Growth Des.*, Vol. 7, (No. 2), pp. 420-424.

Batra, A. K., Stephens, J., Bhat, K., Aggarwal, M. D., Peterson, B. H., Curley, M. & Lal, R. B. (2005). Investigation on the growth kinetics of KDP: LAP and TGS: LAP single crystals, *Proc. of SPIE*, Vol. 5912, pp. 591206-591212.

Batterman, B. W. & Cole, H. (1964). Dynamical diffraction of X-rays by perfect crystals, *Rev. Mod. Phys.*, Vol. 36, (No. 3), pp. 681-717.

Bhagavannarayana, G. (1994). *High resolution X-ray diffraction study of as-grown and BF_2^+ implanted silicon single crystals*, Ph. D. Thesis, University of Delhi, Delhi, India.

Bonse, U. & Hart, M. (1965). Tailless X-ray single crystal reflection curves obtained by multiple reflection, *Appl. Phys. Lett.*, Vol. 7, (No. 9), pp. 238-240.

Boukhris, A., Souhassou, M., Lecomte, C., Wyneke, B. & Thalal, A. (1998). Evolution of the structural and mean square displacement parameters in $(NH_4)_xK_{1-x}H_2PO_4$ solid solutions versus concentration and temperature, *J. Phys.: Condens. Matter*, Vol. 10, (No. 7), pp. 1621-1626.

Davey, R. J. & Mullin, J. W. (1974). Growth of the {100} faces of ammonium dihydrogen phosphate crystals in the presence of ionic species, J. Cryst. Growth, Vol. 26, (No. 1), pp. 45-51.

Dmitriev, V. G., Gurzadyan, G. G. & Nikogosyan, D. N. (1991). *Handbook of nonlinear optical crystals*, Springer, Berlin.

Fu, Y-J., Gao, Z-S., Sun, X., Wang, S-L., Li, Y-P., Zeng, H., Luo, J-P., Duan, A-D. & Wang, J-Y. (2000). Effects of anions on rapid growth and growth habit of KDP crystals, *Prog. Cryst. Growth Charact.*, Vol. 40, (No. 1-4), pp. 211-220.

Ge, W., Liu, H., Zhao, X., Pan, X., He, T., Lin, D., Xu, H. & Luo, H. (2008). Growth and characterization of $Na_{0.5}Bi_{0.5}TiO_3$–$BaTiO_3$ lead-free piezoelectric crystal by the TSSG method, *J. Alloys and Compounds*, Vol. 456, (No. 1-2), pp. 503-507.

Goma, S., Padma, C. M. & Mahadevan, C. K. (2006). Dielectric parameters of KDP single crystals added with urea, *Mater. Lett.*, Vol. 60, (No. 29-30), pp. 3701-3705.

Gunning, M. J., Raab, R. E. & Kucharczyk, W. (2001). Magnitude and nature of the quadratic electro-optic effect in potassium dihydrogen phosphate and ammonium dihydrogen phosphate crystals, *J. Opt. Soc. Am. B.*, Vol. 18, (No. 8), pp. 1092-1098.

Hewat, A. W. (1973). Location of hydrogen atoms in ADP by neutron powder profile refinement, *Nature*, Vol. 246, pp. 90-91.

Kern, R. & Dassonville, R. (1992). Growth inhibitors and promoters exemplified on solution growth of paraffin crystals, *J. Cryst. Growth*, Vol. 116, (No. 1-2), pp. 191-203.

Kubota, N., Fukazawa, J., Yashiro, H. & Mullin, J. W. (1995). Impurity effect of chromium (III) on the growth and dissolution rates of potassium sulfate crystals, *J. Cryst. Growth*, Vol. 149, (No. 1-2), pp. 113-119.

Kumaresan, P., Moorthy Babu, S. & Anbarasan, P. M. (2008). Thermal, dielectric studies on pure and amino acid (L-glutamic acid, L-histidine, L-valine) doped KDP single crystals, *Opt. Mater.*, Vol. 30, (No. 9), pp. 1361-1368.

Kurtz, S. K. & Perry, T. T. (1968). A powder technique for the evaluation of nonlinear optical materials, J. Appl. Phys., Vol. 39, (No. 8), pp. 3798-3813.

Lal, K. & Bhagavannarayana, G. (1989). A high-resolution diffuse X-ray scattering study of defects in dislocation-free silicon crystals grown by the float-zone method and comparison with Czochralski-grown crystals, *J. Appl. Cryst.*, Vol. 22, (No. 3), pp. 209-215.

Li, G., Liping, X., Su, G., Zhuang, X., Li, Z. & He, Y. (2005). Study on the growth and characterization of KDP-type crystals, *J. Cryst. Growth*, Vol. 274, (No. 3-4), pp. 555-562.

Matsushita, E. & Matsubara, T. (1987). The role of hydrogen bonds in antiferroelectricity of $NH_4H_2PO_4$, *J. Phy. Soc. Jpn*, Vol. 56, pp. 200-207.

Meena, M. & Mahadevan, C. K. (2008). Growth and electrical characterization of L-arginine added KDP and ADP single crystals, *Cryst. Res. Technol.*, Vol. 43, (No. 2), pp. 166-172.

Meenakshisundaram, S., Parthiban, S., Madhurambal, G. & Mojumdar, S. C. (2009). Effect of low and high concentrations of KCl dopant on ADP crystal properties, *J. Therm. Anal. Calorimetry*, Vol. 96, pp. 77-80.

Meera, K., Muralidharan, R., Tripathi, A. K. & Ramasamy, P. (2004). Growth and characterisation of L-threonine, DL-threonine and L-methionine admixtured TGS crystals, *J. Cryst. Growth*, Vol. 263, (No. 1-4), pp. 524-531.

Monaco, S. B., Davis, L. E., Velso, S. P., Wang, F. T., Eimerl, D. & Zalkin, A. (1987). Synthesis and characterization of chemical analogs of L-arginine phosphate, *J. Cryst. Growth*, Vol. 85, (No. 1-2), pp. 252-255.

Mullin, J. W. (1993). *Crystallization*, Butterworth-Heinemann, Oxford, London.

Nakatsuka, M., Fujioka, K., Kanabe, T. & Fujita, H. (1997). Rapid growth over 50 mm/day of water-soluble KDP crystal, *J. Cryst. Growth*, Vol. 171, (No. 3-4), pp. 531-537.

Nyvlt, J., Rychly, R., Gottfried, J. & Wurzelova, J. (1970). Metastable zone-width of some aqueous solutions, *J. Cryst. Growth*, Vol. 6, (No. 2), pp. 151-162.

Parikh, K. D., Dave, D. J., Parekh, B. B. & Joshi, M. J. (2007). Thermal, FT-IR and SHG efficiency studies of L-arginine doped KDP crystals, *Bull. Mater. Sci.*, Vol. 30, (No. 2), pp. 105-112.

Podder, J., Ramalingom, S. & Kalkura, S. N. (2001). An investigation on the lattice distortion in urea and KCl doped KDP single crystals by X-ray diffraction studies, *Cryst. Res. Technol.*, Vol. 36, (No. 6), pp. 549-556.

Podder, J. (2002). The study of impurities effect on the growth and nucleation kinetics of potassium dihydrogen phosphate, *J. Cryst. Growth*, Vol. 237-239, (No. 1), pp. 70-75.

Rajesh, N. P., Meera, K., Srinivasan, K., Santhana Raghavan, P. & Ramasamy, P. (2000). Effect of EDTA on the metastable zone width of ADP, *J. Cryst. Growth*, Vol. 213, (No. 3-4), pp. 389-394.

Rajesh, P., Ramasamy, P. & Bhagavannarayana, G. (2009). Effect of ammonium malate on growth rate, crystalline perfection, structural, optical, thermal, mechanical, dielectric and NLO behaviour of ammonium dihydrogen phosphate crystals, *J. Cryst. Growth*, Vol. 311, pp. 4069–4075.

Rak, M., Eremin, N. N., Eremina, T. A., Kzunetsov, V. A., Okhrimenko, T. M., Furmanova, N. G. & Efremova, E. P. (2005). On the mechanism of impurity influence on growth kinetics and surface morphology of KDP crystals-I: defect centres formed by bivalent and trivalent impurity ions incorporated in KDP structure-theoretical study, *J. Cryst. Growth*, Vol. 273, (No. 3-4), pp. 577-585.

Rani, T. J., Loretta, F., Selvarajan, P., Ramalingom, S. & Perumal, S. (2011). Growth Structural and Spectral Studies on L-Proline Added Ammonium Dihydrogen Phosphate Single Crystals, *Recent Research in Science and Technology, Vol. 3, (No. 7), pp. 69-72.*

Razzetti, C., Ardoino, M., Zanotti, L., Zha, M. & Paorici, C. (2002). Solution growth and characterisation of L-alanine single crystals, *Cryst. Res. Technol.*, Vol. 37, (No. 2), pp. 456-465.

Ren, X., Xu, D. & Xue, D. (2008). Crystal growth of KDP, ADP, and KADP, *J. Cryst. Growth*, Vol. 310, (No. 7-9), pp. 2005-2009.

Sangwal, K. (1989). On the estimation of surface entropy factor, interfacial tension, dissolution enthalpy and metastable zone-width for substances crystallizing from solution, *J. Cryst. Growth*, Vol. 97, (No. 2), pp. 393-405.

Sangwal, K. (1996). Effects of impurities on crystal growth processes, Prog. *Cryst. Growth Charact.*, Vol. 32, (No. 1-3), pp. 3-43.

Shantha, K., Philip, S. & Varma, K. B. R. (1997). Effect of KCl addition on the microstructural and dielectric properties of bismuth vanadate ceramics, *Mater. Chem. Phys.*, Vol. 48, (No. 1), pp. 48-51.

Srinivasan, K. (1997). *Growth of device quality KDP family single crystals, fabrication of nonlinear and electro-optical devices out of them and growth of ADP-KDP mixed crystals and their characterization*, Ph. D. Thesis, Alagappa University, Karaikudi, India.

Srinivasan, K., Meera, K. & Ramasamy, P. (1999). Enhancement of metastable zone width for solution growth of potassium acid phthalate, *J. Cryst. Growth*, Vol. 205, (No. 3), pp. 457-459.

Tenzer, L., Frazer, B. C. & Pepinsky, R. (1958). A neutron structure analysis of tetragonal $NH_4(H_2PO_4)$, *Acta Crystallogr.*, Vol. 11, (No. 7), pp. 505-509.

Tukubo, H. & Makita, H. (1989). Refractometric studies of $NH_4H_2PO_4$ and KH_2PO_4 solution growth; experimental setup and refractive index data, *J. Cryst. Growth*, Vol. 94, (No. 2), pp. 469-474.

Ueda, R. (1948). Crystal structure of ammonium dihydrogen phosphate $NH_4H_2PO_4$, *J. Phys. Soc. Jpn.*, Vol.3, pp. 328-333.

Zaitseva, N. P., Rashkovich, L. N. & Bogatyreva, S. V. (1995). Stability of KH_2PO_4 and $K(H_2D)_2PO_4$ solutions at fast crystal growth rates, *J. Cryst. Growth*, Vol. 148, (No. 3), pp. 276-282.

Zaitseva, N. & Carman, L (2001). Rapid growth of KDP-type crystals, *Prog. Cryst. Growth Charact.*, Vol. 43, (No. 1), pp. 1-118.

5

Topochemical Conversion of Inorganic–Organic Hybrid Compounds into Low-Dimensional Inorganic Nanostructures with Smart Control in Crystal-Sizes and Shapes

Deliang Chen
School of Materials Science and Engineering,
Zhengzhou University,
P. R. China

1. Introduction

Nanomaterials with low-dimensional morphologies, including 0D QDs, 1D nanowires (nanotubes and nanoneedles), and 2D nanoplates (nanosheets and nanodisks), have giant potential applications in catalysts, sensors, and medical treatment. Since the end of the last century, there have developed various methods to synthesize low-dimensional nanocrystals. The well-established synthetic methods mainly include CVD, PVD, Sol-Gel processes, soft/hard templating processes, and mechanical balling, most of which have successfully been used to fabricate various nanomaterials with sophisticate morphologies. But at the same time, we usually find a few insurmountable issues when using these methods to synthesize nanomaterials on a large scale: (1) the poor uniformity in crystal-size and shape, (2) the serious agglomeration of small nanoparticles, (3) the high synthetic costs and low outputs, and so on. These issues restrict the extensive applications of low-dimensional nanomaterials to a large extent.

Exfoliation of layered compounds is a common way to synthesize 2D nanocrystals. Colloid nanosheets of titanate, niobate, MnO_2, clays, and layered double hydroxides (LDHs) have been synthesized from totally exfoliating their corresponding layered compounds. The nanosheets obtained by such exfoliation processes are usually extremely thin and tend to form hard aggregates easily when the solvents are removed. At the same time, nanosheets can be used as building blocks to construct novel metastable 3D structures through condensation, thermal decomposition, and recrystallization, allowing fine control over their microscopic or macroscopic structural features.

Recently, we have developed a novel strategy to synthesize low-dimensional nanocrystals on the basis of interaction chemistry. The key points of this method lie in two aspects: (1) synthesis of inorganic-organic hybrid precursors with controlled morphologies and shapes, and (2) topochemical conversion of inorganic-organic hybrid precursors to low-dimensional inorganic nanocrystals by removing the organic species. Fig. 1 shows the schematic process of conversion of tungstate-based inorganic–organic hybrid belts to WO_3 nanoplates via an intermediate product of H_2WO_4 nanoplates. The similar process can be used to synthesize MoO_3 nanoplates. The outstanding advantages of the topochemical process can be described as the followings: (1) The starting materials are easily obtained without captious requirements; (2) The synthetic process is not sensitive to the operation parameters and has high repeatability and stability in product quality; (3) This process is suitable for large-scale and cost-effective synthesis of low-dimensional nanomaterials.

This chapter deals with the synthesis of WO_3 and MoO_3 nanoplates using H_2WO_4, $H_2W_2O_7$, and $MoO_3{\cdot}H_2O$ solid powders as the starting materials.[1-7] The synthesis of carbides and nitrides of tungsten is also introduced using the as-obtained tungstate-based inorganic-organic hybrid compounds as precursors.[8,9] The applications of WO_3 and MoO_3 nanoplates in sensors are also introduced.[6,10,11] The related mechanisms are discussed in some details.[1-11]

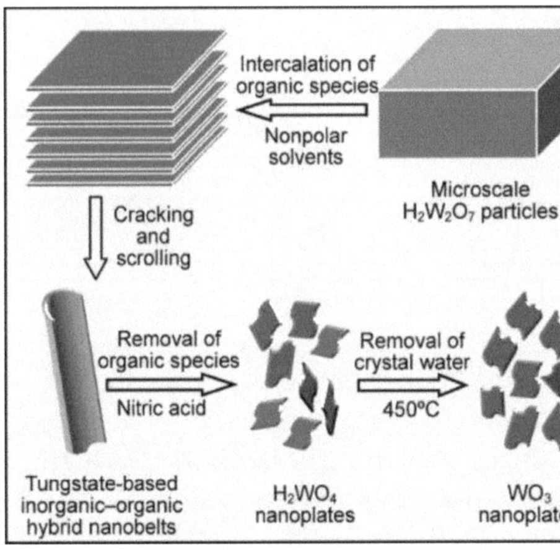

Fig. 1. The schematic representation of the conversion process from microscale $H_2W_2O_7$ particles and tungstate-based inorganic–organic hybrid belts to WO_3 nanoplates via an intermediate product of H_2WO_4 nanoplates. Reprinted with permission from *Small*, 2008, 4(10), 1813-1822. Copyright 2008 Wiley-VCH Verlag GmbH & Co. KGaA.

2. Formation of tungstate-based inorganic–organic hybrid compounds [1-4]

2.1 Material synthesis and characterization

Tungstate-based inorganic-organic hybrid belts were synthesized through reactions of tungstate acids and alkylamines in a nonpolar solvent at ambient conditions. There were two processes to form tungstate-based inorganic-organic hybrid micro-/nanobelts: one was using $H_2W_2O_7{\cdot}xH_2O$ as the host material (i.e., the $H_2W_2O_7{\cdot}xH_2O$ process) and the other was using commercial H_2WO_4 as the host material (i.e., the H_2WO_4 process).

For the $H_2W_2O_7{\cdot}xH_2O$ process, a stoichiometric mixture of Bi_2O_3 and WO_3 was calcined at 800 °C for 2 days with intermittent grinding to synthesize $Bi_2W_2O_9$ powders. Selective leaching of Bi_2O_2 layers from the as-obtained $Bi_2W_2O_9$ by HCl treatment led to the formation of its protonated phase, $H_2W_2O_7{\cdot}xH_2O$. The reactions between $H_2W_2O_7{\cdot}xH_2O$ and *n*-alkylamines were carried out at room temperature under an ambient atmosphere. The same procedure was applied for *n*-alkylamines with various alkyl chain lengths ($C_mH_{2m+1}NH_2$, $4\leq m\leq 14$). The molar ratios of *n*-alkylamine to $H_2W_2O_7{\cdot}xH_2O$ were about 5-30, and the volume ratios of heptane to *n*-alkylamine were maintained at about 2-10. Typically, about 0.3 g of the air-dried $H_2W_2O_7 \cdot xH_2O$ was dispersed in the mixtures of heptane and *n*-alkylamine with magnetic stirring. After reacting for 30 min or 5 days, the products were collected from the suspensions by centrifugation and washed with ethanol. Other solvents were also used as the solvents for the reactions between $H_2W_2O_7 \cdot xH_2O$ and *n*-octylamine under the similar conditions.

For the H_2WO_4 process, 10 g of H_2WO_4 powders was dispersed in a mixture of 0.4 mol of *n*-octylamine and 530 ml of heptane under a constant magnetic stirring at room temperature for 24 h. The molar ratio of *n*-octylamine to H_2WO_4 was 10 and the volume ratio of heptane to *n*-octylamine was 8. After another reaction time of 2 days, the resultant white solids were collected by centrifugation, washed with ethanol, and then dried under a reduced pressure at room temperature for 2 days. The dried sample was tungstate-based inorganic–organic hybrid nanobelts

X-ray diffraction (XRD), scanning electron microscopy (SEM), transmission electron microscopy (TEM), thermogravimetry (TG), CHN analysis and Fourier transform infrared (FT-IR) spectra were used to characterize the microstructures and compositions of the as-obtained hybrid compounds.

2.2 Results and discussion

Fig. 2 shows the typical morphology and microstructure of a tungstate-based inorganic-organic hybrid obtained with a molar ratio of *n*-octylamine to $H_2W_2O_7 \cdot xH_2O$ of about 30 and a volume ratio of heptane to *n*-octylamine of about 2. The low-magnification SEM image (Fig. 2a) indicates that the product possesses a filamentous morphology with lengths of 10–20 μm. A typical FE-SEM image is shown in Fig. 2b. It is clear that the filamentous structures are nanobelts, most of which are scrolled to make nanotubes with apparent diameters of 200–500 nm. The TEM image (Fig. 2c) corroborates that the product obtained exhibits a belt/tubelike morphology. The high-resolution TEM image at point A of Fig. 2c is shown as Fig. 2d, which indicates that the nanobelt/nanotube has a lamellar structure along its length. The thickness of the nanobelt is 20–50 nm. Its XRD pattern suggests that the interlayer distance of the lamellar structure is 2.59(1) nm.

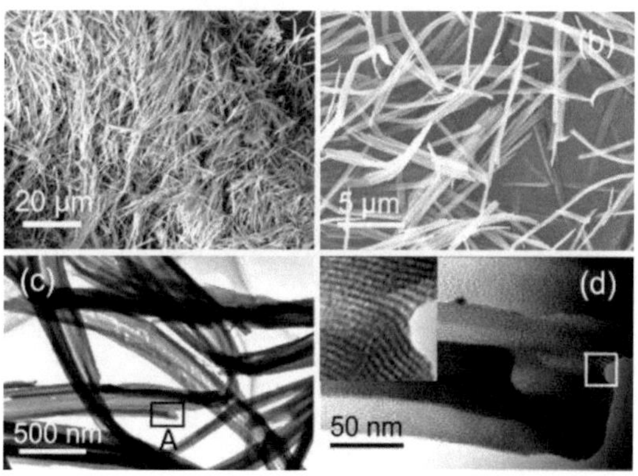

Fig. 2. Typical morphology of the tungstate-based inorganic–organic hybrid accommodating *n*-octylamine with a reaction period of 120 h. Reprinted with permission from *Key Eng. Mater.*, 2007, 352, 85-88. Copyright 2007 Trans Tech Publications Inc.

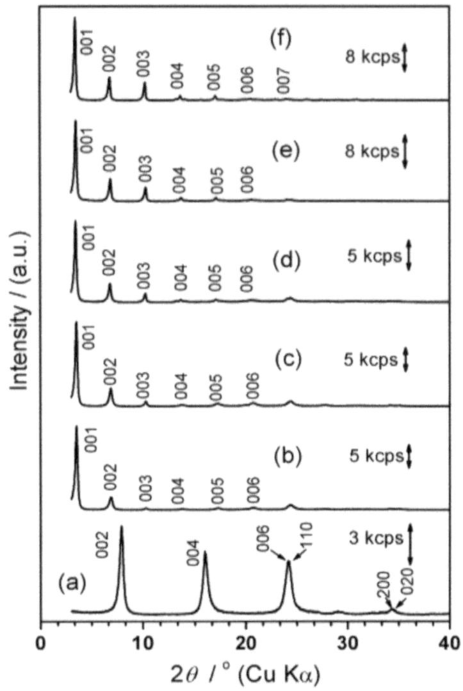

Fig. 3. XRD patterns of the products accommodating *n*-octylamine obtained in a system of heptene/*n*-octylamine/$H_2W_2O_7 \cdot xH_2O$ with various reaction times (*n*-octylamine/ $H_2W_2O_7 \cdot xH_2O$=30): (a) 0 h ($H_2W_2O_7 \cdot xH_2O$), (b) 0.5 h, (c) 1 h, (d) 3 h, (e) 5 h and (f) 8 h. Reprinted with permission from *Chem. Mater.*, 2007, 19(7), 1808-1815. Copyright 2007 American Chemical Society.

The XRD patterns of the products obtained after reaction times of 0, 0.5, 1, 3, 5 and 8 h with an *n*-octylamine/$H_2W_2O_7 \cdot xH_2O$ molar ratio of 30 are shown in Figs. 3a–f, respectively. As Fig. 3b shows, a new series of reflections occur in the low 2θ-angle region for the product with a reaction time of 0.5 h. At the same time, the reflections due to the air-dried $H_2W_2O_7 \cdot xH_2O$ (Fig. 3a) disappear. The sharp reflections in the low-2θ range can be indexed to (00*l*) reflections from a highly ordered lamellar structure. The number of identifiable reflections is up to six, but their intensities are very low for the reflections with $l \geq 3$. The interlayer distance estimated from the (00*l*) reflections is 2.570(9) nm. With increases in reaction time, the intensities of the (00*l*) reflections become stronger, especially for the reflections with $l \geq 3$, as shown in Figs. 3c–f. The interlayer distances of the products with reaction times of 1, 3, 5 and 8 h are 2.56(1), 2.58(1), 2.58(1) and 2.595(9) nm, respectively. There is no more obvious change in the XRD pattern when the reaction time is longer than 5 h, even after several days.

The morphologies of the products obtained at various stages are shown in Fig. 4. The host compound of $H_2W_2O_7 \cdot xH_2O$ takes on a particle-like morphology with distinct cleavage planes, and its particle size ranges from 5 μm to 20 μm (Fig. 4a). The product with a reaction time of 0.5 h (Fig. 4b) shows parallel cracks. The products obtained with longer reaction times (*e.g.* 1 h and 2 h) exhibit denser cracks (Figs. 4c and d). The product after a reaction time of 3 h (Fig. 4e) begins to be transformed into one-dimensional nanostructures. At this stage, the particles and the one-dimensional nanostructures coexist. With the continuation of the reaction to 5 h, most of the particles are transformed into one-dimensional nanostructures, as shown in Fig. 4f. The products obtained with reaction times of 8 and 25 h show similar one-dimensional nanostructures. The FE-SEM (lower left inset of Fig. 4f) and TEM images (Fig. 2c) indicate that the one-dimensional nanostructures are nanobelts or nanotubes with apparent diameters of 200–700 nm and lengths of 5–15 μm. The thicknesses of the nanobelts or the nanotube walls are 20–50 nm.

Fig. 4. SEM images of the products accommodating *n*-octylamine after various reaction times (*n*-octylamine/$H_2W_2O_7 \cdot xH_2O$=30): (a) 0 h ($H_2W_2O_7 \cdot xH_2O$), (b) 0.5 h, (c) 1 h, (d) 2 h, (e) 3 h and (f) 5 h (the lower left inset is an FE-SEM image of a sample with a reaction period of 120 h). Reprinted with permission from *Chem. Mater.*, 2007, 19(7), 1808-1815. Copyright 2007 American Chemical Society.

The TG analyses indicated that the mass-loss-curve profiles of the products with various reaction times are similar. The mass loss between room temperature and 600 °C increases from 21.8% for the product with a reaction time of 0.5 h to 52.9% for the product with a reaction time of 120 h, as summarized in Table 1. The CHN data are also listed in Table 1. The calculated ratios of C:H:N in moles are (7.8-8.0):(19.2-20.1):1, results very close to the composition (8:19:1 for C:H:N) of *n*-octylamine. The typical FT-IR spectra of the products showed a broad band appearing at around 2110 cm^{-1}, which is due to a combination of the asymmetrical bending vibration and torsional oscillation of the $-NH_3^+$ groups interacting with the apical oxygen of the W-O framework, i.e., $R-NH_3^+ \cdots O-W$. When the XRD, TG results and CHN data are taken into account, it can therefore be concluded that the products obtained should be inorganic-organic hybrids with lamellar mesostructures, in which the inorganic W-O layers and the organic species (*n*-octylamine ions) are stacked alternately.

t (h)	M_L (mass %)	C (mass %)	H (mass %)	N (mass %)	R
0.5	21.8	17.68	3.64	2.66	0.563
1	23.7	18.39	3.78	2.73	0.592
3	25.4	19.02	4.03	2.81	0.623
5	36.7	25.52	5.16	3.72	0.973
8	41.3	30.85	6.37	4.51	1.27
25	48.1	35.19	7.31	5.10	1.63
120	52.9	42.48	8.57	6.16	2.16

Table 1. Summary of mass loss (M_L) between room temperature and 873 K, data for CHN analysis, and the molar ratio (R) of *n*-octylamine taken up to W of the products as a function of reaction time (t). Reprinted with permission from *Chem. Mater.*, 2007, 19(7), 1808-1815. Copyright 2007 American Chemical Society.

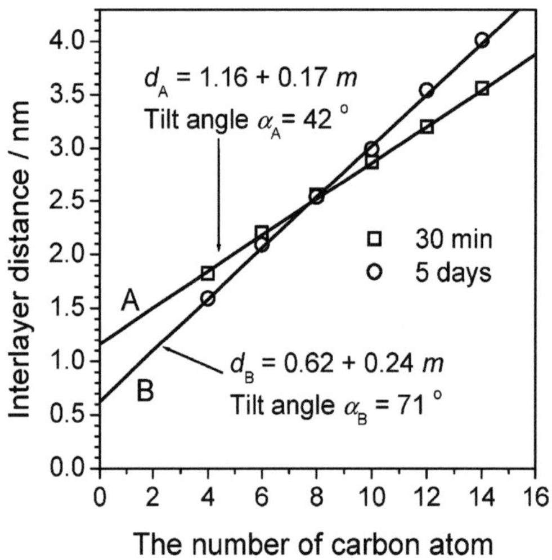

Fig. 5. The plots of the interlayer distance (d) versus the *n*-alkyl chain length (m, $C_mH_{2m+1}NH_2$) for (A) intermediates with a 30-min reaction time and (B) hybrid nanotubes/nanobelts with a reaction time of 5 days. Reprinted with permission from *Chem. Mater.*, 2007, 19(7), 1808-1815. Copyright 2007 American Chemical Society.

The arrangement of the *n*-alkyl chains and the thickness of the inorganic layers can be evaluated by analyzing the variations in interlayer distance versus the alkyl chain length. The relationship of the interlayer distance (d) to the carbon chain length (n_C) can generally be described as d (nm) = $d_0 + kn_C$, where k is the slope and d_0 is the intercept at $n_C = 0$, and that the increment per $-CH_2-$ for a fully extended all-trans alkyl chain is 0.127 nm. The slope k can therefore give useful information about the arrangement of the *n*-alkyl chains. When $k \leq 0.127$, the arrangement should be a monolayer with a tilt angle a ($a = \sin^{-1}(k/0.127)$) or a bilayer with a smaller tilt angle. When $0.127 < k \leq 0.254$, a bilayered arrangement with a tilt angle of $a = \sin^{-1}(k/0.254)$ is usually considered. The intercept d_0 corresponds to the sum of the thickness of the inorganic layer and the spatial separation due to the amine functional groups.

Fig. 5 shows the plots of the interlayer distance (d) versus the *n*-alkyl chain length (m, $C_mH_{2m+1}NH_2$) for intermediates with a 30-min reaction time and hybrids with a reaction time of 5 days. From Fig. 5A, it can be concluded that the *n*-alkyl chains in the intermediates present a bilayered arrangement with a tilt angle a_A = 42 °, and the thickness of the inorganic layers can be calculated to be 1.00 nm, which is very close to the thickness (0.93 nm) of the double-octahedral W-O layers, suggesting that the double-octahedral W-O layers are preserved in the intermediates obtained in an early stage. Similarly, the slope coefficient of k_B = 0.24 suggests that the *n*-alkyl chains in the hybrids take on a bilayered arrangement with a tilt angle of $a_B = \sin^{-1}(0.24/0.254)$ =71° (Fig. 5A). The thickness of the inorganic layers can be estimated to be 0.39 nm, much lower than the thickness (0.931 nm) of the double-octahedral W-O layers, but very similar to the thickness (0.403 nm) of the single-octahedral W-O layers in H_2WO_4 or $H_2WO_4 \cdot H_2O$.

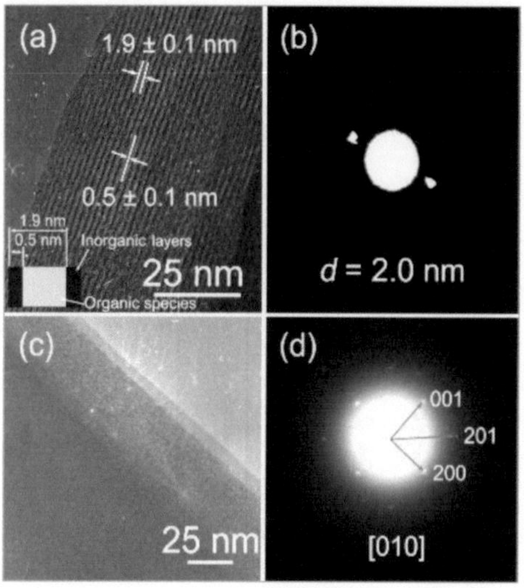

Fig. 6. TEM observation of a tungstate-based organic/inorganic hybrid accommodating *n*-octylamine obtained in heptane with *n*-octylamine/$H_2W_2O_7 \cdot xH_2O$ = 30: (a) a high-resolution TEM image and (b) an SAED pattern obtained with an incident electron beam parallel to the inorganic layer planes; (c) a high-magnification TEM image and (d) an SAED pattern along the [010] zone axis obtained with an incident electron beam perpendicular to the inorganic layer planes. Reprinted with permission from *Chem. Mater.*, 2007, 19(7), 1808-1815. Copyright 2007 American Chemical Society.

Fig. 6 shows the TEM observations of the inorganic layers in the layered inorganic-organic hybrids with a reaction time of 5 days. Figs. 6a and b show the high-resolution TEM image and the corresponding SAED pattern, respectively, of a case in which the incident electron beam is parallel to the inorganic layers. A black-to-white striped structure with a periodical range interval is observed in Fig. 6a. The black stripes belong to inorganic layers and the white ones are organic species. The interlayer distance and the inorganic thickness can be estimated as 1.9 nm and 0.5 nm, respectively. The one-dimensional diffraction lattices shown in Fig. 6b are typically characteristic of an ordered lamellar structure, the interlayer distance of which is about 2.0 nm calculated according to the equation $d = L\lambda / R$. The interlayer distance (2.0 nm) observed by TEM is slightly smaller than that obtained from the XRD results (about 2.5 nm), probably due to redispersing the hybrid in ethanol to make TEM samples. The directly observed thickness of the inorganic layer is much smaller than that of the double octahedral W-O layers but close to the single-octahedral W-O layers. The TEM image and its corresponding SAED pattern obtained with the incident electron beam perpendicular to the inorganic layers are shown in Figs. 6c and d, respectively. The discernible two-dimensional diffractional lattices in the SAED pattern can be used to determine the phase of the inorganic layers. As Fig. 6d shows, it can be indexed to the $H_2WO_4 \cdot H_2O$ phase (JCPDS 18-1420) along the [010] zone axis. The result corroborates that the belt-/tubelike hybrids consist of single-octahedral W-O layers and organic layers.

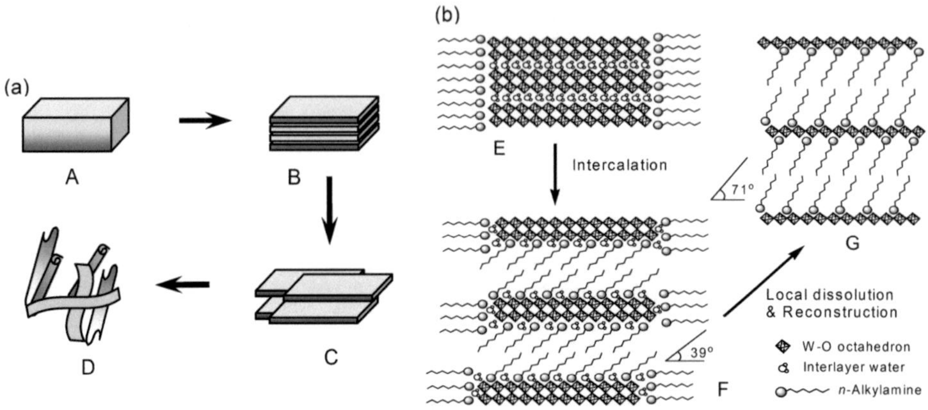

Fig. 7. (a) Macroscopically morphological evolution during the formation of one-dimensional tungstate-based organic/inorganic nanostructures in a system of heptane/*n*-alkylamine/$H_2W_2O_7 \cdot xH_2O$ particles. (b) A formation mechanism of tungstate-based organic/inorganic hybrids with single-octahedral W–O layers from layered $H_2W_2O_7 \cdot xH_2O$ solids with double-octahedral W–O layers. Reprinted with permission from *Chem. Mater.*, 2007, 19(7), 1808-1815. Copyright 2007 American Chemical Society.

Fig. 7a illustrates the possible pathway to form belt/tubelike tungstate-based inorganic-organic hybrids. In the very early stages, the microscale $H_2W_2O_7 \cdot xH_2O$ particles are gradually cleaved into thin plates, which are aggregates of alternately stacked inorganic-organic layers with thicknesses of 200-800 nm (shown as A →B → C). With the continuation of the reaction, the thin plates are dissolved from their edges to form the final filamentous nanobelts, as shown in C → D. Some of the nanobelts with large widths are spontaneously scrolled into nanotubes to minimize the surface energy. The microstructural development of the inorganic W-O layers and the arrangement of *n*-alkyl chains are illustrated in Fig. 7b. The hydrophilic $H_2W_2O_7 \cdot xH_2O$ particles with interlayer water are first dispersed into the nonpolar heptane with the assistance of amphiphilic *n*-alkylamine molecules. The polar heads (-NH_2) of the *n*-alkylamines spread to the surfaces of the inorganic particles, and their alkyl tails spread toward the nonpolar continuous phase, as shown in E. In such reverse-micelle-like media with an excess of *n*-alkylamine, the *n*-alkylamine molecules can directionally diffuse from the nonpolar phase to the interlayer spaces of $H_2W_2O_7 \cdot xH_2O$ particles under the interactions of proton transfer. Consequently, normal intercalation occurs easily. The double-octahedral W-O layers are transitorily preserved in the early stage intermediates, constructed by alternately stacking the double-octahedral W-O layers and the bilayer-arranged *n*-alkylammonium ions with a small tilt angle. A typical microstructural model is shown as F. Upon intercalation of *n*-alkylamine, the interlayer water molecules are released to become "free" water, which then reacts with the surrounding *n*-alkylamine molecules to form highly alkaline solutions in the reverse-micelle-like media. In these space-confined highly alkaline solutions, the double-octahedral W-O layers are therefore dissolved from their edges, and the resultant species subsequently are re-crystallized to form highly ordered lamellar mesostructures with an alternate stacking of single-octahedral W-O layers and bilayer-arranged *n*-alkyl chain arrays with a large tilt angle, as shown in G.

Fig. 8. (a) An SEM image of a hybrid obtained in heptane with an *n*-octylamine-to-$H_2W_2O_7 \cdot xH_2O$ ratio of 2; (b–h) SEM images of hybrids obtained in various solvents with an *n*-octylamine-to-$H_2W_2O_7 \cdot xH_2O$ ratio of 30: (b) ethanol, (c) tetrahydrofuran, (d) 1-octanol, (e) pentane, (f) decane, (g) 2,2,4-trimethylpentane and (h) cyclohexane. Reprinted with permission from *Chem. Mater.* 2007, 19(7), 1808-1815. Copyright 2007 American Chemical Society.

Fig. 8a displays a typical SEM image of the product obtained after a reaction time of 5 days with a molar n-octylamine-to-$H_2W_2O_7 \cdot xH_2O$ ratio of 2. Its major morphology is platelike particles with a diameter of about 10 μm, which differs distinctly from the belt/tubelike morphology of the hybrids obtained with high *n*-octylamine: $H_2W_2O_7 \cdot xH_2O$ molar ratios (e.g., 30). The XRD pattern showed that the product obtained with an *n*-octylamine: $H_2W_2O_7 \cdot xH_2O$ molar ratio of 2 also presented an ordered lamellar structure with an interlayer distance of d = 2.56(2) nm, very close to that of the intermediate obtained after a 30 min reaction, suggesting that the microstructure of the hybrid obtained with a very low molar *n*-octylamine-to-$H_2W_2O_7 \cdot xH_2O$ ratio is very similar to that of the intermediate obtained with a short reaction time.

The solvent types also have important influences on the reaction behavior of $H_2W_2O_7 \cdot xH_2O$ with *n*-alkylamines. No reaction was observed, or the reactions occurred to a very limited extent, in the solvents of 2-propanol, butanol, and acetone. In the cases of ethanol and THF, no belt/tubelike shapes, but instead platelike ones are observed, as shown in Figs. 8b and c. When 1-octanol is used as the reaction solvent, an ordered lamellar hybrid, with beltlike shapes besides a very small fraction of plates, is obtained (Fig. 8d). With the solvents of nonpolar alkanes, on the other hand, including not only heptane but also other alkanes, such as pentane, decane, 2,2,4-trimethylpentane, and cyclohexane, inorganic-organic hybrids with uniform belt/tubelike morphology are readily obtained, as shown in Figs. 8e-h. Their XRD patterns indicated that all these products obtained with other alkanes possess highly ordered lamellar mesostructures.

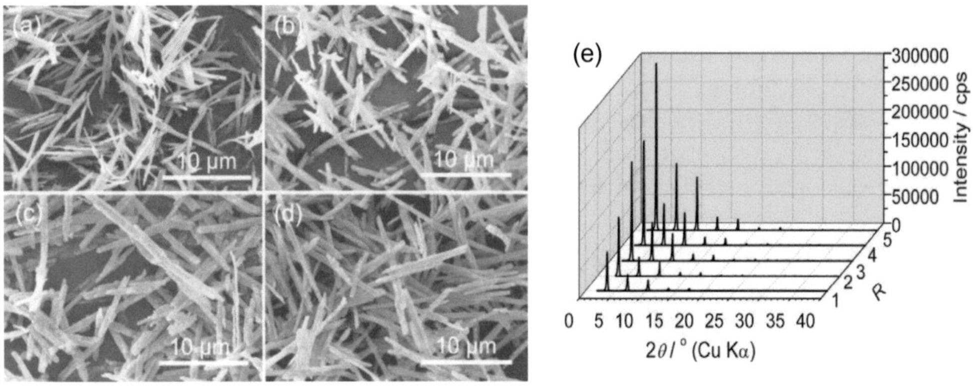

Fig. 9. (a-d) SEM images the products synthesized with various volume ratios (R) of heptane to n-octylamine: (a) 1, (b) 3, (c) 4, and (d) 5; (e) XRD patterns of the products synthesized with various volume ratios (R) of heptane to *n*-octylamine. Reprinted with permission from *Key Eng. Mater.*, 2007, 352, 85-88. Copyright 2007 Trans Tech Publications Inc.

The volume ratio of heptane to *n*-octylamine in the reaction system of $H_2W_2O_7{\cdot}xH_2O$/*n*-octylamine/heptane has an obvious effect on the morphology and microstructure of the final products. Figs. 9(a-d) show typical SEM images of products obtained under similar conditions (molar ratio of *n*-octylamine to $H_2W_2O_7{\cdot}xH_2O$: 30; the reaction time: 120 h; x ~ 3.5), except for the volume ratio of heptane to *n*-octylamine, *R*. As the figure shows, when the *R* value increases from 1 to 5, the products obtained take on a more uniform belt/tubelike morphology in both diameter and length. The corresponding XRD patterns are shown in Fig. 9e. It can be readily observed that the intensities of the reflections of the products become stronger with increases in the *R* value from 1 to 5. These results indicate that the degree of the long-range order of the alternate stacking of the *n*-alkyl chains and the inorganic W–O layers in the products obtained in the diluted reaction systems has obviously been enhanced. The enhancement of both the morphology and the microstructure of the products can be considered to be due to a reduced number of collisions among the filamentous structures formed in the dilute reaction systems.

Fig. 10. SEM images of the products synthesized using $H_2W_2O_7{\cdot}xH_2O$ with various amounts of interlayer water: (a) $x = 4.1$, (b) $x = 2.6$ and (c) $x = 0.85$. Reprinted with permission from *Key Eng. Mater.*, 2007, 352, 85-88. Copyright 2007 Trans Tech Publications Inc.

Fig. 10 shows the SEM images of the products obtained with three kinds of $H_2W_2O_7{\cdot}xH_2O$ with different amounts of interlayer water: x = 4.1, 2.6, and 0.85. All of the products obtained show a very similar belt/tubelike morphology. The rate of the reaction between

$H_2W_2O_7 \cdot xH_2O$ and *n*-octylamine could be qualitatively determined based on the time needed for the reaction system to turn from yellow to white. For $x = 4.1$, a white suspension was obtained after a reaction time of about 1 h, whereas for $x = 0.85$, it took more than 10 h for the reaction system to change to a white suspension. When 120 °C-dried $H_2W_2O_7$ ($x = 0$) was used as a precursor to react with *n*-octylamine under similar conditions, there was no obvious change in color, even after a reaction time of 5 days.

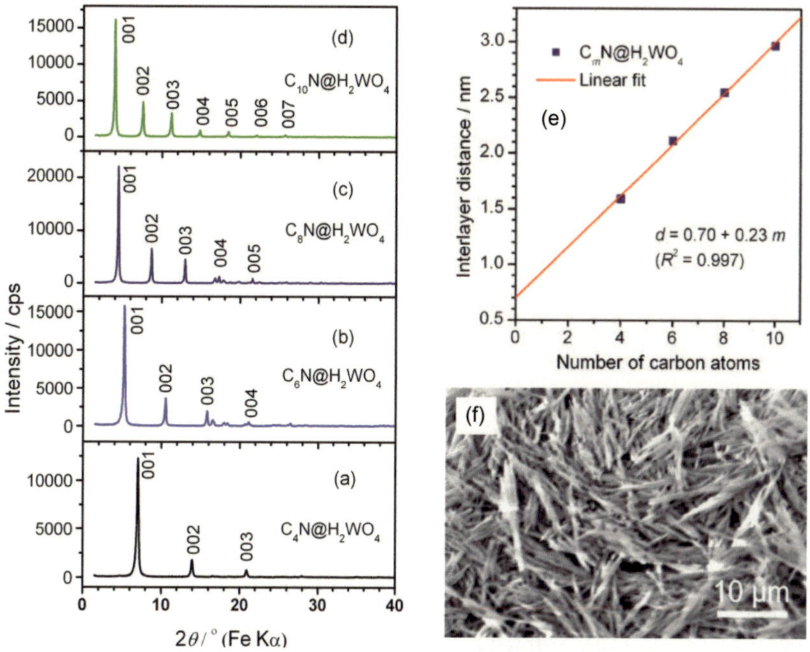

Fig. 11. (a-d) XRD patterns of the tungstate-based hybrid compounds ($C_mN@H_2WO_4$) obtained via reactions between H_2WO_4 and *n*-alkylamines in heptane: (a) $C_4N@H_2WO_4$, (b) $C_6N@H_2WO_4$, (c) $C_8N@H_2WO_4$, and (d) $C_{10}N@H_2WO_4$; (e) A plot of the interlayer distances of the $C_mN@H_2WO_4$ compounds as a function of the carbon numbers of the corresponding *n*-alkylamines; (f) a typical SEM image of the as-obtained $C_8N@H_2WO_4$ hybrid. Reprinted with permission from *Mater. Chem. Phys.*, 2011, 125(3), 838-845. Copyright 2011 Elsevier.

The reaction behavior of commercially available H_2WO_4 with *n*-alkylamines in reverse-microemulsion-like reaction media, *i.e.*, inorganic particles/*n*-alkylamines/heptane is similar to that of $H_2W_2O_7 \cdot xH_2O$ powders. H_2WO_4 powders reacting with *n*-alkylamines at room temperature led to the formation of inorganic-organic hybrid one-dimensional nanobelts, consisting of organic *n*-alkylammonium ions (a bilayered arrangement with a tilt angle of 65°) and inorganic single-octahedral W–O layers, as shown in Fig. 11. The similarity in both compositions and microstructures indicated the reactions of H_2WO_4 and $H_2W_2O_7 \cdot xH_2O$ powders with *n*-alkylamines underwent a similar "dissolution-reorganization" process, where the double-octahedral W–O layers from $H_2W_2O_7 \cdot xH_2O$ particles were firstly decomposed and the decomposed species were then reorganized into ordered lamellar hybrid nanobelts with inorganic single-octahedral W–O layers, as shown in Fig. 12.

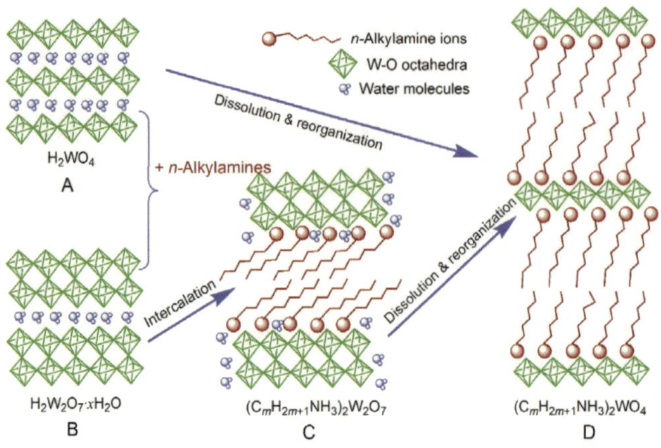

Fig. 12. A schematic of the possible mechanisms for the reactions of $H_2W_2O_7{\cdot}xH_2O$ and H_2WO_4 with *n*-alkylamines in heptane: (A) H_2WO_4 powders with single-octahedral W–O layers, (B) $H_2W_2O_7{\cdot}xH_2O$ powders with double-octahedral W–O layers, (C) intermediate hybrids with double-octahedral W–O layers derived by an intercalation reaction, and (D) tungstate-based inorganic-organic hybrids with single-octahedral W–O layers derived by a dissolution–reorganization process. Reprinted with permission from *Mater. Chem. Phys.*, 2011, 125(3), 838-845. Copyright 2011 Elsevier.

3. Topochemical conversion of tungstate-based inorganic-organic hybrid belts to WO_3 nanoplates [3,5,6]

3.1 Material synthesis and characterization

Typically, 10 g of $H_2W_2O_7{\cdot}xH_2O$ (*ca.* 20 mmol, $x \approx 1.5$) was dispersed in a mixture of 66 mL of *n*-octylamine (400 mmol) and 330 mL of heptane under a constant magnetic stirring at room temperature. After a reaction time of 72 h, the obtained white solids were collected by centrifugation and washed with ethanol for several times, and then dried under a reduced pressure at room temperature for more than 5 h. The obtained product was tungstate-based inorganic–organic hybrid nanobelts, which were then used as the precursors for the synthesis of $WO_3{\cdot}H_2O$ and WO_3 nanoplates. Typically, the obtained hybrid nanobelts (10 g) was dispersed in a mixture of concentrated HNO_3 (60–61 mass %, 200 mL) and distilled H_2O (300 mL) under a stirring condition at room temperature. (***Caution***: *The reaction releases toxic* NO_2 *gas, and has to be carried out in a ventilating cabinet*). A yellow suspension was obtained after a reaction time of more than 2 days. The obtained yellow solids were collected and washed with H_2O and ethanol before air-drying or drying at 120 °C. The air-dried product was $H_2WO_4{\cdot}H_2O$, and the 120 °C-dried product was $WO_3{\cdot}H_2O$ nanoplates. The obtained $WO_3{\cdot}H_2O$ nanoplates (4.2 g) were calcined at 450 °C for 2 h with a heating rate of 2 °C min^{-1} in air, and *ca.* 3.9 g of pale yellow WO_3 nanoplates was obtained.

For the preparation of oriented films from tungsten oxide nanoplates, 0.02 g of WO_3 (or $WO_3{\cdot}H_2O$) nanoplates was dispersed in 20 mL of ethanol, and the obtained suspension was kept stirring for 3–5 h. 100 μL of the WO_3 (or $WO_3{\cdot}H_2O$) suspension was carefully dropped on a pre-washed, horizontally placed XRD glass slice. After the solvent was completely evaporated, another 100 μL of the above suspension was dropped. Such dropping–evaporation process was repeated more than 10 times. The oriented films of WO_3 and $WO_3{\cdot}H_2O$ nanoplates supported by XRD glass slices were obtained.

X-ray diffraction (XRD), scanning electron microscopy (SEM), and transmission electron microscopy (TEM) were used to characterize the products. The nitrigen (N_2) adsorption-desoption technique was also used to characterize the as-obtained products.

3.2 Results and discussion

Fig. 13a shows a typical SEM image of the tungstate-based organic–inorganic hybrid precursor used for the synthesis of $WO_3 \cdot H_2O$ nanoplates. The hybrid takes on a quasi-1D beltlike morphology with a half-rolled microstructure (marked with an arrow). Fig. 13b shows a typical TEM image of the pristine inorganic species without washing or drying, obtained by oxidizing the hybrid nanobelts using nitric acid (*ca.* 6 mol/L). One can find that the pristine inorganic species are soft wrinkly belts, with a length of more than 5 μm, a width of *ca.* 1 μm and a thickness of less than 50 nm. The XRD pattern indicated that the air-dried inorganic species were monoclinic $H_2WO_4 \cdot H_2O$ (JCPDS no. 18–1420). Fig. 13c shows the TEM image of a typical 120 °C-dried sample. The product consists of quadrangular nanoplates lying along the Cu grids. Fig. 13d shows a single quadrangular nanoplate, the area of which is *ca.* 350 nm × 480 nm. Fig. 13e shows its corresponding SAED pattern, indexed to orthorhombic $WO_3 \cdot H_2O$ along the [010] zone axis. The uniform wide ordered diffraction spots indicate that the nanoplate is a whole single crystal. Fig. 13f shows the HRTEM image of the nanoplate. The interplanar distances of *ca.* 0.365 and 0.261 nm can be indexed to (101) (or (10−1)) and (200) crystal planes, respectively.

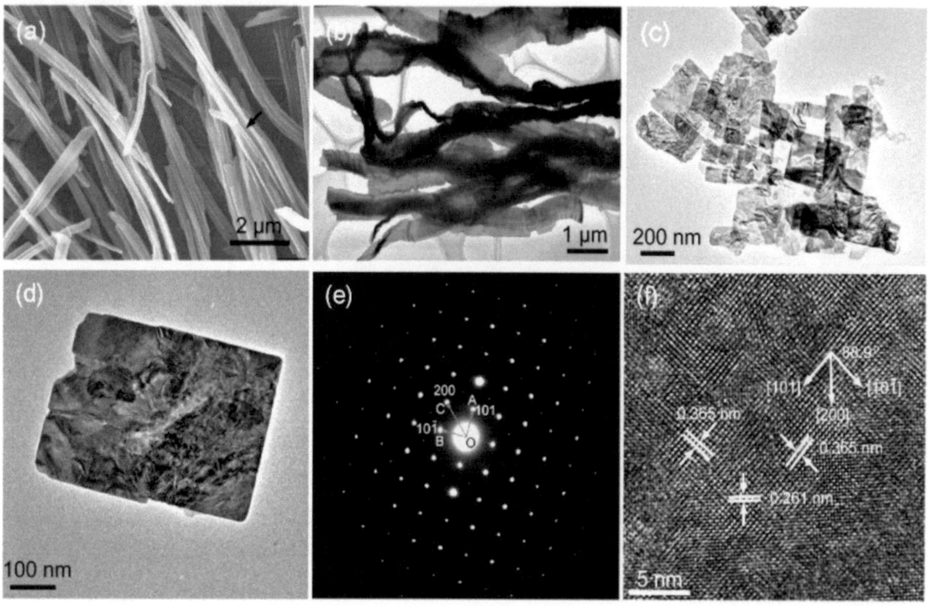

Fig. 13. (a) SEM image of the tungstate-based inorganic–organic hybrid belts; (b) TEM image of $H_2WO_4 \cdot H_2O$ nanobelts without washing or drying treatment; (c) low-magnification TEM image; (d) high-magnification TEM image; (e) SAED pattern along the [010] zone axis; and (f) HRTEM image of H_2WO_4 nanoplates after drying at 120 °C. Reprinted with permission from *Small*, 2008, 4(10), 1813-1822. Copyright 2008 Wiley-VCH Verlag GmbH & Co. KGaA.

Fig. 14 shows the TEM observations of the product derived from $WO_3 \cdot H_2O$ nanoplates by heating them at 450 °C for 2 h with a heating rate of 2 °C min^{-1}. The low-magnification TEM image in Fig. 14a indicates that the calcined product shows a predominant platelike morphology, besides a small fraction of rolled structures (marked with arrows). Fig. 14b shows a single nanoplate with a dimensionality of *ca.* 230 nm × 420 nm, and Fig. 14c shows its corresponding SAED pattern. The uniform, wide, and well-ordered diffraction spots can be assigned to single-crystal monoclinic WO_3, along the [001] zone axis. Fig. 14d shows a typical HRTEM image of the edge of the WO_3 nanoplate. The clear lattice structure corroborates that the obtained nanoplate is single-crystal. The interplanar distances are *ca.* 0.364 and 0.376 nm, assigned to (200) and (020) crystal planes, respectively. Another HRTEM image obtained from the central part of the nanoplate (Fig. 14b) is shown in Fig. 14e. The HRTEM images and SAED pattern indicates that the WO_3 nanoplate derived from the $WO_3 \cdot H_2O$ nanoplate is a whole single crystal. A typical edge dislocation is detected in Fig. 14e (marked with a circle). Fig. 14f shows the EDS spectrum of the WO_3 nanoplate.

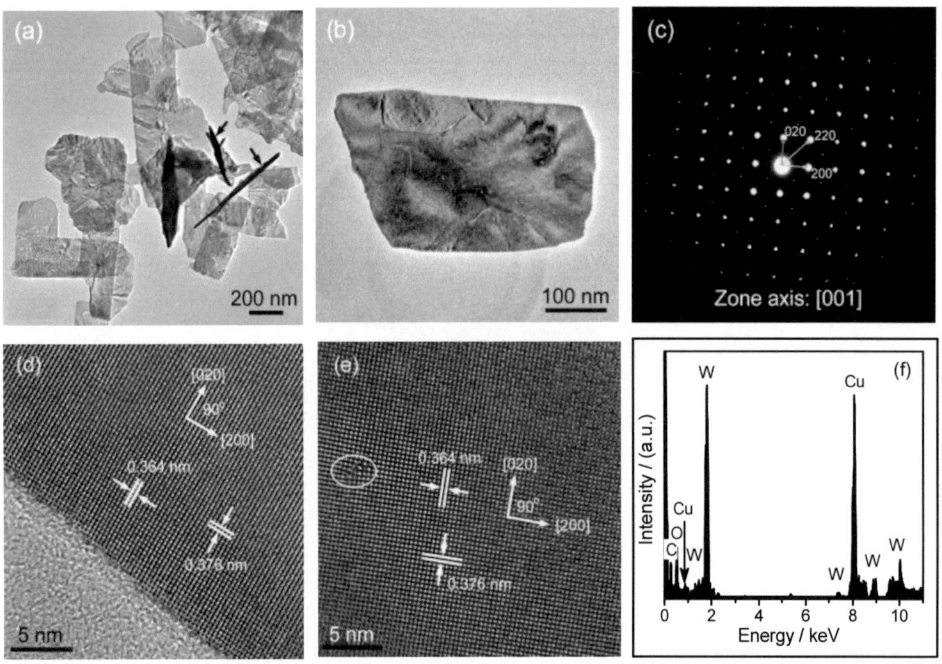

Fig. 14. (a) Low-magnification TEM image; (b) high-magnification image; (c) SAED pattern along the [001] zone axis; (d, e) HRTEM images; and (f) EDS spectrum of the WO_3 nanoplates obtained by calcining H_2WO_4 nanoplates at 450 °C for 2 h with a heating rate of 2 °C min^{-1}. Reprinted with permission from *Small*, 2008, 4(10), 1813-1822. Copyright 2008 Wiley-VCH Verlag GmbH & Co. KGaA.

Fig. 15 shows the SEM images of the obtained $WO_3 \cdot H_2O$ and WO_3 nanocrystals. Both $WO_3 \cdot H_2O$ and WO_3 take on a loose-aggregated-cotton-like morphology. The high-magnification FE-SEM images (Figs. 15b and d) indicate that the cotton-like products consist of thin nanoplates, the thickness of which can be estimated to range from 10 to 30 nm.

Fig. 15. (a, c) Low-magnification SEM images and (b, d) FE-SEM images of (a, b) the H_2WO_4 nanoplates dried at 120 °C and (c, d) WO_3 nanoplates obtained by calcining H_2WO_4 nanoplates at 450 °C for 2 h with a heating rate of 2 °C min^{-1}. Reprinted with permission from *Small*, 2008, 4(10), 1813-1822. Copyright 2008 Wiley-VCH Verlag GmbH & Co. KGaA.

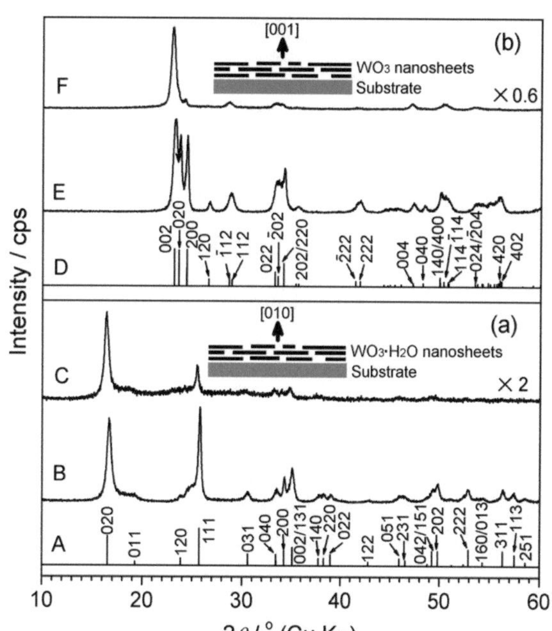

Fig. 16. (a) XRD patterns of (B) the powder sample and (C) the film sample of the obtained $WO_3 \cdot H_2O$ nanoplates (pattern A is the literature data of orthorhombic $WO_3 \cdot H_2O$ (JCPDS no. 43-0679)); (b) XRD patterns of (E) the powder sample and (F) the film sample of the obtained WO_3 nanoplates (pattern D is the literature data of monoclinic WO_3 (JCPDS no. 43-1035).

Fig. 16 shows the XRD patterns of the obtained $WO_3 \cdot H_2O$ and WO_3 nanoplates in the forms of powders and films. Pattern B is the XRD pattern of the powder sample of $WO_3 \cdot H_2O$ nanoplates. It is very close to the literature data of orthorhombic $WO_3 \cdot H_2O$ phase (pattern A, JCPDS no. 43-0679) not only in the peak positions but also in the intensity ratios of the diffraction peaks. The intensity ratios of (020) to (111) reflections (*i.e.*, I_{020}/I_{111}) of the $WO_3 \cdot H_2O$ powder sample and the literature data are 0.88 and 0.80, respectively. This suggests that $WO_3 \cdot H_2O$ nanoplates in the powder sample are essentially a random orientated aggregate. Pattern C shows a typical XRD result of the film of $WO_3 \cdot H_2O$ nanoplates. One can easily find that the (020) reflection becomes the most intense and the (040) reflection are also detectable, but the other reflections are weakened obviously. The intensity ratio of I_{020}/I_{111} is 2.39, which is twice larger than that (0.88) of its powder sample. Therefore, there is a strong preferred orientation along the [020] direction in the film sample of $WO_3 \cdot H_2O$ nanoplates. From the TEM analysis of $WO_3 \cdot H_2O$ nanoplates, it is determined that the inhibited crystal growth direction of $WO_3 \cdot H_2O$ nanoplates is [010]. Due to the large area-to-thickness ratios, the $WO_3 \cdot H_2O$ nanoplates with an inhibited crystal growth direction of [010] tend to lie along the substrate (inset of Fig. 16a). Thus the (0*k*0) crystal planes have the largest probability to diffract and the (0*k*0) reflections have the largest intensities.

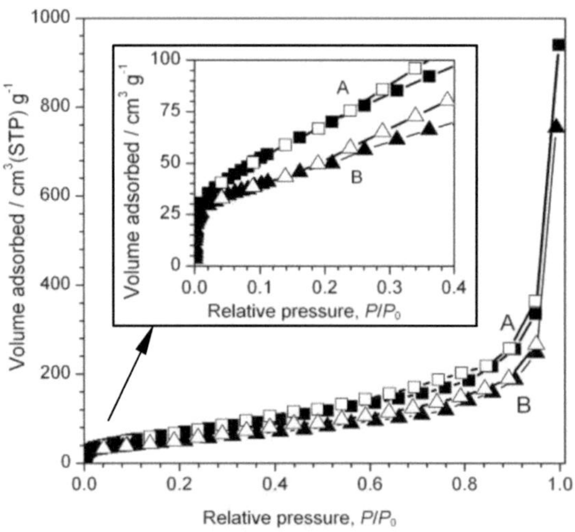

Fig. 17. Nitrogen (N_2) adsorption–desorption isotherms and the corresponding BJH pore-size distribution curves (inset) of the obtained (A) $WO_3 \cdot H_2O$ nanoplates and (B) WO_3 nanoplates (filled: adsorption; open: desorption). Reprinted with permission from *Small*, 2008, 4(10), 1813-1822. Copyright 2008 Wiley-VCH Verlag GmbH & Co. KGaA.

Similarly, the XRD patterns of the powder and film samples of WO_3 nanoplates are shown as patterns E and F in Fig. 16b, respectively. The XRD pattern of the WO_3 nanoplates is consistent with the literature data (pattern D, JCPDS no. 43-1035) of a monoclinic WO_3 phase. The intensity ratios of the (002) to (200) reflection (*i.e.*, I_{002}/I_{200}) of the powder sample of the WO_3 nanoplates and the literature data are 1.21 and 1.01, respectively. The close values in the I_{002}/I_{200} ratios indicate that the powder sample of WO_3 nanoplates shows a very limited preferred orientation and most of the nanoplates aggregate random. But for the

film sample of WO_3 nanoplates, the XRD pattern (pattern F) shows predominant (00*l*) reflections, and the other reflections disappear or weaken obviously. The I_{002}/I_{200} ratio of the film is 7.94, which is 6.5 times larger than that of its powder sample (1.21). As the proposed schematic model (inset of Fig. 16b) indicates, the WO_3 nanoplates with an inhibited crystal growth direction of [001] and high area-to-thickness ratios prefer lying along the substrate surface to form a [001]-direction orientated film.
Fig. 17 shows the nitrogen (N_2) adsorption–desorption isotherms of the $WO_3 \cdot H_2O$ and WO_3 nanoplates obtained. The $WO_3 \cdot H_2O$ and WO_3 nanoplates show a similar type II isotherm of non-porous solids, with sharp knees (inset of Fig. 17) and without any obvious hystereses. The BET surface areas of the $WO_3 \cdot H_2O$ and WO_3 nanoplates are 257 and 180 $m^2\ g^{-1}$, respectively.

4. Formation of MoO_3 nanoplates from molybdate-based inorganic-organic hybrids [7]

4.1 Material synthesis and characterization

$MoO_3 \cdot H_2O$ powders reacted with *n*-octylamine at room temperature to form molybdate-based inorganic–organic hybrid compounds. Typically, 10.3 mL of *n*-octylamine was firstly mixed with 113.7 mL of ethanol in a conical flask under magnetic stirring, and then 5.0 g of $MoO_3 \cdot H_2O$ powders was dispersed into the above mixture to form a white suspension. The as-obtained suspension was kept stirring for 3 days at room temperature in air, and a white mushy mixture was finally obtained. The white solids were collected by centrifugation, followed by washing with ethanol for three times. The as-obtained solids were then air-dried at room temperature in a reduced pressure for more than 3 days, and 8.1 g of white powders, a molybdate-based inorganic–organic hybrid compound, was obtained. The molar ratios of *n*-octylamine to $MoO_3 \cdot H_2O$ were 2–10. The volume ratios of ethanol to *n*-octylamine were higher than 10.
The as-obtained molybdate-based inorganic–organic hybrid compound was used as the precursor to prepare MoO_3 nanoplates. Typically, 2.0 g of the molybdate-based inorganic–organic hybrid compound was placed in an alumina crucible, which was then put into an electric furnace and kept at 550 °C for 1 h. After naturally cooled down to room temperature, about 1.0 g of gray powders was obtained.
X-ray diffraction (XRD) , TEM, HRTEM, TG–DTA, FT–IR, and Raman spectra were used to characterize the molybdate-based inorganic–organic hybrid compounds and the as-obtained MoO_3 nanoplates.

4.2 Results and discussion

Figs. 18a–b show the XRD patterns of the $MoO_3 \cdot H_2O$ and the resultant hybrid compound, respectively. The commercial $MoO_3 \cdot H_2O$ may be a mixture or intermediate phases of molybdic acids with a certain amount of crystal water. Fig. 18b indicate that the the hybrid shows 4 highly intense diffraction peaks with regularly reduced intensities in the low 2θ-angle range of 1.5 – 20°, indicating that the as-obtained product is of a highly ordered layered structure. The peaks at 2θ = 3.836 °, 7.671 °, 11.507 °, and 15.342 ° can be indexed to the reflections from (010), (020), (030) and (040) diffraction planes, respectively, when considering the layered structure of MoO_3 along the *b*-axis direction. The interlayer distance (*d*) can be calculated to be *d* = 2.306(1) nm using a program UnitCell, refined in a cubic system (λ = 1.54055 Å) by minimizing the sum of squares of residuals in 2θ. In the 2θ-angle range of 20°- 40°, there are numerous diffraction peaks with low intensities, as shown in the inset of Fig. 18b. These diffraction peaks are similar to those of molybdic acid (Fig. 18a), and can mainly be attributed to the reflections from the inorganic MoO_6 frames.

Fig. 18(c) shows the XRD pattern of the product obtained by calcining the molybdate-based inorganic–organic hybrid disks at 550 °C for 1 h in air. All the diffraction peaks can be readily indexed to an orthorhombic MoO_3 phase (α-MoO_3, space group: Pbnm (62)) according to JCPDS card No. 05–0508. The calculated cell parameters by refining the XRD pattern are a = 3.964(2) Å, b = 13.862(3) Å and c = 3.6991(8) Å, close to the literature data (a = 3.962 Å, b = 13.858 Å, c = 3.697 Å, JCPDS card No. 05–0508). The $I_{(0k0)}I_{(111)}$ values of the as-obtained α-MoO_3 are obviously larger than those of the literature data, indicating that the as-obtained α-MoO_3 sample is of an obvious preferred orientation growth along the (0k0) planes.

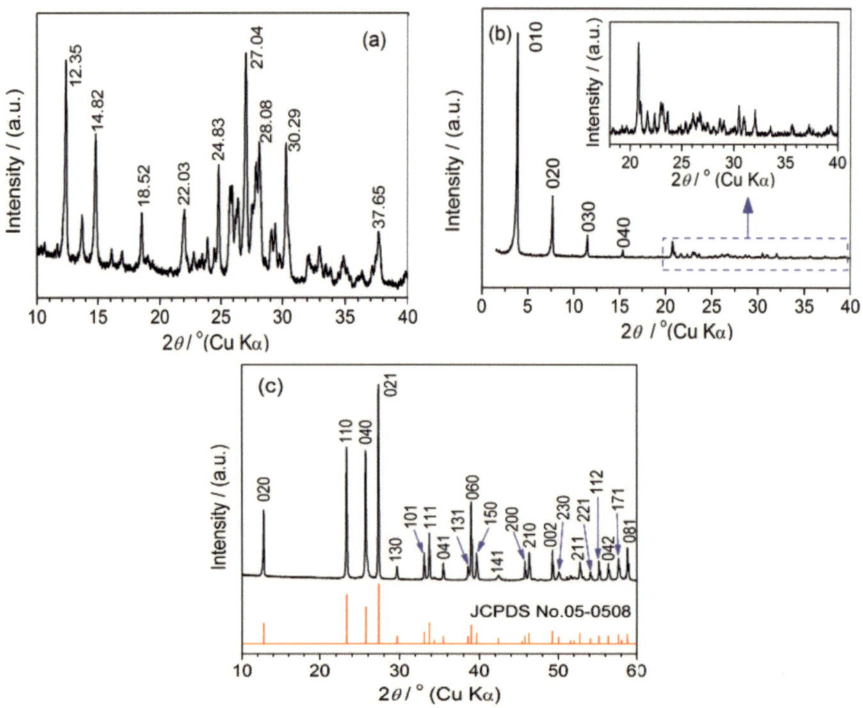

Fig. 18. XRD patterns of (a) commercial molybdic acid; (b) molybdate-based inorganic–organic hybrid compound derived from the reaction of commercial molybdic acid with n-octylamine in ethanol (n_{Mo}:n_N=1:5); (c) MoO_3 nanocrystals derived by calcining the molybdate-based hybrid compound at 550 °C (the inset is the data from JCPDS card No. 05–0508). Reprinted with permission from *J. Mater. Chem.*, 2011, 21(25), 9332-9342. Copyright 2011 Royal Society of Chemistry.

Figs. 19a-d show the SEM images of the commercial molybdic acid and the resultant molybdate-based hybrid compound obtained with n_{Mo}:n_N = 1:5. As Figs. 19a–b show, the commercial molybdic acid consists of large aggregates with sizes of 10 - 15 μm, and the original grains are platelike particles with side sizes of 1 - 3 μm. Figs. 19c–d show the morphology of the hybrid product, consisting of separate disklike particles, with thicknesses of 1.55 ± 0.4 μm and apparent side sizes of 5.2 ± 1.7 μm. The average side-to-thickness ratio of the as-obtained disks is 3.5. Figs. 19e –f show the SEM images of the α-MoO_3 sample derived by calcining the molybdate-based inorganic–organic hybrid disks at 550 °C for 1 h in air. The SEM images with various magnifications suggest that the α-MoO_3 sample is composed of platelike nanocrystals with a good dispersibility in a large view field (Fig. 19e). The platelike particles are of side sizes of 1 - 10 μm and thicknesses of 50 - 150 nm. The platelike morphology of the as-obtained α-MoO_3 sample makes them tend to lie down along a substrate and their large surfaces parallel to the substrate, because of their large side-to-thickness ratios. Considering the preferred growth of the (0*k*0) planes together with their platelike morphology, one can safely conclude that the as-obtained α-MoO_3 nanoplates have a shortest side along the *b*-axis, that is, the thickness of the α-MoO_3 nanoplates is along the *b*-axis (i.e., the [0*k*0] direction).

Fig. 19. SEM images of (a, b) commercial molybdic acid; (c, d) molybdate-based hybrid compound derived from the reaction of commercial molybdic acid with *n*-octylamine in ethanol (n_{Mo}:n_N=1:5); and (e,f) α-MoO_3 nanocrystals derived by calcining the molybdate-based hybrid compound at 550 °C for 1 h in air. Reprinted with permission from *J. Mater. Chem.*, 2011, 21(25), 9332-9342. Copyright 2011 Royal Society of Chemistry.

Fig. 20 shows the TG-DTA curves of the inorganic–organic hybrid disks. The TG curve in Fig. 20a shows that there are four obvious mass-loss stages: (I) 20 - 210 °C with a mass loss of 28%, (II) 210 - 340°C with a mass loss of 11%, (III) 340 - 550 °C with a mass loss of 28 %, and (IV) 700 - 900 °C with a mass loss of 27%. Fig. 20b shows the corresponding DTA result. In stage I, there are two weak endothermic peaks at 136 °C and 185 °C, due to desorption of *n*-octylammonium ions/molecules and the decomposition of crystal water in the Mo–O frames, respectively. The corresponding mass loss of 28% attributes to the removal of the adsorbed *n*-octylammonium ions / molecules and crystal water. In stage II, there is two strong exothermal peaks at around 249 and 280 °C, respectively, attributed to the oxidization of the organic $-(CH_2)_n-$ chains in the hybrid compound. The corresponding mass loss (11%) is due to release of small volatile molecules resulted from the oxidization of long *n*-alkyl chains. The resultant inorganic carbon species formed in stage II then combust at elevated temperatures of 340 - 550 °C, confirmed by the strong exothermal peak at 460 °C, and the mass loss of 28% in stage III. During the temperature range of 550 - 700 °C, there are no detectable changes either in the TG curve or in the DTA curve, indicating that a thermally stable α-MoO_3 phase is formed. When the temperature is higher than 700 °C, *i.e.*, stage IV, there is a weak endothermic peak at around 786 °C, accompanying with a mass loss of 33%. The melting and evaporating behaviors of α-MoO_3 account for the endothermic peak and the large mass loss in stage IV. If the composition of the molybdate-based inorganic–organic hybrid compound can be expressed as $(C_8H_{17}NH_3)_xH_{2-x}MoO_4$, and the product calcined at a temperature higher than 550°C is MoO_3, the x value can be calculated to be about 2.12 according to the total mass loss of 67% at 20 - 520 °C. Therefore, the composition of the molybdate-based inorganic–organic hybrid compound can be described as $(C_8H_{17}NH_3)_2MoO_4$.

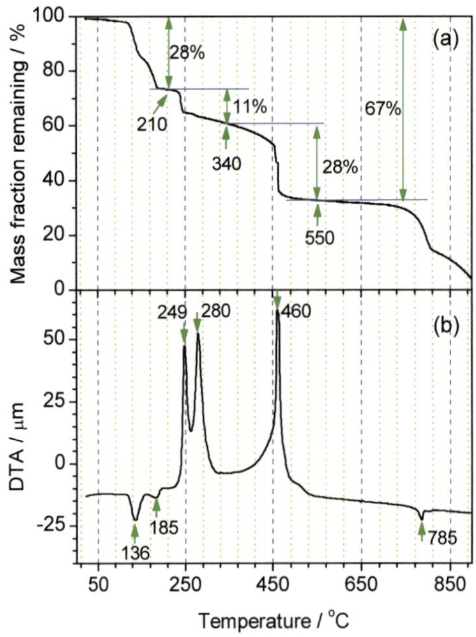

Fig. 20. (a) TG and (b) DTA curves of the molybdate-based hybrid compound obtained from the reaction of commercial molybdic acid with *n*-octylamine in ethanol (n_{Mo}:n_N=1:5). Reprinted with permission from *J. Mater. Chem.*, 2011, 21(25), 9332-9342. Copyright 2011 Royal Society of Chemistry.

Fig. 21 shows the typical TEM observations of the α-MoO_3 nanoplates obtained by calcining the molybdate-based inorganic–organic hybrid disks at 550 °C for 1 h in air. Most of the α-MoO_3 nanoplates take on a quadrilateral platelike shape with a lateral length of 1 - 2 μm

(Fig. 21a), and some α-MoO_3 nanoplates are partially overlapped with each other. Fig. 21b shows a typical individual α-MoO_3 nanoplate with a dimension of 1000 nm × 600 nm and its thickness is very thin (several nanometers) judged by the shallow contrast grade. The typical SAED pattern (Fig. 21c) of the individual α-MoO_3 nanoplate can be indexed to an orthorhombic α-MoO_3 phase with a zone axis along the [010] direction, and indicates that the α-MoO_3 nanoplate is of a single-crystal structure and has a thin and uniform thickness. The HRTEM image (Fig. 21d) of an edge of the α-MoO_3 nanoplate indicates that the α-MoO_3 nanoplate is single-crystal. The distances of lattice stripes are about 0.40 nm and 0.37 nm, corresponding to the (100) and (001) planes of the α-MoO_3 phase, respectively. The as-obtained α-MoO_3 nanoplates may consist of thinner sub-plates because of the formation process on the basis of interaction chemistry. Fig. 21e shows a typical example. The α-MoO_3 nanoplate is composed of three super-thin subplates, which overlap loosely with each other. Fig. 21f shows an HRTEM image of a tip of one of the subplates, and the clear lattice stripes indicates the sub-plate is also of a well-defined single-crystal structure.

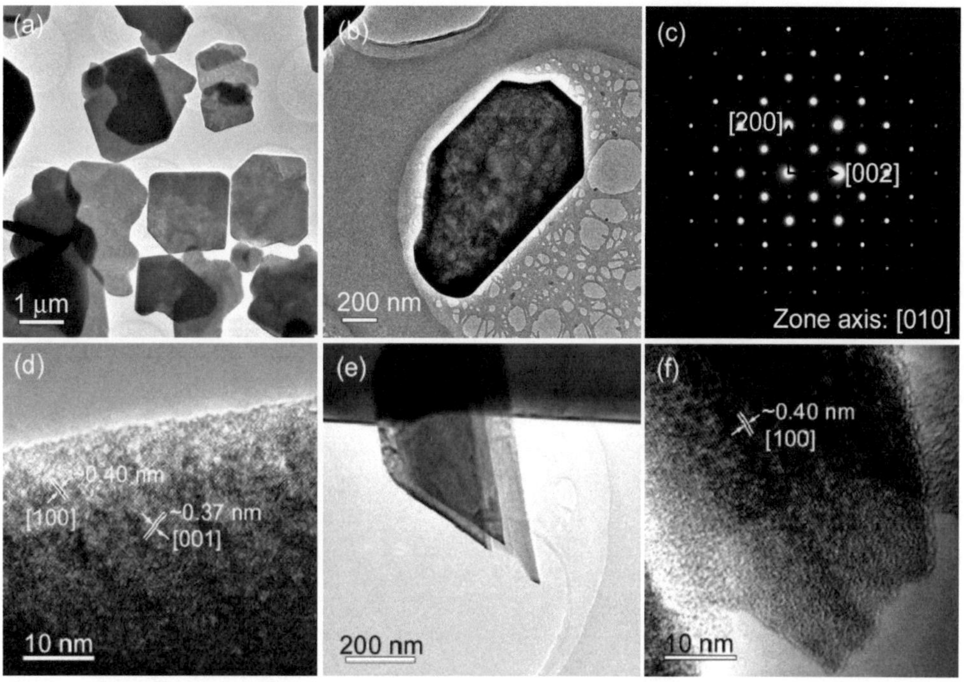

Fig. 21. (a) A low-magnification TEM image of α-MoO_3 nanoplates; (b) a TEM image, (c) an SAED pattern and (d) an HRTEM image of an individual α-MoO_3 nanoplate; (d) a TEM image of laminated α-MoO_3 nanoplates, and (f) an HRTEM image of the tip of an α-MoO_3 nanoplate. Reprinted with permission from *J. Mater. Chem.*, 2011, 21(25), 9332-9342. Copyright 2011 Royal Society of Chemistry.

Fig. 22 shows a typical Raman spectrum of the α-MoO_3 nanoplates. The vibration modes appearing in the frequency ranges of 1000 - 600 cm^{-1} and 600 - 200 cm^{-1} correspond to the stretching and deformation modes, respectively. The narrow band at 994 cm^{-1} is assignable to the antisymmetric $\nu(Mo{=}O_1)$ stretching (A_g), in which the bonding aligns along the *b* axis direction. The strong band at 819 cm^{-1} represents the symmetric ν (Mo-O_3-Mo) stretching (A_g) with the bonding aligning along the *a*-axis direction. The weak and broad bands at 666 and 470 cm^{-1} are ascribable to the antisymmetric ν(Mo-O_2-Mo) stretching (B_{2g}) and bending (A_g), respectively. The bands at 377 and 364 cm^{-1} correspond to the δ(O_2=Mo=O_2) scissor (B_{1g} and A_g modes). The band at 336 cm^{-1} is characteristic of the δ (O_3-Mo-O_3) deformation (B_{1g}). The band at 282 cm^{-1} and a weak shoulder centered at 289 cm^{-1} correspond to the δ (O_1=Mo=O_1) wagging (B_{2g} and B_{3g}, respectively). The bands at 242 and 215 cm^{-1} correspond to the δ(O_2-Mo-O_2) scissor (B_{3g} and A_g modes, respectively).

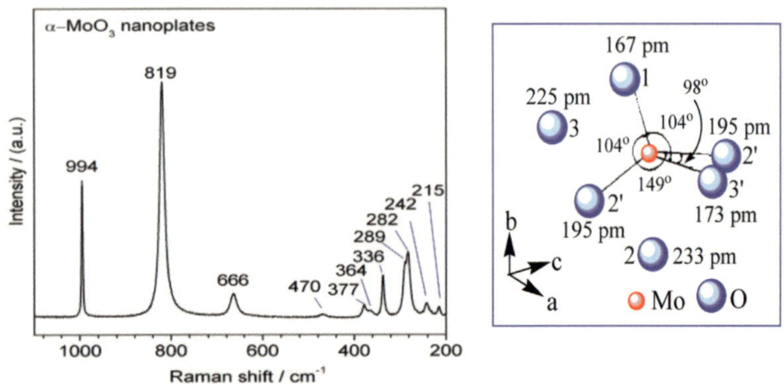

Fig. 22. A Raman spectrum of α-MoO_3 nanoplates derived from molybdate-based inorganic-organic hybrid disks. Reprinted with permission from *J. Mater. Chem.*, 2011, 21(25), 9332-9342. Copyright 2011 Royal Society of Chemistry.

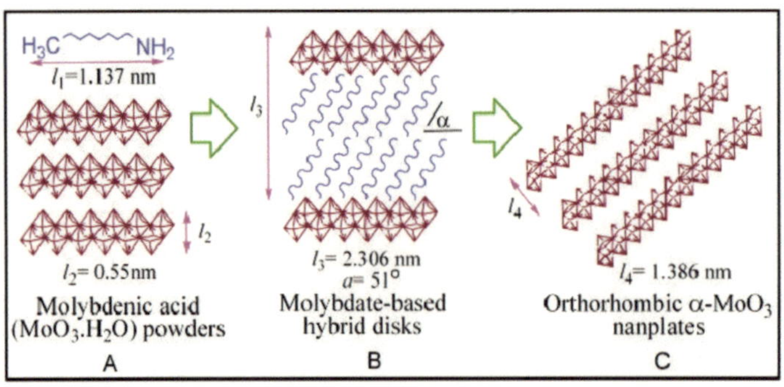

Fig. 23. A schematic of the formation mechanism for molybdate-based inorganic-organic hybrid compound and the resultant α-MoO_3 nanoplates. Reprinted with permission from *J. Mater. Chem.*, 2011, 21(25), 9332-9342. Copyright 2011 Royal Society of Chemistry.

Fig. 23 shows the schematic of the formation mechanisms for the molybdate-based inorganic-organic hybrid disks and the α-MoO_3 nanoplates. In the stage of A→B, molybdic acid reacts with *n*-octylamine via an acid-base mechanism, where an intercalation and reorganization process is undergone. The length of an *n*-octylamine molecule can be evaluated to be l_1=1.137 nm, according to the length (0.248 nm) of a CH_3NH_2 molecule and the length (0.127 nm) per -CH_2-. The contribution (l_0 = 1.756 nm) of the organic species to the interlayer distance can be calculated by subtracting the thickness (l_2 = 0.55 nm) of the double MoO_6 layer from the interlayer distance (l_3 = 2.306 nm). The thickness (l_0) of the organic species is larger than the length (l_1) of an *n*-octylamine molecule (or an *n*-octylammonium ion), and the *n*-octyl chains, therefore, take a double-layer arrangement with a tilt angle of $\alpha = \sin^{-1}(l_0/2l_1) = 51°$ (Fig. 23B), which is in good agreement with the previously reported result (50.8°). The organic layers of molybdate-based inorganic–organic hybrid disks are removed by a thermal treating process at 550 °C and the platelike α-MoO_3 nanocrystals are obtained, as shown as the stage of B→C. The synthesis of α-MoO_3 nanoplates can be seen as a pseudo-topochemical transformation, where the platelike morphology is from the precursors to the final products. The possible chemical reactions can be described as Eqs. 1-2.

$$MoO_3 \cdot H_2O\,(s) + C_8H_{17}NH_2(l) \xrightarrow{\text{r.t. \& in ethanol}} (C_8H_{17}NH_3)_2MoO_4(s,\text{disks}) \qquad (1)$$

$$\begin{aligned}&(C_8H_{17}NH_3)_2MoO_4\ (s,\text{disks}) \xrightarrow{550\ ^\circ\text{C in air}} \\ &MoO_3\ (s,\text{nanoplates}) + CO_2\ (g) + H_2O\ (g) + NH_3(g)\end{aligned} \qquad (2)$$

Fig. 24 shows a typical SEM image of the as-obtained α-MoO_3 product obtained by calcining commercially available $MoO_3 \cdot H_2O$ powders at 550 °C for 1 h in air. The particulate shapes are the dominant morphology for the α-MoO_3 sample obtained directly from $MoO_3 \cdot H_2O$ powders, and most of the α-MoO_3 particles are conglutinated to form large agglomerates. This comparative investigation indicates that molybdate-based inorganic–organic hybrid disks are necessary in the formation of singel-crystal α-MoO_3 nanoplates with a large diameter-to-thickness ratio.

Fig. 24. An SEM image of particulate MoO_3 nanocrystals derived by calcining commercial molybdic acid ($MoO_3 \cdot H_2O$) at 550 °C for 1 h in air. Reprinted with permission from *J. Mater. Chem.*, 2011, 21(25), 9332-9342. Copyright 2011 Royal Society of Chemistry.

5. Topochemical conversion of tungstate-based inorganic–organic hybrid belts to β-W_2N nanoplates [8]

5.1 Material synthesis and characterization

Typically, 0.5 g of tungstate-based inorganic–organic hybrid nanobelts was placed in a semi-cylindrical Al_2O_3 crucible, and the crucible was then inserted to the center of a tubular

quartz furnace (ϕ500 mm × 1000 mm) in a lying manner. The gas-out end of the quartz tube was loosely stuffed with a porous cylindrical alumina tile. The quartz tube was first purged with a high-pure Ar gas (100 mL/min) for 1 h. After the Ar gas was closed and the NH_3 gas was simultaneously opened with a flux of 80 mL/min, the furnace was rapidly heated to 500 °C with a heating rate of 25 °C /min, and then slowly heated from 500 °C to 650–800 °C with a heating rate of 2 °C /min. After kept at 650–800 °C for 2 h, the furnace naturally cooled down to room temperature under a NH_3-gas condition.
X-ray diffraction (XRD), SEM, TEM, TG-DTA and FT-IR were used to characterize the as-obtained products.

5.2 Results and discussion

A typical XRD pattern of the sample obtained by treating tungstate-based inorganic–organic hybrid nanobelts at 750 °C for 2 h in a NH_3 flow is shown in Fig. 25a. There are four obvious diffraction peaks at 36.4°, 45.1°, 64.0° and 76.2°, which can readily be indexed to the reflections from (111), (200), (220) and (311) planes of the cubic β-W_2N phase, respectively, according to the literature data (JCPDS card No. 25-1257; space group: Pm3m [221], a = 0.4126 nm). The calculated cell parameters according to the XRD data and the minimum square method are $a = b = c = 0.4133(8)$ nm, close to the literature data (a = 0.4126 nm). As Fig. 25a shows, the diffraction peaks are obviously broadened, indicating that the product consists of nanocrystals with very small crystal sizes. According to the Debye-Scherrer equation ($d=0.89\lambda/B\cdot\cos\theta$), λ=0.15406 nm, $B\cdot\cos\theta$ = 0.0428 for the (111) reflection, and the calculated average size of the crystalline β-W_2N grains is about 3.2 nm.

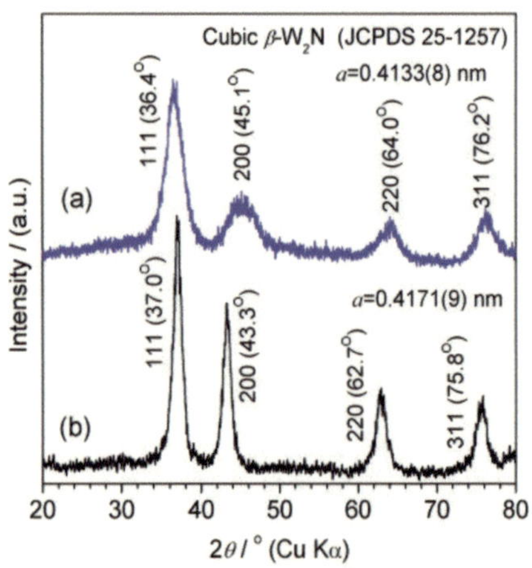

Fig. 25. Typical XRD patterns of (a) cubic β-W_2N nanoplates derived from the tungstate-based inorganic–organic hybrid nanobelts, and (b) cubic β-W_2N nanocrystals derived from commercial H_2WO_4 powders. Reprinted with permission from *J. Solid State Chem.*, 2011, 184(2), 455-462. Copyright 2011 Elsevier.

Fig. 26 shows the typical TEM observations of the β-W_2N nanocrystals derived from tungstate-based inorganic–organic hybrid nanobelts. Fig. 26a shows a low-magnification TEM image, indicating the β-W_2N product consists of well-dispersed nanoplates. Most of the nanoplates are of sharp profiles, including triangles, trapezoids and rectangles. The enlarged TEM image shown in Fig. 26b corroborates that the β-W_2N nanoplates obtained are not nanorods, but thin nanoplates, according to the shallow contrasts. The contrast from different parts of a nanoplate suggests that some β-W_2N nanoplates are curly or partly scrolled. Fig. 26c shows an individual partly-scrolled β-W_2N nanoplate with an apparent dimension of about 200 nm× 500 nm. The enlarged TEM image in Fig. 26d shows a β-W_2N nanoplate with a dimension of about 130 nm× 300 nm. Its contrast indicates the nanoplate is composed of nanoparticles with sizes of several nanometers. The corresponding SAED pattern of the β-W_2N nanoplate (Fig. 26d) is shown in Fig. 26e. The SAED pattern consists of a series of concentric diffraction rings superimposed a series of separate diffraction spots, indicating that the β-W_2N nanoplate is composed of small crystalline nanoparticles. The concentric diffraction rings can be readily indexed to the reflections from the cubic β-W_2N phase. The TEM observations are consistent to the XRD results. The grain sizes (about 3.2 nm) of the β-W_2N product calculated from the XRD result are much smaller than the sizes of the plate-like particles from the TEM observations, suggesting that an individual β-W_2N plate-like particle is not a single crystal, but is composed of a great number of small β-W_2N nanocrystals. This point is corroborated by the corresponding SAED pattern with a series of concentric diffraction rings.

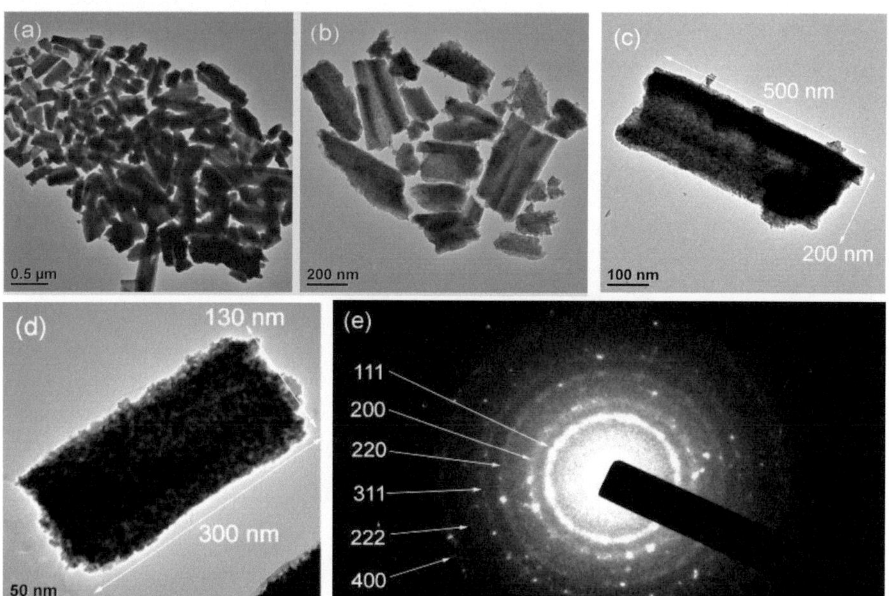

Fig. 26. (a–d) TEM images and (e) SAED pattern of the β-W_2N nanoplates derived from the tungstate-based inorganic–organic hybrid nanobelts. Reprinted with permission from *J. Solid State Chem.*, 2011, 184(2), 455-462. Copyright 2011 Elsevier.

For comparison, tungsten nitride was also synthesized by directly nitridizing commercially available H_2WO_4 powders. Fig. 25b shows the XRD pattern of the as-obtained product by nitridizing H_2WO_4 powders in an NH_3 flow. It can be indexed to be cubic β-W_2N with a calculated cell parameter of a=0.4171(9) nm, a little larger than that [a=0.4133(8) nm] of β-W_2N nanoplates. Fig. 27 shows a typical SEM image of the β-W_2N sample, consisting of shape-irregular particles, with a large size-distribution (1-5 μm), similar to their precursor of commercial H_2WO_4 powders. The comparative experimental results corroborate that the plate-like morphology of β-W_2N nanoplates is inherited from the tungstate-based inorganic-organic hybrid nanobelts.

Fig. 27. A typical SEM image of cubic β-W_2N sample derived from commercial H_2WO_4 powders. Reprinted with permission from *J. Solid State Chem.*,2011,184(2),455-462. Copyright 2011 Elsevier.

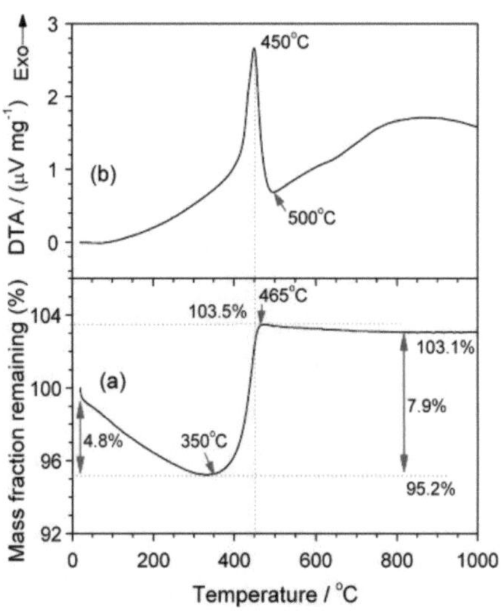

Fig. 28. TG–DTA curves of the β-W_2N nanoplates derived from the tungstate-based inorganic–organic hybrid nanobelts. Reprinted with permission from *J. Solid State Chem.*, 2011, 184(2), 455-462. Copyright 2011 Elsevier.

Fig. 28 shows typical TG–DTA curves of the β-W_2N nanoplates derived from tungstate-based inorganic–organic hybrid nanobelts. The TG curve (Fig. 28a) shows that there is a mass loss of 4.8% from room temperature to 350 °C, followed by a sharp mass gain of 8.3% from 350 °C to 465 °C. There is a small mass loss of 0.4% at 465–800 °C. Fig. 28b shows the corresponding DTA curve, with a sharp exothermal peak between 400 - 500 °C, and weak and broadened exothermal peaks in the temperature regions of 100 - 400 ^oC and 500 - 800 ^oC. Considering the TG result together with its corresponding DTA curve, the large mass loss with a sharp exothermal peak at 350 - 465 ^oC is mainly due to the oxidation of β-W_2N to WO_3, and the slow mass losses and their corresponding weak exothermal peaks at 100 - 400 ^oC and 500 - 800 ^oC should be due to the oxidation of the residual C species. Supposing that β-W_2N and residual C are completely oxidized to be WO_3 and CO_2, respectively, the content of the β-W_2N phase in the sample can be calculated to be 91.7% according to the total mass gain of 3.1% between 100 ^oC and 800 ^oC. It is noted that the oxidation temperature region of the as-obtained β-W_2N nanoplates is 350–500 ^oC.

6. Topochemical conversion of tungstate-based inorganic–organic discs to hierarchical tungsten carbide micro-/nanocrystals [9]

6.1 Material synthesis and characterization

The tungstate-based inorganic–organic hybrids were synthesized by a reaction between H_2WO_4 and *n*-octylamine in heptane. Typically, 10 g of H_2WO_4 powders was moistened by 5 mL of distilled water, and the H_2O-moisted H_2WO_4 was then dispersed in the mixture of 0.4 mol of *n*-octylamine and 530 mL of heptane under a constant magnetic stirring at room temperature. The molar ratio of H_2WO_4 to *n*-octylamine was about 1:10 and the volume ratio of *n*-octylamine to heptane was 1:8. After a reaction time of more than 48 h, the resulting white solid was collected by centrifugation, washed with ethanol, and then dried under a reduced pressure at room temperature for 30 h. The dried product, a tungstate-based inorganic-organic hybrid compound, was used as the precursor for the synthesis of tungsten carbide micro-/nanocrystals.

Typically, 0.3 g of the hybrid precursor was placed in the bottom of the quartz tube with a diameter of 10 mm and a length of 25 cm. The quartz tube was then vacuumized to 2×10^{-3} Pa and sealed. The sealed quartz tube was placed in an electrical resistance furnace, and thermally treated at 750~1050^oC with a heating rate of 5^oC/min for 2~10 h. After a given reaction time, the sealed tube was cooled naturally to room temperature. The black powders at the bottom of the quartz tube were carefully collected.

X-ray diffraction (XRD), SEM and FT-IR were used to characterize the as-obtained intermediate and final products.

6.2 Results and discussion

Fig. 29 shows the XRD patterns of the products obtained by thermally treating the tungstate-based inorganic-organic hybrid compound at 750~1050 ^oC for 2~10 h in a sealed quartz tube. The product obtained at 750 ^oC for 5 h, Fig. 29a, can be readily indexed to monoclinic WO_2 (JCPDS card No. 32-1393), and no other phases are observed. The product obtained at 800 ^oC for 5 h, Fig. 29b, has a major phase of monoclinic WO_2 and minor phases of cubic β-$W_{40.9}N_{9.1}$ (JCPDS No. 03-1031) and hexagonal WC (JCPDS No. 25-1047). When the thermal treating temperature increases to 850 ^oC, the phases of WO_2 and β-$W_{40.9}N_{9.1}$ disappear, but a new phase of hexagonal α-W_2C (JCPDS No. 35-0776) occurs, as shown in Fig. 29c. With increases in thermal treating time from 5 h to 10 h, the hexagonal WC becomes the dominant phase, and a very small amount of α-W_2C and cubic W (JCPDS No. 04-0806) coexist, as shown as Figs. 29c-d. The product obtained at 950 ^oC for 5h mainly consists of hexagonal WC, coexisting with a very small amount of α-W_2C (Fig. 29e). An elevated treating temperature and shortened treating time, e.g., at 1000ºC for 2h, can also achieve a product with major compositions of hexagonal WC and α-W_2C, as shown in Fig. 29f. The product obtained at 1050 ºC for 2 h (Fig. 29g) is composed of a pure hexagonal WC phase.

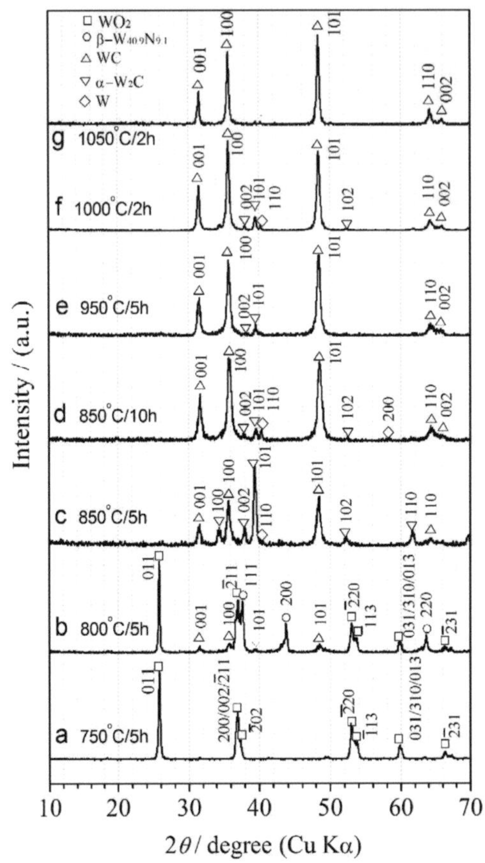

Fig. 29. XRD patterns of the products obtained by thermally treating the tungstate-based inorganic-organic hybrid compound in the sealed quartz tube under various conditions: (a) 750ºC for 5 h, (b) 800ºC for 5 h, (c) 850ºC for 5 h, (d) 850ºC for 10h, (e) 950ºC for 5h, (f) 1000ºC for 2h and (g) 1050ºC for 2h. (Δ) WC (JCPDS No.25-1047), (∇) α-W_2C (JCPDS No.35-0776), (◊) W (JCPDS No.04-0806), (○) β-$W_{40.9}N_{9.1}$ (JCPDS No.03-1031) and (□) WO_2 (JCPDS No.32-1393). Reproduced with permission from *J. Am. Ceram. Soc.*, 2010, 93(12), 3997-4000. Copyright 2010 John Wiley and Sons.

Fig. 30 shows the typical SEM images of the $C_8N@H_2WO_4$ hybrid (Fig. 30a) and the corresponding WC product obtained at 1050 ºC for 2 h with various magnifications (Figs. 30b-d). The hybrid compound takes on a plate-like morphology, with side-sizes of about 1 μm, as shown in Fig. 30a. According to the FT-IR spectrum (the inset of Fig. 30a), the tungstate-based inorganic-organic hybrid compound consists of inorganic W-O layers and organic ammonium ions. The TG-DTA curves suggested the mass remaining is 65 % after thermal treating at 600 ºC in an air flow. A low-magnification SEM image in Fig. 30b indicates the WC product consists of spherical microparticles, which are well dispersed in a large view field. Fig. 30c shows an enlarged SEM image, indicating that the spherical microparticles are porous aggregates with an apparent size range of 5~15 μm. A high-magnification SEM image, as shown in Fig. 30d, indicates that the porous microparticles are loose aggregates of WC nanoparticles with a particle size of 100~250 nm. The result of laser particle size analysis indicated that the apparent particle sizes of the WC powders obtained at 1050 °C for 2 h range in 4.0~18.0 μm, with a peak size of 8.6 μm. The above result is consistent with the SEM observations (Figs. 30b & c). The specific surface area of the WC powders obtained at 1050 °C for 2 h is 1.7 m^2/g. Their average diameter estimated according to the specific surface area is about 200 nm, which is very close to the high-resolution SEM observation (Fig. 30d). The molar ratio of W to C of the WC sample obtained at 1050 °C for 2 h is close to 1:1 according to the EDS spectra.

Fig. 30. (a) A typical SEM image of the as-obtained $C_8N@H_2WO_4$ hybrid derived from a reaction between H_2O-moisted H_2WO_4 powders and *n*-octylamine (the inset is its FT-IR spectrum); (b-d) SEM images with various magnifications of the WC powders obtained by treating the $C_8N@H_2WO_4$ hybrid at 1050 °C for 2h. Reproduced with permission from *J. Am. Ceram. Soc.*, 2010, 93(12), 3997-4000. Copyright 2010 John Wiley and Sons.

The formation process of hexagonal WC particles undergoes the following steps according to the XRD results. (1) The pyrolysis of the tungstate-based inorganic-organic hybrid precursor leads to formation tungsten oxides (e.g., WO_2) and some species containing carbon and nitrogen in the sealed quartz tube. (2) The species containing carbon and nitrogen react with the tungsten oxides, and form β-$W_{40.9}N_{9.1}$, α-W_2C, W and WC. (3) The phases of β-$W_{40.9}N_{9.1}$, α-W_2C and W further react with the C-containing species, and finally form a hexagonal WC phase by elevating the reaction temperature and prolonging the reaction time.

In the sealed quartz tube, the pyrolyzed organic ammonium ions emit excess C-containing species relative to the amount of W, and this is helpful for the formation of a pure WC phase. The excess C forms carbon films and carbon beads in the other end wall of the quartz tube. The above results are validated by EDS examination on the samples collected from different parts of the quartz tube. The possible chemical reactions can be described as Eqs.3–6.

$$(C_8H_{17}NH_3)_2WO_4 \xrightarrow{C/CO/C_mH_n,\leq 750^oC} WO_2 + C + CO + NH_3 + C_mH_n \quad (3)$$

$$WO_2 \xrightarrow{C/CO/C_mH_n,\geq 800^oC} W + \alpha - W_2C + WC \quad (4)$$

$$WO_2 \xrightarrow{NH_3,750\text{-}800^oC} \beta - W_{40.9}N_{9.1} \xrightarrow{C/CO/C_mH_n,800\text{-}850^oC} W + \alpha - W_2C + WC \quad (5)$$

$$W + \alpha - W_2C \xrightarrow{C/CO/C_mH_n,\geq 850^oC} WC \quad (6)$$

7. Typical applications of WO_3 and MoO_3 nanoplates

7.1 Photocatalysts for water splitting [6]

Fig. 31 shows the UV/Vis spectra of the as-obtained WO_3 nanoplates and commercially available WO_3 powders. The on-set value of the absorption band of the synthesized WO_3 nanoplates was *ca.* 466 nm (spectrum A), by which the energy band gap was estimated as *ca.* 2.67 eV. Although the energy gap of the WO_3 nanoplates is a little larger than that of the bulk WO_3 (2.58 eV) powders (spectrum B) due to the quantum size effect, the WO_3 nanoplates are still able to absorb a majority of solar energy and can be an efficient photocatalyst for the visible-light utilization.

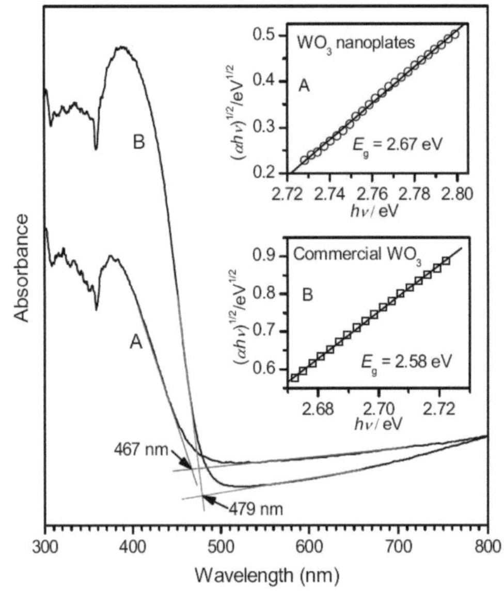

Fig. 31. UV/Vis absorption spectra of A) as-obtained WO_3 nanoplates and B) commercially available WO_3 powders. The insets are the plots of $(ah\nu)^{1/2}$ versus $h\nu$, by which the values of the energy band gaps of the as-obtained WO_3 nanoplates and the commercial WO_3 powders are determined as 2.67 and 2.58 eV, respectively. Reprinted with permission from *Small*, 2008, 4(10), 1813-1822. Copyright 2008 Wiley-VCH Verlag GmbH & Co. KGaA.

Fig. 32 shows the results of the visible-light-induced water splitting for oxygen (O_2) generation using the WO_3 nanoplates as the photocatalyst. For comparison, the commercial WO_3 powders were also used as the photocatalyst for water splitting under the same experimental conditions. As Fig. 32 shows, the amount of O_2 generated using WO_3 nanoplates as the photocatalyst is larger than the case of commercial WO_3 powders by an order of magnitude. The enhanced photocatalytic properties should be attributed to the superhigh specific surface areas and high crystallinity of the synthesized single-crystal WO_3 nanoplates.

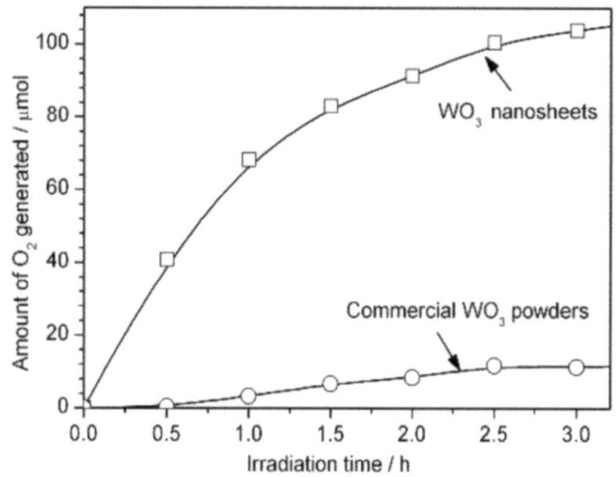

Fig. 32. Plots of the amount of O_2 generated *vs.* the irradiation time (0.6 g of WO_3 in 300 mL of 0.05 M $AgNO_3$ aqueous solution; visible light irradiation from an 18 W straight-tube fluorescence lamp). Reprinted with permission from *Small*, 2008, 4(10), 1813-1822. Copyright 2008 Wiley-VCH Verlag GmbH & Co. KGaA.

7.2 Sensitive materials for chemical sensors [10,11]

7.2.1 Fabrication of sensors

WO_3 (or α-MoO_3) nanoplates (or nanoparticles) were mixed with a small amount of de-ionized H_2O to form WO_3 (α-MoO_3) pastes in a glass dish. The as-obtained pastes were then brush-coated onto the surfaces of an Al_2O_3 microtube with four Pt electrodes (Fig. 33a). After the WO_3 coating was air-dried, the coating process was repeated until a complete coating was formed. The Al_2O_3 microtubes coated with WO_3 (α-MoO_3) nanoplates were then fixed to a special pedestal with 6 poles (Fig. 33c) by welding the four Pt electrodes to 4 poles of the pedestal, respectively. A heating coil (Fig. 33b) was then inserted through the Al_2O_3 microtube and its two ends were welded to the other two poles of the pedestal. A photograph of the as-obtained WO_3 (α-MoO_3) sensor was shown in Fig. 33d.

7.2.2 Gas-sensing test system

The sensing properties of WO_3 (α-MoO_3) sensors were measured using a commercial computer-controlled HW-30A system under a static testing condition. The sensors, integrated in a large circuit board with 32 inlet-sites, were encased in a transparent glass chamber with a volume of 13.8 L. The testing system was placed in a ventilating cabinet with a large draught capacity. Various vapors of volatile organic liquids were used as the target gases to characterize the sensing performance of the WO_3 (α-MoO_3) sensors. Volatile organic liquids were sampled using syringe-like samplers with ranges of 1–10 μL. The

operating temperatures were r.t–400 °C, controlled by an electric heating system (applied voltages: 4.2–5.0 V). The relative humidity (RH) of the environment was 35–50%.
The concentrations (ppm) were calculated according to the densities of volatile organic liquids and the volume of the chamber using the following equation:

$$V_t = \frac{10^{-9} V_0 \cdot M \cdot C_t}{22.4 \rho \cdot p} \tag{7}$$

Here, V_t is the required volume of the target liquid (μL), V_0 is the volume of the chamber (V_0=13.8 L), ρ is the density of volatile organic liquids (g cm^{-3}), M is the mole mass (g mol^{-1}) of volatile organic liquids, p is the rate of purity of volatile organic liquids, and C_t is the concentration (ppm) of volatile organic liquids.
An equivalent circuit of the gas-sensing testing sytem is shown in Fig. 33e. the sensor (R) is connected in series with a load resistor (R_0) with a known resistance (22–1000 KΩ), and a source voltage (U_0) of 5 V is loaded on the circuit. The system measures the voltages (U) loaded on the resistor R_0, and the resistances (R) of the WO_3 sensors can therefore be calculated according to Eq. 8.

$$R = \frac{U_0 - U}{U} \times R_0 \tag{8}$$

For reducing gases of alcohols and n-type semiconducting sensors, the sensitivity (S_r) is defined as Eq. 9, where R_a and R_g are the resistances of the sensor in air ambient and in alcohol ambient, respectively.

$$S_r = R_a / R_g \tag{9}$$

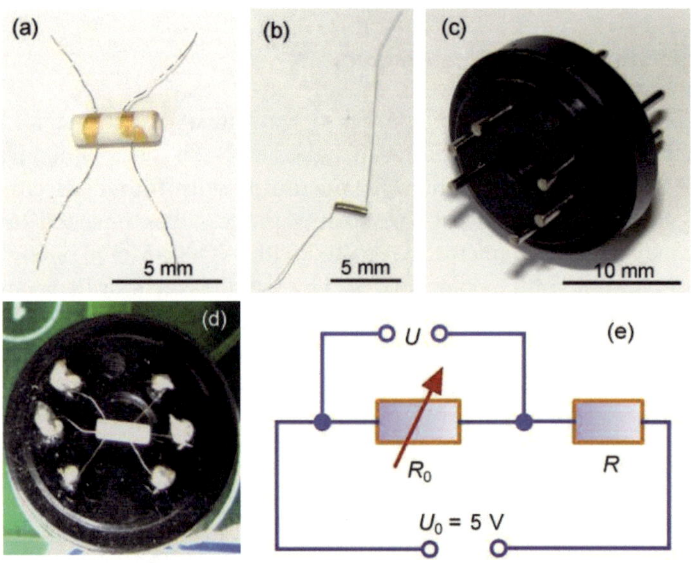

Fig. 33. Components of a sensor and its gas-sensing testing principle: (a) an Al_2O_3 microtube with 4 Pt-electrodes; (b) a heating coil; (c) a pedestal with 6 poles; (d) a photograph of a complete sensor; (e) an equivalent circuit of the alcohol-sensing testing system. Reprinted with permission from *Nanotechnology*, 2010, 21, 035501. Copyright 2010 IOP Publishing Ltd.

The response time (T_{res}) is defined as the time required for the sensor to reach 90% of the stabilized value of its resistance in the presence of the test gas. Similarly, the recovery time (T_{rec}) is defined as the time required for the sensor to reach 10% of the initial steady state value of its resistance after the gas was removed.

7.2.3 Results and discussion

Fig. 34 shows the changing trend of the sensitivities of WO_3 nanoplate sensors as the concentrations of alcohols increase from several ppm to several hundred ppm at an operating temperature of 300 °C. One can find that the sensitivities increase with increases in the concentration of alcohols, including methanol, ethanol, isopropanol and butanol. For methanol, the sensitivity increases from 6 at 10 ppm to 33 at 300 ppm (Fig. 34a). The sensitivity for ethanol increases from 8 at 10 ppm to 38 at 200 ppm (Fig. 34b). For the case of isopropanol, the sensitivity increases from 12 to 75 as its concentration increases from 10 ppm to 200 ppm (Fig. 34c). The sensitivity of WO_3 nanoplate sensors to butanol increases from 31 at 2 ppm to 161 at 100 ppm (Fig. 34d), much higher than the sensitivities to methanol, ethanol or isopropanol. There is a linear relationship between the sensitivity and the concentration for all the tested alcohols. The solid lines in Fig. 34 are the linear fitting results and their linear correlation coefficients (R) are not less than 0.96. When compared the slope coefficients of the fitting equations (inlets in Fig. 34), one can find that the increase rate in the sensitivity to butanol (1.24 per ppm) is much higher than those to isopropanol (0.33 per ppm) and ethanol (0.15 per ppm), whereas the sensitivity to methanol shows a lowest increase rate (0.09 per ppm).

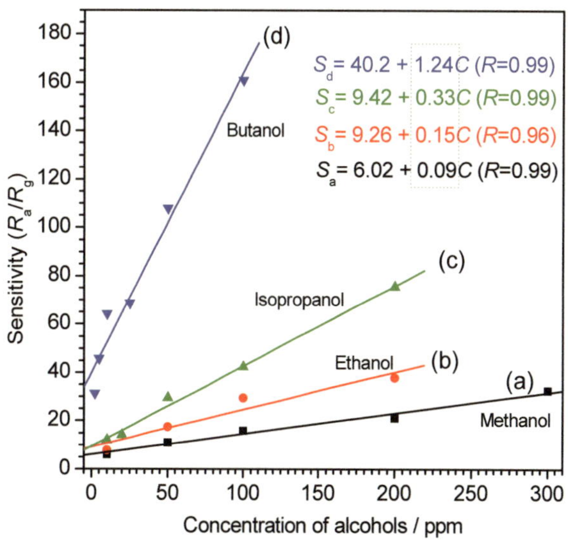

Fig. 34. Sensitivities (R_a/R_g) of the WO_3 nanoplate sensors as a function of alcohol concentration with an operating temperature of 300 °C: (a) methanol, (b) ethanol, (c) isopropanol and (d) butanol. Reprinted with permission from *Nanotechnology*, 2010, 21, 035501. Copyright 2010 IOP Publishing Ltd.

Fig. 35 shows the response and recovery times of the WO_3 nanoplate sensors upon exposure to various alcohol vaopors with different concentration levels at an operating temperature of

300 ºC. As Fig. 35a shows, the response times of methanol are close to their recovery times at a concentration range of 10–300 ppm, and their values are about 10–14 s. For ethanol, as shown in Fig. 35b, the response times are less than 7 s for the concentration range of 10–200 ppm, but their recovery times are about 10 s, longer than their corresponding response times. The response and recovery times of isopropanol are shown in Fig. 35c. One can find that the response time (less than 10 s) is obviously shorter than its corresponding recover time (about 15 s). But for the case of butanol, the response time (10–15 s) is longer than its corresponding recovery time (9–10 s), especially in the low concentration range of 2–10 ppm, as shown in Fig. 35d. It is very different from the cases of methanol, ethanol and isopropanol, which have longer recovery times than their corresponding response times. The above-mentioned difference suggests that the semiconductor time response is strongly correlated with the length of alcohol alkyl tails, which have a decreased vapour tension from methanol to butanol.

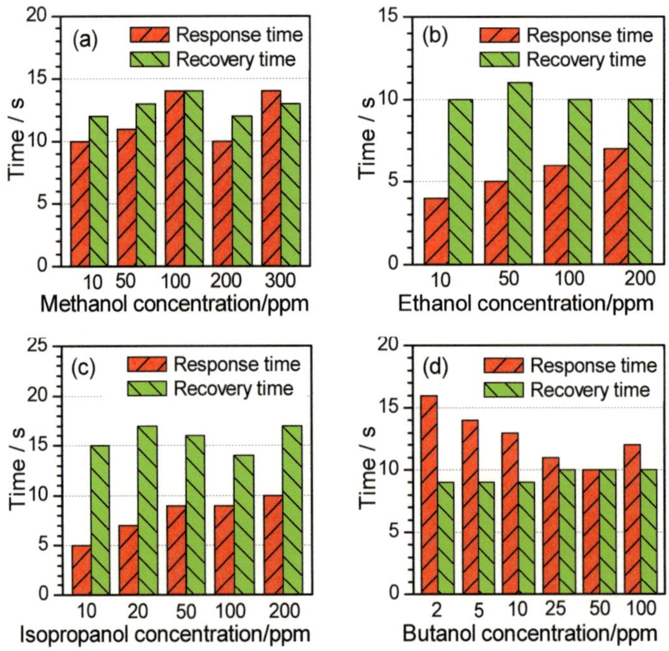

Fig. 35. Response and recovery times of the WO_3 nanoplate sensors operating at 300 ºC under various alcohol concentrations: (a) methanol, (b) ethanol, (c) isopropanol and (d) butanol. Reprinted with permission from *Nanotechnology*, 2010, 21, 035501. Copyright 2010 IOP Publishing Ltd.

Fig. 36 shows the acetone-sensing response profiles of the sensors made using the as-obtained WO_3 nanoplates as the sensitive material. Fig. 36a shows a typical response profile of the WO_3 nanoplate sensors operating at 300 ºC to acetone vapors with various concentrations from 2 ppm to 1000 ppm. One can find that there are sharp rises and drops in *U* values when the acetone vapors are injected and discharged, respectively, which indicates that the WO_3 nanoplate sensors are of fast response and recovery speeds to acetone vapors. Fig. 36b shows a similar rapid acetone-sensitive response of the WO_3 nanoplate sensors at an operating temperature of 250 °C. Fig. 36c shows typical response results of the WO_3 nanoplate sensors operating at 200 °C, and Fig. 36d shows the response curve operating at 100 °C. One can find that when the operating temperature decreases to 100 °C, the change amounts in *U* obviously decrease, the response speed and the detectable limits decrease, and the response profiles become instable, as shown in Figs. 36a–d.

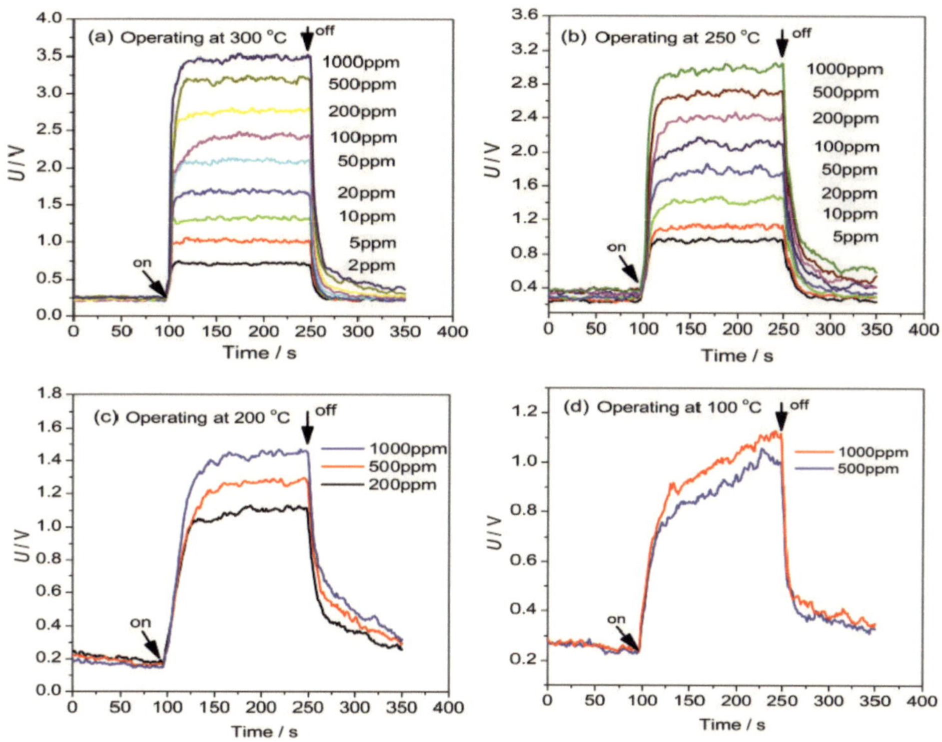

Fig. 36. Acetone-sensing response profiles of the WO_3 nanoplate sensors operating at various temperatures and various acetone vapor concentrations. The R_0 values are 22 kΩ, 100 kΩ, 100 kΩ and 470 kΩ, respectively, for the measurements operating at (a) 300 °C, (b) 250 °C, (c) 200 °C and (d) 100 °C. Reprinted with permission from *Sensor. Actuat. B-Chem.*, 2011, 153(2), 373-381. Copyright 2011 Elsevier.

Fig. 37a presents the sensitivities (R_a/R_g) of the WO_3 nanoplate sensors operating at various temperatures and acetone concentrations. As curve A shows, the sensitivities of the WO_3 nanoplate sensors decrease as the operating temperature decreases in the range of 100–300 °C under the same acetone concentrations in the range of 2–1000 ppm. Also, we can find that the sensitivity increases with an increase in the acetone concentration at the same operating temperature. At an operating temperature of 300 °C, the WO_3 nanoplate sensor has a sensitivity as high as 42 for a 1000 ppm acetone vapor, and it has a detectable limit as low as 2 ppm of acetone vapor with a sensitivity of about 4, as shown as curve A. At a low operating temperature of 100 °C, the sensitivities of the WO_3 nanoplate sensors are about 3 for 100–500 ppm acetone vapors (curve D in Fig. 37a). Figs. 37b-c show the response and recovery times of the WO_3 nanoplate sensors under various concentrations of acetone vapors at operating temperatures of 250–300 °C. The response times of the WO_3 nanoplate sensors operating at 300 °C are 3–10 s in the vapor concentration range of 2–1000 ppm, and their corresponding recovery times are 6–13 s. For the case operating at 250 °C, the response times lie in a range of 8–15 s, and their recovery times are 15–31 s. Both the response times and recovery times increase with the increase in operating temperature.

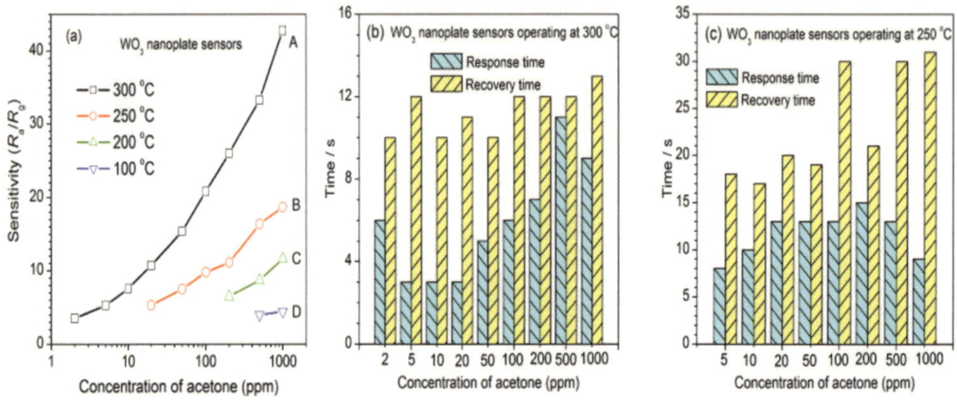

Fig. 37. (a) Sensitivities of the WO_3 nanoplate sensors as a function of the concentration of acetone vapor at various operating temperatures, and (b-c) response and recovery times of the WO_3 nanoplate sensors to acetone vapor with various concentrations operating at (b) 300°C and (c) 250 °C. Reprinted with permission from *Sensor. Actuat. B-Chem.*, 2011, 153(2), 373-381. Copyright 2011 Elsevier.

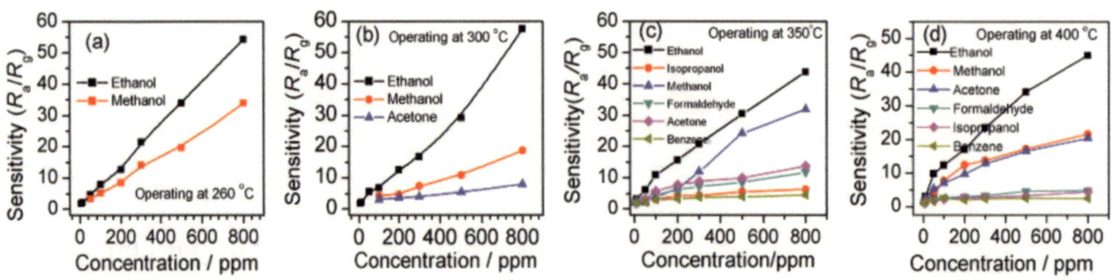

Fig. 38. The plots of the sensitivities of the α-MoO_3 nanoplate sensors as a function of the concentrations of different reagent vapors operating at (a) 260 °C (R_0 = 20 MΩ), (b) 300 °C (R_0 = 20 MΩ), (c) 350 °C (R_0 = 20 MΩ), and (d) 400 °C (R_0 = 4.7 MΩ). Reprinted with permission from *J. Mater. Chem.*, 2011, 21(25), 9332-9342. Copyright 2011 Royal Society of Chemistry.

Fig. 38 shows the plots of the sensitivities of the α-MoO_3 nanoplate sensors as a function of the concentrations of different reagent vapors operating at 260 - 400 °C. The sensitivities to ethanol vapors present a linearly increasing trend from 2 to 54 as the concentration of the ethanol vapor increases from 5 to 800 ppm, and a similar linear relationship is also seen for methanol vapors (4 - 34 for 50 - 800 ppm), when the operating temperature is 260 °C (Fig. 38a). When operated at 300 °C, the α-MoO_3 nanoplate sensors show similar high sensitivities to ethanol vapors (2 - 58 for 5 - 800 ppm), whereas their sensitivities to methanol vapors are 4 - 19 for 50 - 800 ppm, lower than those (4 - 34 for 50 - 800 ppm) operated at 260 °C, as shown in Fig. 38b. The sensitivities of the α-MoO_3 nanoplate sensors to acetone vapors shows a good linear increase with the increase in concentrations, but their sensitivities are less than 10 even at 800 ppm. For ethanol vapors, the sensitivities show a similar change trend when the operating temperature increases from 350 to 400 °C, and the sensitivity values increase from about 2 to 44 when the concentrations of ethanol vapors increase from 5 to 800 ppm. When the operating temperature is 400 °C, the sensitivities of isopropanol, formaldehyde and benzene vapors are very low (less than 5) in the concentration range of 5-800 ppm, whereas the sensitivities are about 1.2 - 22 for 5 - 800 ppm of methanol and acetone vapors.

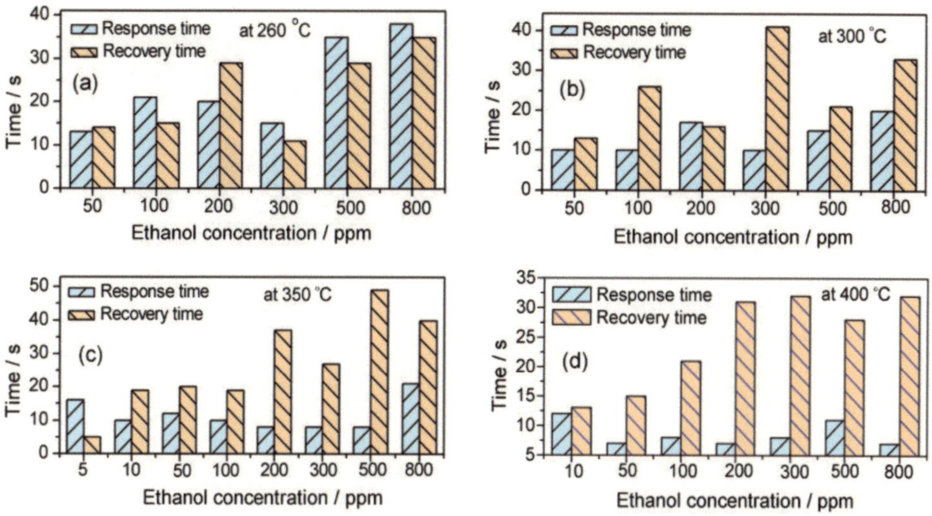

Fig. 39. The response and recovery times of the α-MoO_3 nanoplate sensors as a function of the ethanol concentrations operating at 400 °C (R_0 = 4.7 MΩ). Reprinted with permission from *J. Mater. Chem.*, 2011, 21(25), 9332-9342. Copyright 2011 Royal Society of Chemistry.

Fig. 39 shows the response and recovery times of the α-MoO_3 nanoplate sensors to ethanol vapors with various concentrations. Generally, the response times of α-MoO_3 nanoplate sensor decrease with the increase in operating temperature, from 15 - 38 s at 260 °C to 7 - 14 s at 400 °C. However, the recovery times (i.e., 10 - 40 s) almost show no obviously change in the operating temperature range of 260 - 400 °C. The recovery time slightly increases with an increase in the concentration of ethanol vapors (Figs. 39c-d).

The space-charge layer model has often been applied to explain the possible gas-sensing mechanism of a semiconducting metal oxide sensor. WO_3 and MoO_3 are typical *n*-type metal oxide semiconductors, and the space-charge layer model is suitable to WO_3 and MoO_3 nnaoplate sensors. When the nanoplate sensor is exposed to air, O_2 molecules will adsorb on the surfaces of WO_3 nanoplates. The O_2 molecules adsorbed then transform to be oxygen ions (*e.g.*, O^-, O^{2-}, or O_2^-) by capturing free electrons from the conductance band of nanoplates. The electron-capture process leads to a depletion region in semiconductor nanoplates, which reduces the conductive regions, and thus a high-resistance state is formed, as shown in Fig. 40a. When the nanoplate sensors are exposed to reducing gases (i.e., alcohol, and acetone vapors), the molecules adsorbed on the surfaces of the semiconductor nanoplates can provide electrons to reduce oxygen ions, and then release free electons back into nanoplates. This process decreases the depletion region and form a conducting channel, which results in a low-resistance state, as shown in Fig. 40b.

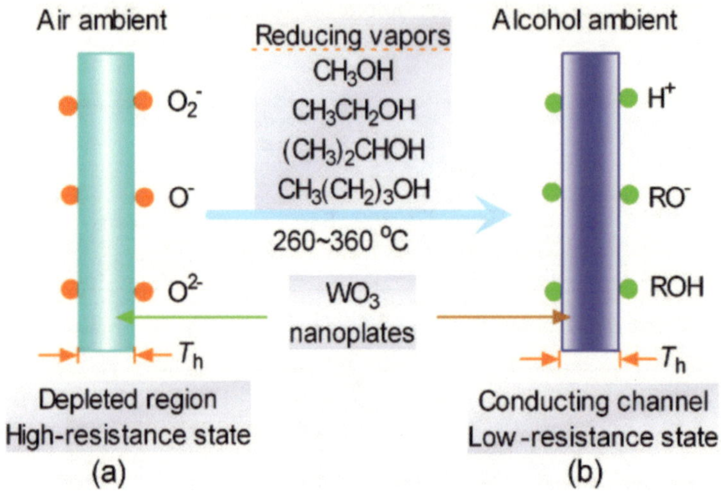

Fig. 40. A schematic representation of the alcohol sensing mechanism for WO_3 nanoplate sensors. Reprinted with permission from *Nanotechnology*, 2010, 21, 035501. Copyright 2010 IOP Publishing Ltd.

The sizes and shapes of the semiconductor nanocrystals, together with their configurartion, are the key factors influencing the gas-sensing properties. Reducing the sizes of the active materials down to several nanometers can enhance the gas-sensing performance. However, the paticles with nanoscale sizes tend to form aggregates, as shown in Fig. 41c. The long diffusion length and the sluggish diffusion of a target gas into the inner parts of the secondary aggregates is inefficient to improve the gas-sensing property. Only the resistance of the primary nanoparticles near the surfaces of the aggregates is affected by the target gas molecules, and thus the sensitivity is low and the response time is long, as shown in Fig. 41d. But for the ultrathin 2D nanoplates, they usually form a loose assembly containing a great number of gaps due to the steric effect of the platelike morphology, as shown in Fig. 41a. The gaps between nanoplates not only enhance the effective surface to adsorb target gases, but also provide capacious channels for target gases to diffuse. The large surface areas for the effective adsorption and the loosely assembling structures for the rapid diffusion of target gases are helpful to enhance the sensitivities and shorten the response times of the 2D nanoplate sensors (Fig. 41b).

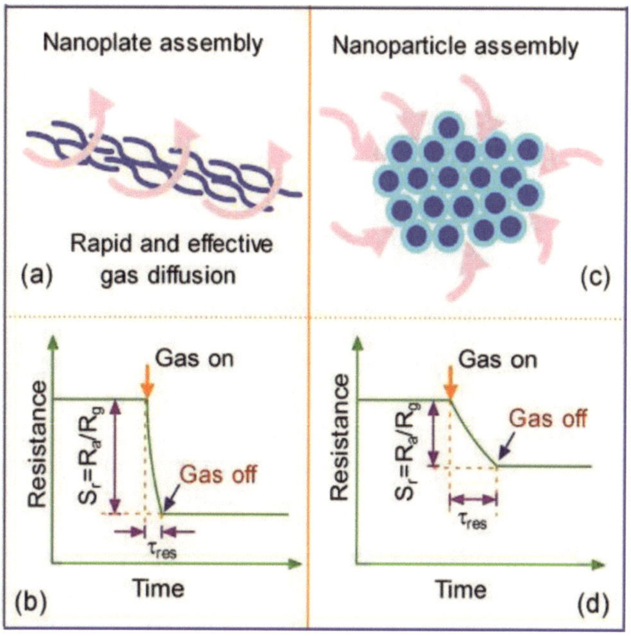

Fig. 41. A schematic representation of the configuration effect of the semiconductor nanoplates and nanoparticles on their alcohol sensing performance. Reprinted with permission from *Nanotechnology*, 2010, 21, 035501. Copyright 2010 IOP Publishing Ltd.

8. Conclusions

Topochemical conversion of inorganic-organic hybrid compounds into low-dimensional inorganic nanostructures is an efficient strategy to achieve advanced functional materials with smart control in crystal size and shape. The inorganic-organic hybrid compounds can be synthesized via intercalation chemical reactions between layered inorganic host solids and suitable organic guest molecules. The nanostructures obtained by this topochemical conversion route usually inherit the morphologies and sizes of their corresponding hybrid precursors, and using this method one can readily synthesize some special low-dimensional materials, e.g. carbides and nitrides. The outstanding advantages of the topochemical conversion route are the ready control in dimensions and the ability of cost-effective and large-scale synthesis. The as-obtained nanocrystals can have wide applications in photocatalysts, sensors, and energy conversion.

9. Acknowledgment

This work was supported by the National Natural Science Foundation of China (Grant No. 50802090, 51172211), the China Postdoctoral Science Foundation (Grant No. 20090450094, Grant No. 201003397), the Opening Project of State Key Laboratory of High Performance Ceramics and Superfine Microstructure (Grant No. SKL200905SIC), and the Young Scientist Exchange Program between the Republic of Korea and the People's Republic of China. The author thanks Prof. Yoshiyuki Sugahara (Waseda University) for his valuable discussion in the formation mechanism of tungstate-based inorganic-organic hybrid compounds. The author also thanks Prof. Lian Gao, Prof. Rui Zhang, Prof. Jing Sun, Prof. Xinjian Li, Prof. Hongxia Lu, Prof. Hongliang Xu, Prof. Daoyuan Yang, Prof. Hailong Wang, Dr. Bingbing Fan, Mr. Hejing Wen, Mr. Xianxiang Hou, Ms Zhongsheng Liu, Ms Li Yin, Mr. Tao Li, Ms Mina Liu, Mr. Yi Zhang and Mr. Haitao Zhai for their kind help in some of the experiments.

10. References

[1] D. L. Chen and Y. Sugahara, Tungstate-based inorganic-organic hybrid nanobelts/nanotubes with lamellar mesostructures: synthesis, characterization, and formation mechanism. *Chem. Mater.*, 2007, 19(7), 1808-1815.

[2] D. L. Chen, and Y. Sugahara, Investigation of factors influencing the formation of tungstate-based inorganic-organic hybrid nanobelts/nanotubes. *Key Eng. Mater.*, 2007, 352, 85-88.

[3] D. L. Chen, H. L. Wang, R. Zhang, S. K. Guan, H. X. Lu, H. L. Xu, D. Y. Yang, Y. Sugahara and L. Gao, Synthesis, characterization and formation mechanism of single-crystal WO_3 nanosheets via an intercalation-chemistry-based route. *Chem. J. Chinese U.*, 2008, 29(7), 1325-1330.

[4] D. L. Chen, T. Li, L. Yin, X. Hou, X. Yu, Y. Zhang, B. Fan, H. Wang, X. Li, R. Zhang, T. Hou, H. Lu, H. Xu, J. Sun and L. Gao, A comparative study on reactions of n-alkylamines with tungstic acids with various W-O octahedral layers: Novel evidence for the "dissolution-reorganization" mechanism. *Mater. Chem. Phys.*, 2011, 125(3), 838-845.

[5] D. L. Chen, H. L. Wang, R. Zhang, L. Gao, Y. Sugahara and A. Yasumori, Single-crystalline tungsten oxide nanoplates. *J. Ceram. Process. Res.*, 2008, 9(6), 596-600.

[6] D. L. Chen, L. Gao, A. Yasumori, K. Kuroda and Y. Sugahara, Size- and shape-controlled conversion of tungstate-based inorganic-organic hybrid belts to WO_3 nanoplates with high specific surface areas. *Small*, 2008, 4(10), 1813-1822.

[7] D. L. Chen, M. Liu, L. Yin, T. Li, Z. Yang, X. Li, B. Fan, H. Wang, R. Zhang, Z. Li, H. Xu, H. Lu, D. Yang, J. Sun and L. Gao, Single-crystalline MoO_3 nanoplates: topochemical synthesis and enhanced ethanol-sensing performance. *J. Mater. Chem.*, 2011, 21(25), 9332-9342.

[8] D. L. Chen, H. Wen, T. Li, L. Yin, B. Fan, H. Wang, R. Zhang, X. Li, H. Xu, H. Lu, D. Yang, J. Sun and L. Gao, Novel pseudo-morphotactic synthesis and characterization of tungsten nitride nanoplates. *J. Solid State Chem.*, 2011, 184(2), 455-462.

[9] D. L. Chen, H. Wen, H. Zhai, H. Wang, X. Li, R. Zhang, J. Sun and L. Gao, Novel synthesis of hierarchical tungsten carbide micro-/nanocrystals from a single-source precursor. *J. Am. Ceram. Soc.*, 2010, 93(12), 3997-4000.

[10] D. L. Chen, X. Hou, H. Wen, Y. Wang, H. Wang, X. Li, R. Zhang, H. Lu, H. Xu, S. Guan, J. Sun and L. Gao, The enhanced alcohol-sensing response of ultrathin WO_3 nanoplates. *Nanotechnology*, 2010, 21, 035501.

[11] D. L. Chen, X. Hou, T. Li, L. Yin, B. Fan, H. Wang, X. Li, H. Xu, H. Lu, R. Zhang and J. Sun, Effects of morphologies on acetone-sensing properties of tungsten trioxide nanocrystals. *Sensor. Actuat. B Chem.*, 2011, 153(2), 373-381.

6

Synthesis of Cobalt-Zinc Phosphates Templated by Polyamines

Yue Ding, Niu Li, Daiping Li, Ailing Lu,
Naijia Guan and Shouhe Xiang
Nankai University
P.R. China

1. Introduction

Open-framework structures of metal phosphates, especially those zinc and cobalt phosphates, have exhibited many fascinating structural features and potential applications in catalysis, separation processes, and as photoluminantphosphors (Rao et al., 2000; Chang et al., 2004). The feasibility of zinc and cobalt to tetrahedrally coordinate their phosphates makes them to be an important group of transition metal phosphates (Mandal & Natarajan, 2002; Harrison et al., 1991; Neeraj et al., 1999; Fan et al., 2000). Usually, these materials have been synthesized by employing hydrothermal or solvothermal conditions in the presence of organic amines as the structure-directing agents (SDA). Studies have shown that the important role of host–guest charge matching between the inorganic frameworks and the organic amines makes linear polyamines the preferred candidates for the structure-directing agents to form the higher negative charge framework of zinc and cobalt phosphates (Bu et al., 1997, 1998; Feng et al., 1997). Moreover, diethylenetriamine (DETA), triethylenetetramine (TETA) and tetraethylenepentamine (TEPA) become typical amines and have directed the formation of a large number of open-framework structures of zinc (or cobalt) phosphates by varying their concentrations accompanying with the changes in inorganic compositions (Choudhury et al., 2000).

Recently, a new synthesis route involving alkylformamide as a template precursors has been reported (Lakiss et al., 2005; Vidal et al., 2000). It seems that organic amines generated in situ by the decomposition of alkylformamide show a special effect on the formation of metal phosphates with a novel structure. In our studies aimed to synthesize transition metal phosphates, more than one cobalt-zinc phosphates with different framework structures have been obtained from a single gel using diethylenetriamine as the organic basis (Lu et al., 2008; Li et al., 2009). It is noticed that the guest species encapsulated in their structures are not diethylenetriamine, but some smaller amines decomposed from it. Controlling the in situ decomposition of these linear polyamines plays the key role for directing the synthesis of pure-phase cobalt-zinc phosphates. According to these matters, large amines can also be used as a source for smaller amines as they will provide hydrolysis decomposition in the reaction mixture during crystallization. In the synthesis a temperature of crystallization is an important factor for controlling the hydrolysis of linear polyamines, and higher temperature makes decomposition of polyamines easier. On the other hand, adding small amines, such as propylamine, butylamine, can control the decomposition of the chain-type polyamines.

2. Three open-framework cobalt-zinc phosphates synthesized in the presence of linear polyamines as structure-directing agents (SDA)

In this section, three unique frameworks of cobalt-zinc phosphate have been synthesized under hydrothermal or solvothermal conditions. They were named as $CoZnPO_4$-IV, $CoZnPO_4$-V and $CoZnPO_4$-VI, respectively. $CoZnPO_4$-IV with cross-linked 10- and 8-ring channels was synthesized in the presence of diethylenetriamine. $CoZnPO_4$-V with cross-linked 16-, 12- and 10-ring channels was synthesized using diethylenetriamine as a structure-directing agent. $CoZnPO_4$-VI has 16-ring one-dimensional channels. It was synthesized in the presence of triethylenetetramine. Polyamines did not decompose during the synthesis process (Fig.13d, 13e, and 13f). Extra-large-pore structures ($CoZnPO_4$-V and $CoZnPO_4$-VI) with 16-ring channels have been synthesized in the presence of linear polyamines as the structure-directing agents (SDA).

2.1 Synthesis and initial characterization

The main reactants are $Zn(CH_3COO)_2$ (denoted as $ZnAc_2$), $Co(CH_3COO)_2$ (denoted as $CoAc_2$), H_3PO_4, linear polyamines, and ethylene glycol. In a typical synthetic procedure, solution A was prepared by mixing $ZnAc_2 \cdot 2H_2O$ with 85% H_3PO_4 and ethylene glycol (EG), the mixture was stirred at room temperature until $ZnAc_2 \cdot 2H_2O$ was dissolved. Solution B was prepared by mixing $CoAc_2 \cdot 4H_2O$, distilled water, and 85% H_3PO_4. Solution A and a kind of linear polyamine, such as diethylenetriamine were added to solution B. Then the pH value of the mixture was adjusted to 8.0 by adding an assistant amine. Finally the mixture was transferred to a stainless steel autoclave and heated at 140°C~160°C for 3-10 days. Specific gel compositions and crystallization conditions of three cobalt-zinc phosphates are shown in Table.1. (In the synthesis of $CoZnPO_4$-VI, drips of HF(40%) were added to the gel).

Sample	Reactant composition						Crystallization condition	
	ZnO	CoO	H_3PO_4	polyamine	H_2O	EG	Temp/ °C	Time/day
$CoZnPO_4$-IV	2.0	1.0	10.5	1.8DETA	300	0	160	3
$CoZnPO_4$-V	2.0	1.0	9.0	1.8DETA	95.4	9.6	140	10
$CoZnPO_4$-VI	2.4	0.6	4.0	2.0TETA	400	0	140	4.5

Table 1. Gel compositions and crystallization conditions of three cobalt-zinc phosphates.

Three large vivid blue crystal ($CoZnPO_4$-IV, $CoZnPO_4$-V and $CoZnPO_4$-VI) have been synthesized hydrothermally. Scanning electron micrograph (SEM) images show their morphologies being distinct from each other as rectangular-like, rectangular column, thin plate-like, respectively (Fig.1). XRD patterns of the three compounds also show their phases being distinct from each other (Fig.1).

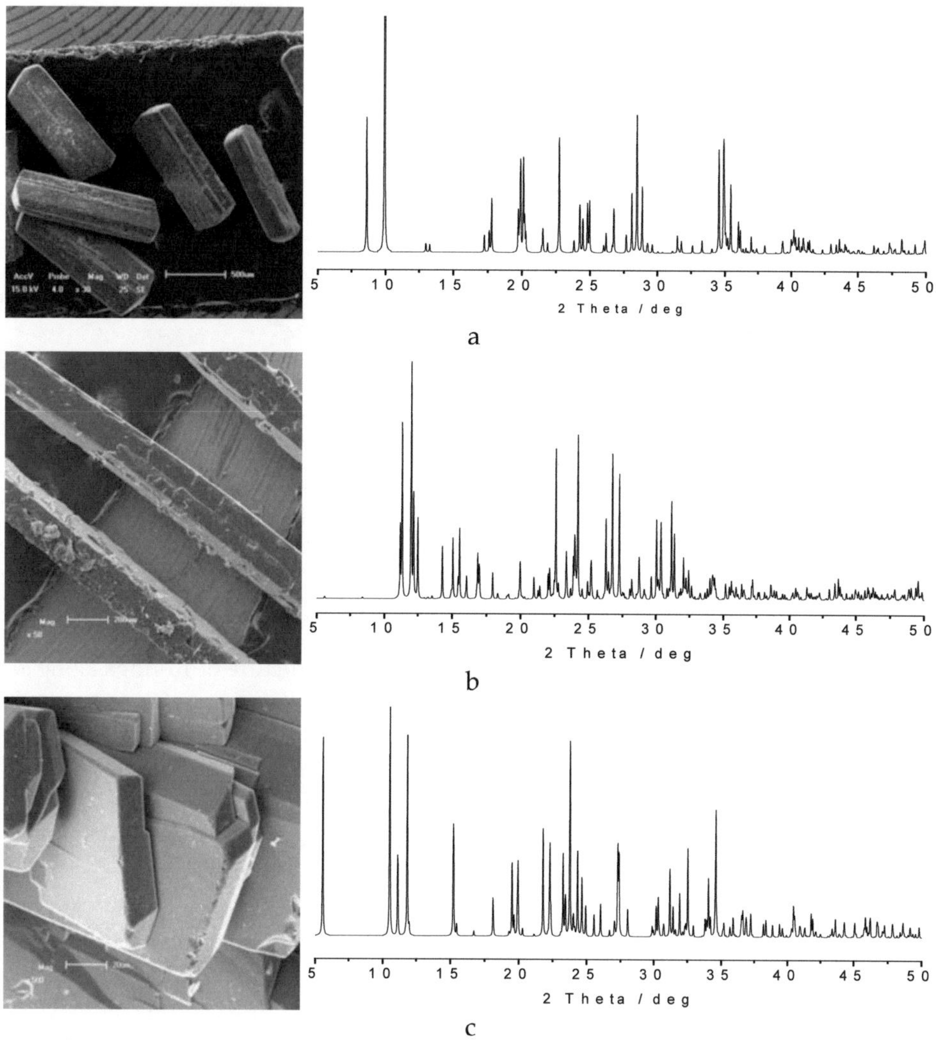

Fig. 1. Scanning electron micrograph (SEM) images and XRD patterns of the three compounds(a) $CoZnPO_4$-IV; (b) $CoZnPO_4$-V; (c) $CoZnPO_4$-VI.

2.2 Crystal structure of $CoZnPO_4$-IV

Single crystal analysis and Inductively Coupled Plasma atomic emission spectroscopy (ICP) analysis reveal that $CoZnPO_4$-IV is composed of $C_4N_3H_{15}Zn_{4.12}Co_{0.88}$ P_4O_{16}; it crystallizes in the monoclinic system, space group *Cc* (No. 9), with a = 26.926(7) Å, b = 5.1912(10) Å, c = 17.834(6) Å, $\alpha = \gamma = 90°$, β = 130.25(2)°, V = 1902.6(9) Å^3, and z = 4.

The secondary structural unit of $CoZnPO_4$-IV is shown in Fig.2a, and it is composed of 3-rings and 4-rings. Secondary structural units link with each other through sharing oxygen atoms to form the layers along *ac*-plane. Then neighboring layers link with each other along *b*-axis, resulting in cross-linked 10- and 8-ring channels in $CoZnPO_4$-IV (Fig.2b). The 8-ring channels locate among the layers (Fig.2c).

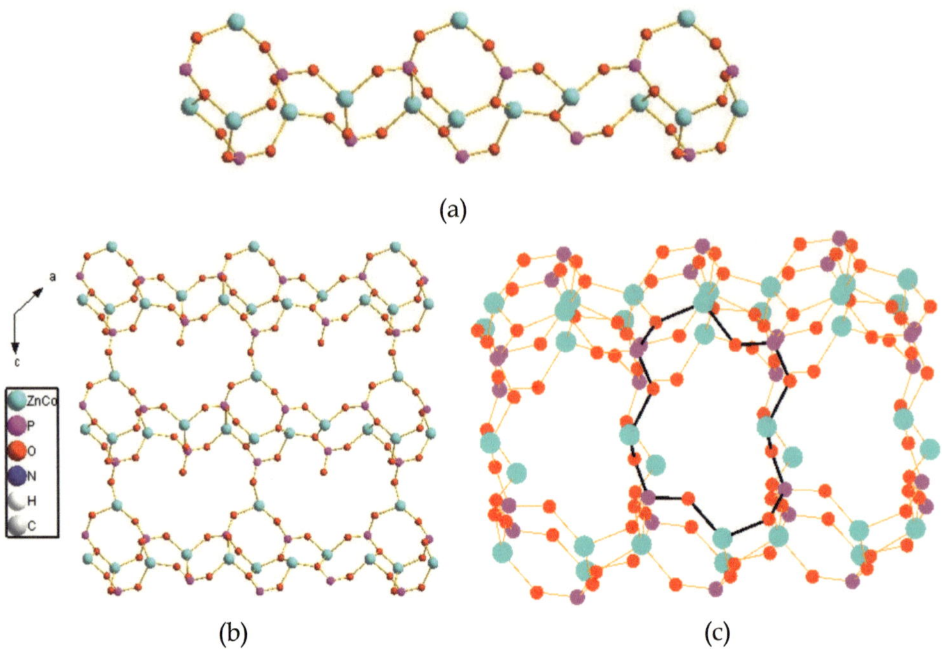

Fig. 2. (a) The secondary structural unit of $CoZnPO_4$-IV; (b) Structure of 10-rings along *ac*-plane in $CoZnPO_4$- IV; (c) Structure of 8-ring channel in $CoZnPO_4$- IV.

2.3 Crystal structure of $CoZnPO_4$-V

Single crystal analysis and Inductively Coupled Plasma atomic emission spectroscopy (ICP) analysis reveal that $CoZnPO_4$-V is composed of $C_{12}N_9H_{52}Zn_{4.5}Co_{1.5}P_8O_{32}$; it crystallizes in the monoclinic system, space group $P2_1/c$ (No. 14), with a = 31.936(3) Å, b = 8.3775(7) Å, c = 15.7874(13) Å, $\alpha = \gamma = 90°$, β = 97.0530(10)°, V = 4191.8(6) Å^3, and z = 4.

Three edge-shared 4-ring units of Zn-O and P-O tetrahedral groups could be the building unit, which was named as 4^3 unit (Fig. 3a). The neighboring units link with each other through 4-ring to form an infinite zigzag chain along the *b*-axis direction (Fig. 3b). 2D (4, 10) nets are conformably formed of the chains of 4-ring units from the corner-shared 4^3 units (Fig. 3b).

Viewed along the plane parallel to *b*-axis, the neighboring 4^3 units link with each other in a corner- shared way to form another infinite chain (Fig. 3c). The neighboring chains link together from 4-ring every two 4^3 units to form a layer. Therefore, 16-ring channels are constructed (Fig. 3c). On the other hand, the wall of each 16-ring channel can be described as the coiling of a net composed of 10-rings and infinite 4-ring zigzag chain (Fig. 3b).

2.4 Crystal structure of $CoZnPO_4$-VI

Single crystal analysis and Inductively Coupled Plasma atomic emission spectroscopy (ICP) analysis reveal that $CoZnPO_4$-VI is composed of $C_6N_4H_{22}Co_{0.16}Zn_{5.84}(PO_4)_4(HPO_4)_2$; it crystallizes in the triclinic system, space group $P\bar{1}$ (No. 2), with a = 5.1919(10) Å, b = 8.7263(17) Å, c = 16.000(3) Å, α =89.07(3)°, β = 83.45(3)°, γ = 74.34(3)°, V = 693.4(2) Å^3, and z = 1.

The three-dimensional framework of $CoZnPO_4$-VI can be described by with two secondary structural units (SBU-a and SBU-b). SBU-a (Fig.4a) is a special structure composed of two 6-rings ladder linked face to face; SBU-b (Fig.4b) is a common 4-ring ladder in the molecular sieves. The linkage of SBU-a conducted by sharing 4-rings form the layers along *ab*-plane. SBU-a and SBU-b connect with one sharing oxygen atom in the structure, forming the three-dimensional framework of $CoZnPO_4$-VI and resulting in 16-ring large channels along the direction of *c*-axis (Fig.4c). The arms of channels are constituted mainly by SBU-a and SBU-b.

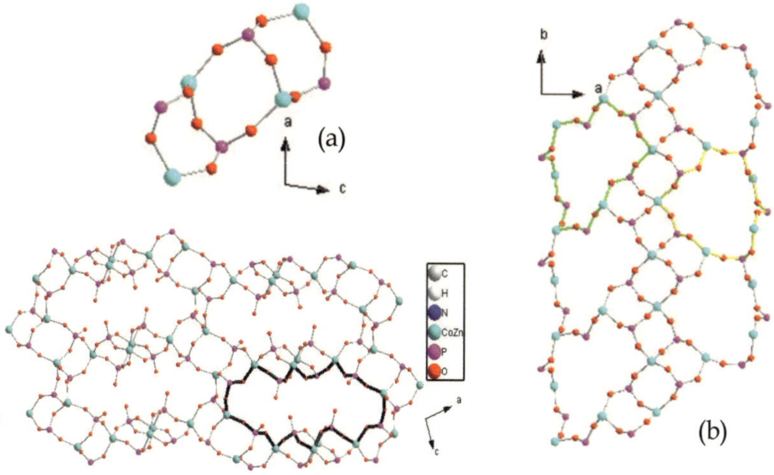

Fig. 3. (a) 4^3 unit along b-axis in $CoZnPO_4$-V; (b) Structure of 10-rings along *ab*-plane in $CoZnPO_4$-V. (c) Structure of 16-rings along *ac*-plane in $CoZnPO_4$-V.

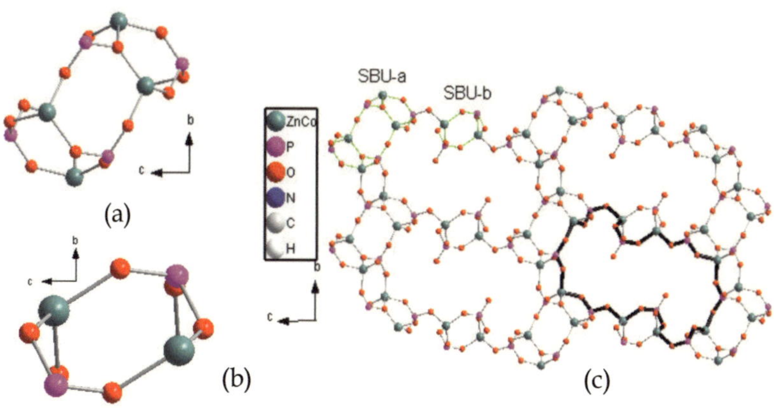

Fig. 4. (a) SUB-a; (b) SUB-b; (c) Structure of 16-rings along *bc*-plane in $CoZnPO_4$-VI.

3. Controlling the in situ decomposition of linear polyamines to synthesize open-framework cobalt-zinc phosphates

As the organic templates, linear polyamines are easy to decompose at high reaction temperature (higher than 160°C) in the hydrothermal condition to synthesize microporous metal phosphate materials. In this section, three open-framework cobalt-zinc phosphates,

$CoZnPO_4$-I, $CoZnPO_4$-II, and $CoZnPO_4$-III, have been synthesized from a single gel at 200 °C. Organic amine in the starting mixture is diethylenetriamine. Structural analysis reveals that species encapsulated in the structures of $CoZnPO_4$-I, $CoZnPO_4$-II, and $CoZnPO_4$-III are ammonium, ethylenediaminum, and diethylenetriaminum, respectively, which can be attributed to the partly in situ decomposition of diethylenetriamine molecules during the crystallization procedure (Fig.13a, 13b, and 13c). When a second small organic amine, such as propylamine, or butylamine, was added to reacting mixture, single phase of $CoZnPO_4$-I, $CoZnPO_4$-II, or $CoZnPO_4$-III can be obtained respectively, according to the kind or concentration of small organic amine, which plays an important role on controlling the in situ decomposition of linear polyamines.

3.1 Synthesis and initial characterization

The main reactants are $ZnAc_2$, $CoAc_2$, H_3PO_4, DETA, and ethylene glycol. In a typical synthesis procedure, solution A was prepared by mixing $ZnAc_2$ $2H_2O$ (2.217 g) with 85% H_3PO_4 (0.98 mL) and ethylene glycol (EG) (2.7 mL), and then stirring at room temperature until $ZnAc_2$ $2H_2O$ was dissolved. Solution B was prepared by mixing $CoAc_2$ $4H_2O$ (1.252 g), distilled water (8.6 mL), and 85% H_3PO_4 (2.6 mL). Solution A and DETA (0.97 mL) were added to solution B. The pH value of the mixture was adjusted to pH 8.0 by the addition of *n*-propylamine (4.5 mL). The final molar ratio of the mixture was $2ZnAc_2 \cdot 2H_2O$:$1CoAc_2 \cdot 4H_2O$:9.6EG:10.5H_3PO_4:95.4H_2O:1.8DETA. The mixture was then transferred to a stainless steel autoclave and heated at 200 °C for 6 days. Three types of large blue crystal were filtered and washed with deionized water and dried in air.

a b c

Fig. 5. SEM images of the large crystals of three cobalt-zinc phosphates : (a) crystal of $CoZnPO_4$-I with elongated hexagonal-dipyramid morphology;(b) $CoZnPO_4$-II crystal with square bifrustum morphology; (c) $CoZnPO_4$-III crystal with cubical-like morphology.

Three open-framework cobalt-zinc phosphates have been synthesized hydrothermally from a single gel at 200°C by using DETA as the SDA with *n*-propylamine as the assistant organic amine. They are all large vivid blue crystals with sizes more than 400 μm. Scanning electron micrograph (SEM) images show their morphologies being distinct from each other, shaped as elongated hexagonal-dipyramid, square bifrustum, and cubical, respectively (Fig.5). XRD patterns of $CoZnPO_4$-I, -II, and -III (Fig.6) have shown that they are three distinguishing phases with different diffraction peaks.

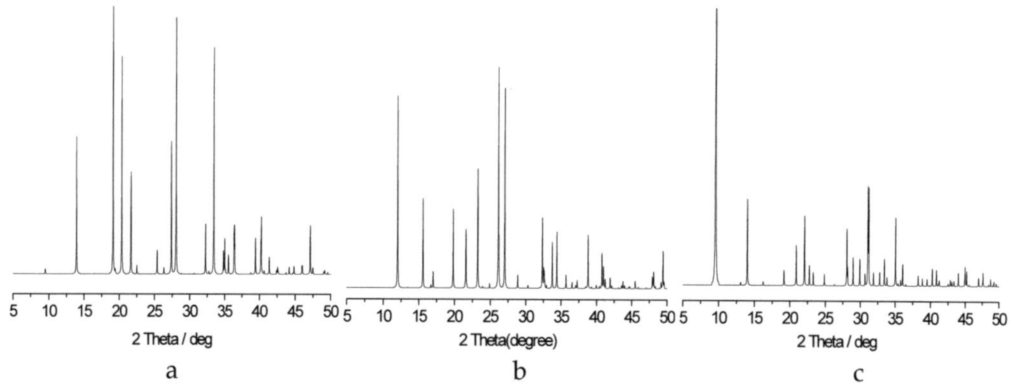

Fig. 6. XRD patterns of three cobalt-zinc phosphates: (a) $CoZnPO_4$-I; (b) $CoZnPO_4$-II; (c) $CoZnPO_4$-III.

3.2 Crystal structures of three compounds

It is clear that there are only diethylenetriamine and *n*-propylamine as the SDAs present in the starting materials. However, organic species encapsulated in the structure of $CoZnPO_4$-I, -II, and -III are ammonium, ethylenediaminum, and diethylenetriaminum cations, respectively (Fig.13a, 13b, 13c). The appearance of ethylenediaminum and ammonium may be attributed to the in situ decomposition of diethylenetriamine during the crystallization.

Single crystal analysis reveals that $CoZnPO_4$-I is a novel cobalt-zinc phosphate. It is composed of $(NH_4)_2Co_{0.34}Zn_{1.66}(PO_4)_2$, which was modeled as $(NH_4)_2Co_2(PO_4)_2$ in the single crystal data and crystallizes in the hexagonal system, space group $P6_3$ (No. 173), with $a=b=10.7207(5)$Å, $c=8.7241(8)$ Å, $\alpha=\beta=90°$, $\lambda=120°$, V=868.36(10) Å^3, and $z=4$. Viewing along the plane parallel to *c*-axis, a net consisting of 6-ring channels has been found. The 6-ring channels with regularly alternating Zn(Co)- and P-centered tetrahedral can be considered as the building units of $CoZnPO_4$-I (Fig. 7a). J. V. Smith determined a simple hexagonal net, which has only one type of 3-connected node and is described as the Schläfli symbol 6^3 because each node lies between three circuits of 6 nodes (Smith, J. V. *Am. Mineral.* 1977).For any node in a simple hexagonal net in horizational position, an additional perpendicular linkage pointing either upward (U) or downward (D) may result in the linkage joined to a node of another simple hexagonal net lying either above or below the first hexagonal net. Eight ways of the sequence of U and D linkages around each 6-ring has been used to enumerate 4-connected, 3-dimensional nets and classification of framework of silicates and aluminophosphates.

The linking sequences of UDUDUD around each 6-ring produces a dense tridymite framework, whereas the UUUDDD results in an open framework ABW containing 8-ring channels. However, infinity of frameworks can be produced if more than one sequence occurs in the hexagons of a 6^3 net. This has just happened in the structure of $CoZnPO_4$-I. In its framework, the 6-rings of tetrahedral are present in two different conformations based on the more than one type circuit as the following sequence of up and down tetrahedral: UDUDUD and UUUDDD in the proportion 1:3. A second building unit, which is called a 6^94^6cage, has been formed (Fig. 13a).

On the other hand, it can be found along the (010) plane that double 4-rings and one 6-ring linked in turn by sharing edges to form an infinite chain (Fig. 7b). These chains are cross-linked along the *c*-axis direction to the similar chains from the linkage Zn-O-P or P-O-Zn to form the sheet structure with edgesharing infinite 10-ring chains (Fig. 7b). Adjacent sheets are cross-linked with each other along the vertical direction of (010) plane via the Zn-O-P bonds displaced by +1/2 in the [001] direction to form the three-dimensional framework of $CoZnPO_4$-I (Fig.7c). Each 10-ring is blocked by double 4- and 6-rings around it. So the tendency to construct 10-ring channels is interdicted with only clathrate $6^9 4^6$ cages to be formed.

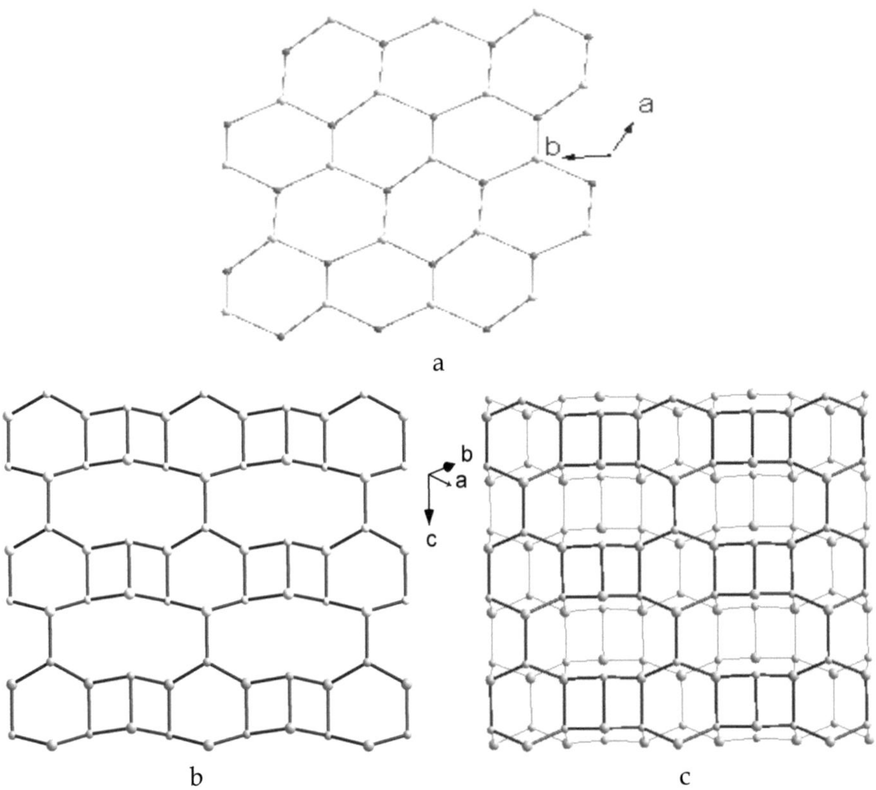

Fig. 7. (a) The crystal structure of $CoZnPO_4$-I consisted of 6-rings viewed down [001]; (b) A sheet extends along the (010) plane consisted of 4-, 6-, and 10-rings;(c) The sheets stacked along the direction vertical to the (010) plane (each node is a T-atom, oxygen atoms are omitted for clarity).

$CoZnPO_4$-II and -III are isostructural with DTF topology and $C_6N_4H_{22}Co_7(PO_4)_6$, respectively. $CoZnPO_4$-II crystallizes in the tetragonal system, space group $P4_2bc$ (No. 106), with $a = b = 14.694(2)$ Å, $c = 8.936(2)$ Å, $\alpha = \beta = \lambda = 90°$, $V = 1929.4(6)$ Å^3, and $z = 8$. $CoZnPO_4$-III crystallizes in the trigonal system, space group $R\bar{3}$ (No. 148), with $a = b = 13.5393(8)$ Å, $c = 15.0443(10)$ Å, $\alpha = \beta = 90°$, $\lambda = 120°$, $V = 2388.3(3)$ Å^3, and $z = 3$. The three-dimensional framework of $CoZnPO_4$-II can be described as the stacking of 4.8^2 nets along the [001] direction connecting through the UUDDUUDD linkages of each tetrahedral atom around the 8-ring channels with protonated ethylenediame cations accommodated in the 8-ring channel (Fig.8).

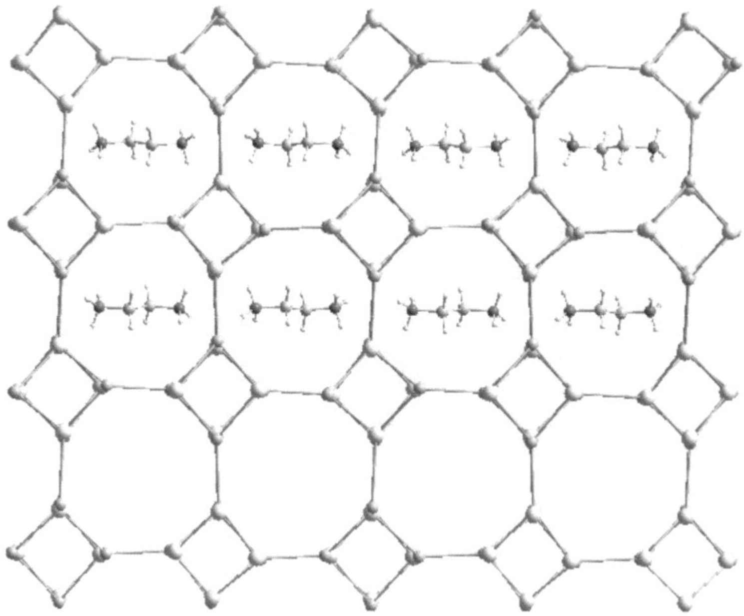

Fig. 8. $CoZnPO_4$-II seen along [001] direction showing the stacked 4.8^2 nets to form 3D framework and 8-ring with diprotonated ethylenediame cations accommodated in the channel (each node is a T-atom, oxygen atoms are omitted for clarity).

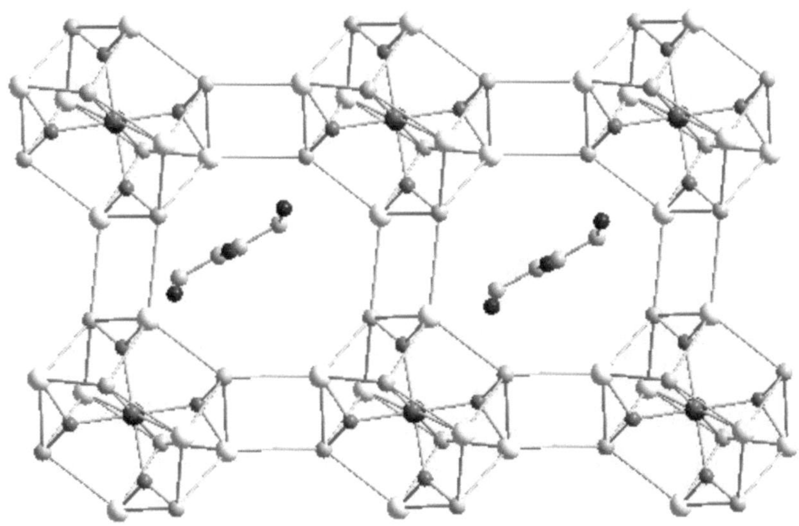

Fig. 9. $CoZnPO_4$-III I seen along [100] direction showing the 8-ring channels with diethylenetriaminum cations accommodated in the channel.

$CoZnPO_4$-III possesses a three-dimensional architecture similar to $C_6N_4H_{22}Co_7(PO_4)_6$, but with the composition of $[C_4N_3H_{16}]_{1.33}[Co_{2.41}Zn_{3.59}(PO_4)_6]$. Its structure is made of CoO_6, ZnO_4 (or CoO_4), and PO_4 polyhedra. The cobalt octahedron is surrounded by six zinc tetrahedra to form the CoZn(or Co)$_6$O$_6$ cluster which are linked via six PO_4 tetrahedra to other clusters to form its framework. The clusters are arranged in such a manner that each cluster is displaced by half its *a*-distance from its neighbors, forming a honeycomb layer. The next layer of clusters is identical to the first but is displaced along the *a*-axis by half the unit cell so that the honeycomb channels are capped. The framework of AB-type stacking results in the formation of 8-ring channels at an angle to *a*-axis (Fig. 9).

4. The in situ framework transformation of cobalt-zinc phosphates in the presence of linear polyamines

Although linear polyamines cannot decompose at low crystallization temperature, experiments have shown the crystallization time also affects the crystal structure transformation. Another typical result that we have found is the transformation of two cobalt-zinc phosphates, $CoZnPO_4$-VI and $CoZnPO_4$-VII. $CoZnPO_4$-VI was synthesized by using TETA as the template, which can be transformed into $CoZnPO_4$- VII with the increase of crystallization time in 140°C.

4.1 Synthesis and initial characterization

$CoZnPO_4$-VI and $CoZnPO_4$-VII were synthesized hydrothermally from a mixture of ZnO, $2CoCO_3 \cdot 3Co(OH)_2 \cdot H_2O$ (47.5%), H_3PO_4 (85%), TETA, HF (40%), H_2O with a typical molar ratio being 2.4ZnO:0.6CoO:4H_3PO_4:2TETA:4HF:400H_2O, and the experiment was similar to the previous synthesis. When the crystallization period was 4.5 days, prevailingly light blue sheet-like crystals (named $CoZnPO_4$-VI) and a less number of dark blue block crystals (named $CoZnPO_4$-VII) can be obtained. When the crystallization period was extended to 15 days, sheet-like crystals become significantly fewer and block crystals increase. It can be seen that $CoZnPO_4$-VI can be gradually transformed into $CoZnPO_4$-VII with the increase of crystallization time.

Scanning electron micrograph (SEM) image (Fig.10) shows the morphology of $CoZnPO_4$-VII is as a relatively large irregular block, it is distinct from $CoZnPO_4$-VI (Fig.1c). XRD patterns of $CoZnPO_4$-VI and $CoZnPO_4$-VII have also shown they are two distinguishing phases with different diffraction peaks (Fig.10 and Fig.1c).

Fig. 10. SEM image and XRD pattern of $CoZnPO_4$-VII.

4.2 Crystal structure of $CoZnPO_4$-VII

Single crystal analysis and Inductively Coupled Plasma atomic emission spectroscopy (ICP) analysis reveal that $CoZnPO_4$-VII is composed of $C_6H_{22}Co_0N_4O_{16}P_4Zn_4$; it crystallizes in the triclinic system, space group $P\bar{1}$ (No. 2), with a = 8.0871(16) Å, b = 8.4237(17) Å, c = 14.961(3) Å, α =89.57(3)°, β = 88.29(3)°, γ =76.23(3)°, V = 989.4(3) Å^3, and z = 2.

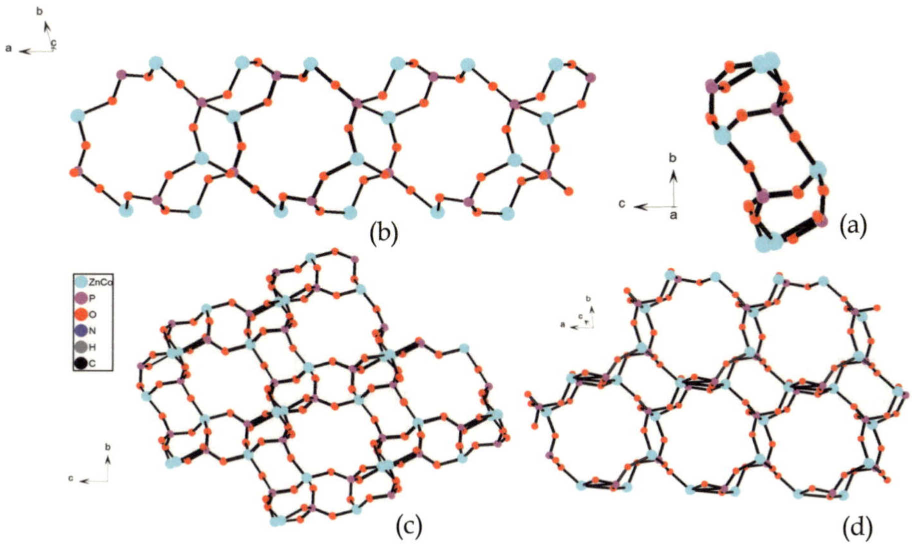

Fig. 11. (a) SUB-c; (b) SUB-c along a-axis; (c) Structure of 8-rings along a-axis; (d) Structure of 8-rings along c-axis.

The biggest feature in the structure of $CoZnPO_4$-VII is that there are two cross-linked 8-ring channels in the structure. The three-dimensional framework of $CoZnPO_4$-VII can be also described as the stack of a secondary structural unit (SBU-c) (Fig.11a). In the direction of a-axis, the connection of 8-rings by three 4-rings forms a ladder (SBU-c) (Fig.11b); in the direction of b-axis ladders are joined by sharing edges, spreading the layers running ab-plane. Then adjacent layers are connected with 4-rings. All described above spread the three-dimensional framework of $CoZnPO_4$-VII and exhibit a 8-ring channel system running along the direction of a-axis (Fig.11c). The arms of channels are constituted mainly by SBU-c. At the same time they form another 8-ring channel system running the direction of c-axis (Fig.11d).

4.3 Relationship between $CoZnPO_4$-VI and $CoZnPO_4$-VII

The similarities between $CoZnPO_4$-VI and $CoZnPO_4$-VII provide the possibility of transformation. On one hand, both of them are composed of Co,Zn,P,O elements, and all Zn (Co) atoms and P atoms in their structures are tetrahedrally coordinated by oxygens, which have been seen from their asymmetric unit structures (Fig.12). On the other hand, the templates filled in their inorganic frameworks are TETA, and both of them crystallize in the triclinic system, space group $P\bar{1}$ (No. 2). All of these result in the transformation of $CoZnPO_4$-VI and $CoZnPO_4$-VII with the increase of crystallization time.

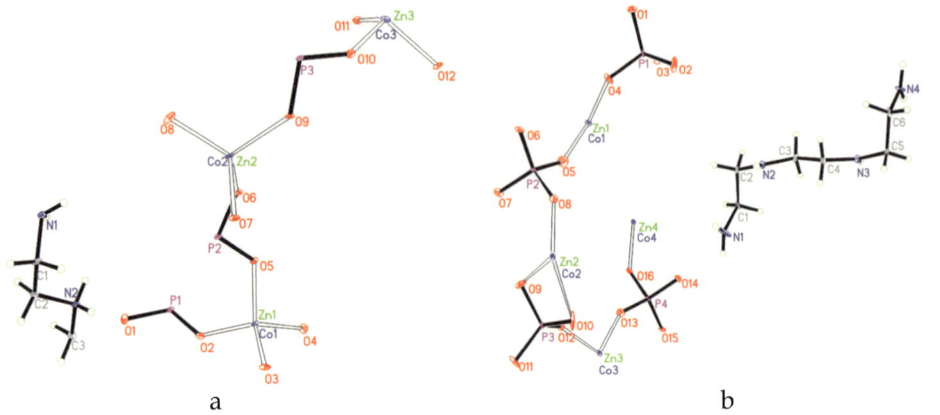

Fig. 12. The asymmetric unit structures of two compounds (a) $CoZnPO_4$-VI; (b)$CoZnPO_4$-VII.

5. Investigation of the in situ decomposition of linear polyamines during hydrothermal synthesis of cobalt-zinc phosphates

In this section, polyamines such as diethylenetriamine, triethylenetetramine, and tetraethylenepentamine were used as structure directing agents to investigate the in situ decomposition of linear polyamines during the crystallization of microporous zinc-cobalt phosphate materials. Mixtures of water and ethylene glycol were used as solvents. Monoamines, such as n-propylamine, n-butylamine, n-dipropylamine, n-tripropylamine and ethanolamine were used as assistant amines.

5.1 Experimental methods

The important basis for determination of polyamines' decomposition is consideration of the template molecules in the framework of all crystals. From the summarized above, we know that the template molecules in the framework of $CoZnPO_4$-I, $CoZnPO_4$-II, $CoZnPO_4$-III, $CoZnPO_4$-IV, $CoZnPO_4$-V and $CoZnPO_4$-VI are NH_4^+, en, DETA, DETA, DETA and TETA, respectively (Fig.13).

Previous studies have shown preliminary that temperature is an important factor for controlling the hydrolysis of linear polyamines. Other, the addition of different assistant templates has different effects on decomposition of propylamines while using the straight-chain polyamines as a precursor of the template in the synthesis of cobalt-zinc phosphates. In order to investigate the decomposition of polyamines, the same ratio of gel should react for a certain time at different temperatures, and the pH value of gel should be adjusted to pH (7.0-9.0) by the addition of different assistant templates. The selected molar ratio of gel is $2ZnAc_2 \bullet 2H_2O$:$1CoAc_2 \bullet 4H_2O$:9.6EG:$10.5H_3PO_4$:$95.4H_2O$:1.8R (R is DETA, TETA, TEPA respectively).

5.2 Discussions of factors affecting decomposition of polyamine

The results are shown in Table. 2 when crystallizing at different temperatures for 6 days.

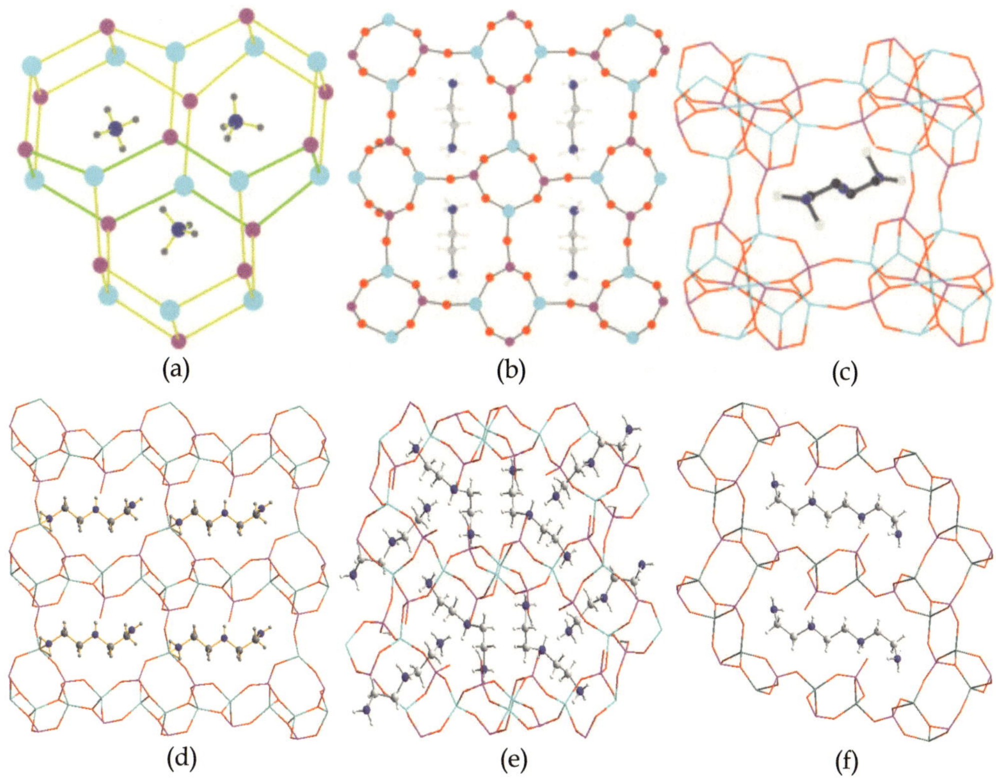

Fig. 13. The structures of six compounds:(a) $CoZnPO_4$-I;(b) $CoZnPO_4$-II;(c) $CoZnPO_4$-III; (d) $CoZnPO_4$-IV;(e) $CoZnPO_4$-V;(f) $CoZnPO_4$-VI.

5.2.1 Temperature effect on decomposition of polyamines

From Table.2, it can be found that high temperature favors the hydrolysis of linear polyamines, prior to adding the same case of secondary amines (*n*-propylamine). For DETA, it is special that it can be decomposed completely into NH_4^+ in 200 °C, still partially decomposed. It is decomposed completely into en in 180 °C, and cannot break down in 160 °C. For TETA, it can be decomposed completely into DETA in 200 °C, and can not break down lower than 160 °C. For TEPA, it also can be decomposed completely into NH_4^+ in 200 °C, and decomposed completely into DETA in 180 °C and 160 °C. Therefore, pure phase is easy to be synthesizes at low temperature.

5.2.2 Assistant template on decomposition of polyamines

We only study the products synthesized at the temperature of 200°C. In the present investigation when crystallizing the mixture of 2:1:9.6:10.5:95.4:1.8 $ZnAc_2$:$CoAc_2$:EG:H_3PO_4:H_2O:DETA at 200 °C for 6 days (assistant amines have not been added), only the pure phase of $CoZnPO_4$-I has been obtained with only ammonium in the structure. It means the complete decomposition of DETA. Meanwhile, the case using triethylenetetramine or tetraethylenepentamine as the SDA also gives the same results, namely, crystallizing at 200 °C has also resulted in the formation of $CoZnPO_4$-I as the only product

However, when an assistant amine, such as *n*-propylamine, was added to the mixture to increase the pH value to 7–9, even if the mixture is crystallized at 200 °C, three metal phosphates ($CoZnPO_4$-I, -II, -III) have been obtained with diethylenetriamine still encapsulated in the pores of $CoZnPO_4$-III. It indicates that the presence of *n*-propylamine can restrain the hydrolysis of diethylenetriamine during the crystallization. It can be found from Table 2 that, apart from *n*-propylamine, other assistant amines, such as *n*-butylamine, dipropylamine, and tripropylamine, can also restrain the decomposition of diethylenetriamine to some extent. The influence of second organic amine may be attributed to the effect of the pH on hydrolysis of the linear polyamines. In the case of triethylenetetramine or tetraethylenepentamine, the controlling effects of assistant amine are less prominent than for diethylenetriamine. $CoZnPO_4$-II and -III have only been obtained as the pure phases by varying assistant amine or crystallizing temperature.

$CoZnPO_4$	-I	-II	-III	-IV	-V	-VI
Diethylenetriamine (Crystallizing at 200 °C)						
without assistant amine	y	n	n	N	n	N
n-propylamine	y	y	y	N	n	N
n-propylamine(180 °C)	n	y	n	N	n	N
n-propylamine(160 °C)	n	n	y	N	n	N
n-butylamine	y	y	n	N	n	N
dipropylamine	y	y	y	Y	n	N
tripropylamine	n	y	n	N	n	N
Triethylenetetramine (Crystallizing at 200 °C)						
without assistant amine	y	n	n	N	n	N
n-propylamine	n	n	y	Y	n	N
n-propylamine(160 °C)	n	n	n	Y	n	Y
n-propylamine(140 °C)	n	n	n	N	n	Y
Tetraethylenepentamine (Crystallizing at 200 °C)						
without assistant amine	y	n	n	N	n	N
n-propylamine	y	n	n	N	n	N
n-propylamine(180 °C)	n	n	y	N	n	N
n-propylamine(160 °C)	n	n	y	N	n	N

[a] The reacting mixture with the molar ratio of 2:1:9.6:10.5:95.4:1.8 $ZnAc_2$:$CoAc_2$:EG:H_3PO_4:H_2O:R was crystallized for 6 days; y = yes, compound has been obtained; n = no,compound has not been obtained.

Table 2. Products obtained by adding various assistant amines to adjust the pH value of the mixture to 7-9[a]

6. Conclusions

Synthesis of zinc-cobalt phosphate materials have been investigated by using linear polyamines, such as tetraethylene-pentamine, triethylenetetramine or diethylenetriamine as SDA. During the investigation of decomposition of polyamines seven cobalt-substituted zinc phosphates were obtained: $CoZnPO_4$-I-VII.

$CoZnPO_4$-I, $CoZnPO_4$-II, and $CoZnPO_4$-III, have been synthesized from a single gel at 200 °C in the presence of diethylenetriamine as the structure-directing agent. Structural analysis reveals that species encapsulated in the structures of $CoZnPO_4$-I, $CoZnPO_4$-II, and

$CoZnPO_4$-III are ammonium, ethylenediaminum, and diethylenetriaminum, respectively, which can be attributed to the partly in situ decomposition of diethylenetriamine molecules during the crystallization procedure. $CoZnPO_4$-IV, $CoZnPO_4$-V and $CoZnPO_4$-VI with differently unique 3D framework topologies have been synthesized under hydrothermal or solvothermal conditions in the presence of linear polyamines as structure directing agents (SDA). $CoZnPO_4$-VI and $CoZnPO_4$-VII, which were synthesized by using TETA as the template, are symbiotic and $CoZnPO_4$-VI can be transformed into $CoZnPO_4$-VII with the increase of crystallization time in 140°C.

In the synthesis of zinc-cobalt phosphates, temperature is an important factor, favoring the hydrolysis of linear polyamines at higher temperature. Investigations have also found that the in situ decomposition of tetraethylene-pentamine, triethylenetetramine, or diethylenetriamine can be controlled by using second organic amines, such as propylamine as the assistant organic additive. The presence of an assistant amine can restrain the hydrolysis of diethylenetriamine during the crystallization. If synthesis conditions can be controlled, different structures can be transformed into each other.

7. Acknowledgment

This work was funded by the Natural Science Foundation of China (20873069) and State Basic Research Project (2009CB623502).

8. References

Bu, X.; Feng, P. & Stucky, G. D. (1880). Large-Cage Zeolite Structures with Multidimensional 12-Ring Channels. *Science,*Vol. 278, No. 5346, (December1997), pp. 2080-2085, ISSN 0036-8075

Bu, X.; Feng, P.; Gier, T. E.; Zhao, D. & Stucky, G. D.(1879). Hydrothermal Synthesis and Structural Characterization of Zeolite-like Structures Based on Gallium and Aluminum Germanates *J. Am. Chem.Soc.*, Vol.120, No.51, (December 1998), pp. 13389-13397, ISSN 0002-7863

Chang, W.; Chiang, R.; Jiang, Y.; Wang, S.; Lee, S. & Lii, K.(1962). Metamagnetism in Cobalt Phosphates with Pillared Layer Structures:[Co_3(pyz)$(HPO_4)_2F_2$] and [Co_3(4,4'-bpy) $(HPO_4)_2F_2$] xH_2O. *Inorg.Chem.*, Vol.43, No.8, (March 2004), pp.2564-2568, ISSN 0020-1669

Choudhury, A.; Natarajan, S. & Rao, C. N. R.(1962). Formation of One-, Two-, and Three-Dimensional Open-Framework Zinc Phosphates in the Presence of a Tetramine. *Inorg. Chem.*, Vol.39, No.19, (August 2000), pp. 4295-4304, ISSN 0020-1669

Fan, J. & Hanson, B. E.(1962). Novel Zinc Phosphate Topologies Defined by Organic Ligands. *Inorg. Chem.*, Vol.44, No.20, (September 2000), pp.6998-7008, ISSN 0020-1669

Feng, P.; Bu, X. & Stucky, G. D.(1869). Hydrothermal Synthesis and Structure Characterization of Zeolite Analogue Compounds based on Cobalt Phosphate. *Nature*, Vol.14, (August 1997), pp.735-740, ISSN 0028-0836

Harrison, W. T. A. & Phillips, M. L. F.(1989). Hydrothermal Syntheses and Single-Crystal Structures of Some Novel Guanidinium-Zinc-Phosphates. *Chem. Mater.*, Vol.9, No.8, (September 1991), pp.1837-1846, ISSN 0897-4756

Lakiss, L.; Simon-Masseron, A. & Patarin, J.(1981) New simplified synthesis method of TREN–GaPO using ethylamine in situ generated by decomposition of the ethylformamide. *Microporous Mesoporous Mater.*, Vol.84, (June 2005), pp. 50-58, ISSN 1387-1811

Li, D.; Song, H.; Wang, H.; Li, N.; Guan, N. & Xiang, S.(1950). Small organic amine assisted synthesis of an extra-large pore containing open-framework zinc-cobalt phosphate templated by a chain-type polyamine. *Science in China Series B: Chemistry.*, Vol.52, No.9, (September 2009), pp.1485-1489, ISSN 1006-9291

Lu, A.; Li, N.; Ma, Y.; Song, H.; Li, D.; Guan, N.; Wang, H. & Xiang, S.(2001). Controlling the In situ Decomposition of Chain-Type Polyamines To Direct the Crystal Growth of Cobalt zinc Phosphates. *Cryst. Growth Des.*, Vol.8, No.7, (June 2008), pp.2377-2383, ISSN 0020-1669

Mandal, S. & Natarajan, S.(2001). Hydrothermal Synthesis and Structures of Three-Dimensional Zinc Phosphates Built-Up from Two-Dimensional Layers and One Dimensional Chains and Ladders. *Cryst. Growth Des.*, Vol.2, No.6, (June 2002), pp.665-673, ISSN 0020-1669

Neeraj, S.; Natarajan, S. & Rao, C. N. R. (1989). A Zinc Phosphate Possessing Ladder-like Layers Made Up of Three- and Four-Membered Rings and Infinite Zn-O-Zn Chains. *Chem. Mater.*, Vol.11, No.5, (April 1999), pp.1390-1395, ISSN 0897-4756

Rao, C. N. R.; Natarajan, S. & Neeraj, S.(1879). Exploration of a Simple Universal Route to the Myriad of Open-Framework Metal Phosphates. *J. Am. Chem. Soc.*, Vol.122, No.12, (March 2000), pp.2810-2817, ISSN 0002-7863

Smith, J. V. Enumeration of 4-connected 3-dimensional nets and classification of framework silicates. I. Perpendicular linkage from simple hexagonal net. *Am. Mineral,* Vol.62, (1977), pp.703-709, ISSN 0003-004X

Vidal, L.; Pray, C. & Patarin, J.(1981). $AlPO_4$-41 aluminophosphate molecular sieve: a new reproducible synthesis route involving diisopropylformamid as template precursor. *Microporous Mesoporous Mater.*,Vol.39, (February 2000), pp. 113-116, ISSN 1387-1811

Part 3

Biomineralization

7

Unusual Crystal Formation in Organisms - Exceptions that Confirm Biomineralization Rules

Davorin Medaković[1] and Stanko Popović[2]
[1]*Ruđer Bošković Institute, Center for Marine Research, Rovinj,*
[2]*Department of Physics, Faculty of Sciences, University of Zagreb,*
Croatian Academy of Science and Arts, Zagreb,
Croatia

The universe is full of things that care of themselves.
Atoms spontanousely assemble into molecules.
Mountains steadily rise out of the sea.
Stars continously cluster into galaxies.
Yet nowhere is self-organization more impressive than in crystals.
(Fagan, J.F. & Ward, M.D. 1992)

1. Introduction

Biomineralization is a complex combination of biochemical and physiological processes, depending on endogenous activity of an organism and the exogenous environmental influence. In marine organisms (bacteria, calcareous algae, molluscs, crustaceans, fish and mammals) biomineralization starts at early stages of development and continues during the entire life. These processes take place in different tissues, shells and skeletons, and are manifested in calcification, decalcification and in polymorph transitions of calcium carbonate in particular shell or skeleton layers (Lowenstam, 1981; McConnaughey, 1989). Biomineralization processes depend on a number of conditions of which the main ones, that influence the type and structure of the biomineral, are the following: supersaturation of the medium with calcium and other ions, the nature of the critical nucleus, characteristics of the organic matrix, promotors and inhibitors of phase transitions, environmental temperature, salinity concentration and pH ranges (Wilbur & Saleuddin, 1983; Addadi & Weiner, 1992; Mann, 1991, 2001; Bauerlein, 2004; Veis, 2008). The research related to the biomineralization processes is interdisciplinary and the main aim of the present review is to express and interpret results obtained by different methods and approaches. In this review we address three major topics.

First, we turn back to the period before the modern biomineralization theory, which presents several innovative hypotheses and proposes mechanisms fundamental in understanding crystal formation in the living organisms. These advanced ideas were randomly or deliberately neglected and minimized in recent biomineralization literature, and they deserve attention.

Next, we take up the role of different research methodologies that directly promote and address some of the pressing profiles in explanation or acceptance of unusual-usual biomineralization processes.

Third, we turn to areas that require a closer attention and a more intense effort in presenting the recent synthesis and generalization of the mineralization of the organisms. One of these is well confirmed and accepted knowledge and techniques underlying the environmentally induced biomineralization and the influence of different extrinsic conditions on crystal formation in adult invertebrates. The area is apparently much intricate and faces an increasing difficulty in well-trained understanding of the biomineralization related to starting-initial biomineralization processes, the first crystals formation and growth during early embryogenesis up to the juvenile and adult organisms. Major achievements and examples of fundamental results in recent years will be presented. It depends of the personal scientific judgement whether the distinction between the basic mechanisms of the initial crystal formation and evolutionary role of these processes on the general aspects of the biomineralization represents a real difference.

2. Usual-unusual biomineralization

Biomineralization is recognized like a separate scientific discipline comprising and interacting within natural and technical scientific fields in the late sixties - early seventies of the previous century. According to Bauerlein (2000), the modern biomineralization era began by Lowenstam's synthesized review entitled "Minerals formed by Organisms" (Lowenstam, 1981), and continued in numerous books, book serials and conference volumes as well as extraordinary reviews up to recent. The biomineralization research could be presented in few graduated steps as:

- structural and morphological recognition –differentiation - at the cell, tissue or organism level,
- physiological function determination - propositions on the tissue, organism, species or family level,
- possible mechanisms used in appropriate synthesis of different species,
- application/correcting deviations and production of new materials with unexpected properties.

The level of reliable recognition and evaluation of contribution of biomineralization to the general knowledge in life sciences is dependent on the methodology applied and exploited in the research as well as on spatial and temporal approach in explanation of unusual to usual biomineralization.

Definition of UNUSUAL: not usual: uncommon, rare (Webster Dictionary). Related to UNUSUAL - Synonyms: curious, extraordinary, funny, odd, offbeat, out-of-the-way, peculiar, queer, rare, singular, strange, unaccustomed, uncommon, uncustomary, unique, weird

Related Words: bizarre, eccentric, far-out, kooky (also kookie), oddball, outlandish, outré, way-out; aberrant, abnormal, atypical, exceptional, irregular; newsworthy, notable, noteworthy, noticeable, particular, remarkable, special

Increasing interests of diverse groups of scientists for inclusion in the biomineralization research and application of new instrumental methods resulted in exponential growth of the general knowledge and in numerous crucial studies and research that enormously broaden understanding in the field. Remarkable literature data could explain the difference and gradation in our understanding from *unusual* to *usual* biomineralization. Furthermore, only several basic studies and experimental data that improve step by step growing importance in the biomineralization field could be enumerated. Odum (1951) described strontium content of sea water celestite radiolarian and strontianite snail shells. Vinogradov (1953) presented chemical series of skeletal formation in marine organisms. Beside expected carbonate polymorphs calcite, aragonite and vaterite he emphasized importance of $CaCO_3$ crystallohydrates (amorphous $CaCO_3$ in further literature - ACC), and its role in biomineral formation. Amorphous $CaCO_3$ is found frequently in rather dispersed state in young Mollusca and Crustacea and later it changes into calcite or aragonite. He also observed that formation of aragonite in marine organisms is stimulated by temperature, pH and "the presence of a crystalline seed of isomorphic matter," especially concentration of magnesium, phosphate and carbonate salts. Both inspired findings of Vinogradov (1953) were confirmed in further basic biomineralization research and interpreted as crucial for starting crystals formation in the organisms, ACC as a template and nucleation centre as promotors of the biomineralization. Lowenstam (1954) compared environmental conditions on the modification of carbonate secretion in several marine invertebrates. Stenzel finding of aragonite in the resilium of oysters (1962), and different calcite and aragonite regions in the adult oysters *Crassostrea virginica* (1963), deserves attention and it was "unusual" only during a short period. In the next years, using X-ray diffraction, Stenzel (1964) found that the larval shell of the common North American oyster, *Crassostrea virginica,* in a strait-hinge stage to the latest umbo stage (before metamorphosis), was composed of mineral aragonite and that no calcite was indicated. He also stated that it was most likely that all, or nearly all, Bivalvia had aragonitic larval shells because most of them had aragonitic adult shells. According to Stenzel one might assume that the larvae of oysters conform to the general pattern in the Bivalvia. Larval oyster shells are aragonitic because their ancestors had aragonitic shells, and there was, and is, no adaptive need for the free-swimming larvae to have shells of a composition other than aragonitic. Many further morphological studies and research related to bivalve larval shells are based on and support the Stenzel findings and hypothesis (Taylor et al., 1969; Carriker & Palmer, 1979; Carriker et al., 1980; Waller, 1981; Watabe, 1988). Hare (1963) analyzed amino acids in the proteins from aragonite and calcite layers in the shells of the mussels *Mytilus californianus* and expressed importance of the organic matrix in the biomineralization at organisms' level and contributed to general interpretation of the invertebrate biomineralization. The molluscs produce a shell composed mostly of biogenic calcium carbonate, $CaCO_3$. The most common precipitating biominerals are either calcite or aragonite. Vaterite, the third polymorph of calcium carbonate, was registered in the freshwater gastropod *Viviparus intertextus* and in newly formed layers of bivalve shells (Wilbur, 1964; Taylor et al., 1969; Watabe, 1988). Also, some authors showed that, beside calcium carbonates, other minerals could be incorporated in the structure of mollusc shells. In the larval shell of the pearl oyster *Pinctada martensi,* Watabe (1988) observed mineral dahlite-carbonate hydroxylapatite, $Ca_5(PO_4,CO_3)_3$ (OH). Barite, $BaSO_4$, which is usually mineralised only by two protozoan organisms, was recorded in the shell of field-collected freshwater Asian bivalve *Corbicula fluminea* (Fritz et al., 1990). In laboratory experiments the same species co-precipitated barite crystals with aragonite in the inner shell surface (Fritz et al., 1992). Carriker et al., (1980 and 1991) demonstrated a remarkable capacity of oysters to incorporate other elements in their shells. Fritz et al., (1990) suggested that minerals of elemental composition similar to that of calcium carbonate could also be mineralised by molluscs.

3. Environmentaly induced biomineralization

Many recent reports and ecological studies have correlated harmful environmental factors (lack of oxygen, toxic algae, heavy metals, organic pollutants, antifouling paints) with disease, deformation, mutagenicity or genotoxicity of biomineralization, calcification processes and mineral composition of marine organisms. The calcareous skeletons of marine organisms act as concentrators of dissolved minor polluting elements in the seawater, permitting easier analysis and more reliable measurements than those carried out directly on water or on soft tissues. The minor elements, including heavy metals, nanoparticles and organic molecules could be incorporated into the crystal structure through isomorphism, where remain stabilized in the shell and skeleton structures and thus can be used as bioindicators of the different contamination in the sea.

Marine organisms have been commonly used as bio-indicators of changes in marine environment. Bioaccumulation of metals and lipophilic organic compounds through extraction from sea water is typically confirmed by chemical analysis of soft tissue. However, the analysis of the hard structures consisting of biominerals is rarely applied, despite the possibility that it could provide additional information on toxic elements concentration, understanding of biomineralization process and environmental evaluation.

In this review unusual biominerals influenced by unexpected environmental events, and caused by anthropogenic activity in several groups of marine organisms, freshwater and subterranean snails will be presented. In order to explain how complex and sentinel mechanisms regulate environmental and physiological functions in the mollusks, a concise report of outstanding study related to the shell microgrowth patterns of bivalve *Ruditapes philippinarum* performed by Kanazawa and Sato (2008) will be presented. They found that each microgrowth increment in the shell corresponds to one lunar day and the local tidal periodicity could be easily determined from it (Figure 1). However, spawning season and spawning breaks of some specimens could be also recognized and distinguished from the shell microgrowth patterns. Also, it may be possible to estimate the age of sexual maturity and the number of spawning events that occurred per year in fossil bivalves.

3.1 Shell crystals of marine bivalve as a site of pollutant deposition

3.1.1 Pink coloured nacreous layer of the Mediterranean mussels *Mytilus galloprovincialis*

This chapter reports on the unusual crystal morphology and color found in parts of the inner nacreous layer of the Mediterranean mussels *Mytilus galloprovincialis* during intensive anoxic conditions caused by intensive phytoplancton bloom in the Northern Adriatic in the years 1989-1991.

Molluscs in general are believed to exert only a minimal vital effect over their isotopic composition (Jones, 1985), and thus their isotopic values are representative of environmental water conditions. However, calcium carbonate may not always be precipitated in equilibrium with the environment, and in such cases stable isotope analysis of the shell carbonate may be an unreliable technique for environmental reconstruction. Isotopic disequilibrium may be in part due to metabolic effects or kinetic effects that are inherent in fast-growing shells, or areas of the shell (Mitchel et al., 1994). Many molluscs use aragonite to build their shells but calcitic and those with mixed mineralogy also are abudant. In shells with mixed aragonite-calcite composition, low temperatures are favourable for deposition of calcite, while warm waters favour formation of aragonite (Schifano & Censi, 1986).

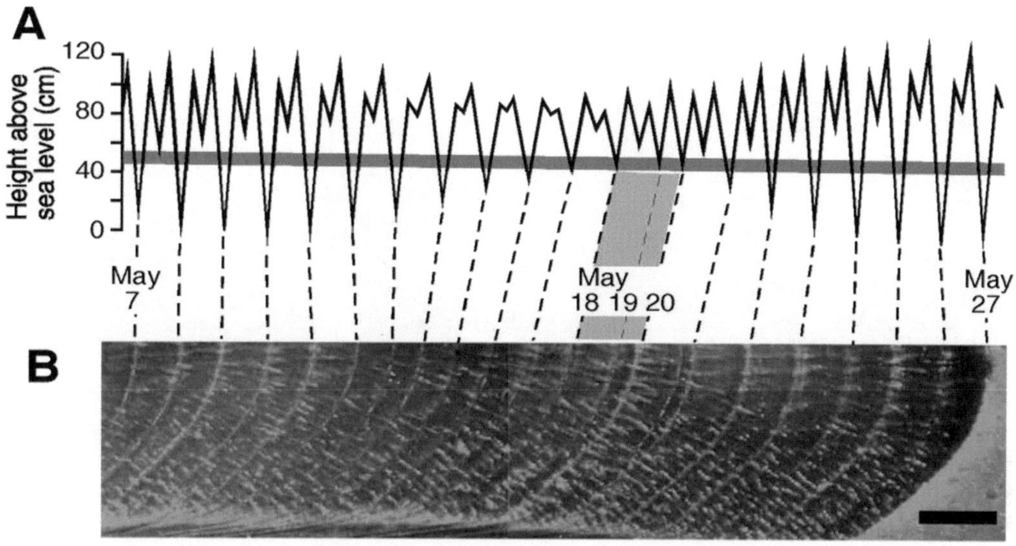

A. Local tidal periodicity, as determined using TIDE for WIN. The bold grey line indicates the hight above sea level of the sampling site. B. Shell microgrowth pattern formed from 7 to 27 May 2005 for the individual (shell length, 29 mm; age, 3) of *Ruditapes philippinarum* collected from Matsukawa-ura on 27 May 2005. Each microgrowth line is respective to immersion event in each day. The grey area indicates shell microgrowth increments in 1.5 lunar day during a neap tide, from 18 to 20 May 2005. Scale bar = 100 μm.

Fig. 1. Copy from: Kanazawa, T., & Sato, S. (2008). *Journal of Molluscan Studies*, 74, 89-95; doi:10.1093/mollus/eym049.

The impact of environmental factors such as marine anoxia in the coastal zones of the Northern Adriatic on the growth of Mediterranean mussels *Mytilus galloprovincialis* shell is indicated by changes in the growth structure of the outer part of the nacreous aragonitic layer, which was pink in colour and out of isotopic equilibrium with oxygen and carbon from the ambient seawater (Figure 2a). The inner nacreous shell layer was roughly divided into two sublayers, and according to scanning electrone microscope (SEM) analysis, the white uncoloured part of the nacreous layer consisted of flattened aragonite prisms (Figure 2b). The outer pink coloured sublayer was up to 900 μm thick and composed of irregular aragonite grains showing a massive crystal structure. These aragonite crystals are precipitated during a prolonged stress condition and deposited more rapidly than ucoloured ones. The comparison of results obtained by energy dispersive spectroscopy (EDS) and stable isotope analysis showed that the pink coloured aragonite contains more carbon than oxygen relative to the uncoloured aragonite, thus pink aragonite layer indicates anoxic conditions during its precipitation. Isotopic disequilibrium most probably resulted due to a faster precipitation of aragonite during the summer period, as well as due to a lack of oxygen and higher concentrations of isotopically light organic derived CO_2 in the anoxic environment. The influence of higher concentrations of ambiental CO_2 is also reflected in a higher C/O ratio of pink coloured aragonite as compared to the uncoloured one (Dolenec et al., 2000).

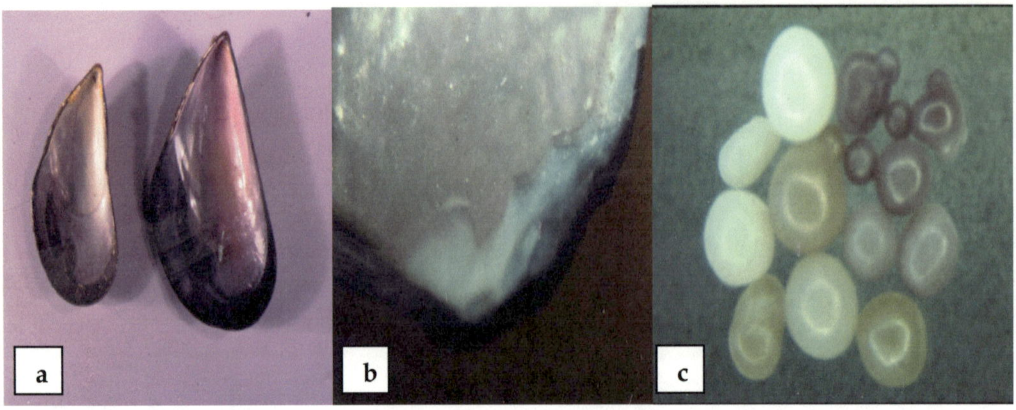

Fig. 2. a) Right valves of mussel *Mytilus galloprovincialis* - uncoloured and Pink colored sampled during anoxic conditions; b) transsection of the shell valve – upper part of pink colored nacreous layer; middle white part of the "normal" nacreous layer; down black colored outside prismatic calcite layer; c) white mussel pearls collected in "normal" environmental condition and colored pearls separated from the shells during anoxic period.

The inside pink layer of the shell was carefully extracted and powdered and analysed by X-ray diffraction. The results were compared by those for the white inside layer from the shell collected in the year that preceded the apearance of the pink shells. Diffraction patterns showed that both samples consisted of dominant aragonite fraction and some calcite crystals. As it was confirmed by SEM analysis the pink layer contained approximately 10 % of calcite. The diffraction lines of calcite in the sample of the pink layer were broadened. Further, the electron spin resonance (ESR) spectroscopy of Mn^{2+} ions that are paramagnetic substitutes for Ca^{2+} in the host calcite was also performed in order to obtain more details of the structural properties of calcium carbonate in the shell (Brečević at al., 1996). The ESR analysis confirmed that the pink layer contained a smaller fraction of calcite than the white nacreous part, and showed that the symmetry of the nearest environment of Mn^{2+} (Ca^{2+}) was lower in the pink layer than in the white. Also, the ESR spectrum of the pink layer indicated the presence of other paramagnetic ions, besides Mn^{2+}, that appeared in an unevenly dispersed clusters. On the other hand, the standard chemical analysis and X-ray spectroscopy could not find any significant difference in the composition of the examined pink and white layers from the mussel shells.

A long term monitoring of the pink mussels showed that in 1990, in locally more polluted zones the number of pink mussels ranged between 40 and 100%, while in the Mirna River estuary and some other areas under the influence of freshwater inflow, the number was much lower and ranged from 8 to 14%. The analyzed mussel samples indicated a high number of pink mussels among specimens longer than 40 mm. The smallest pink colored mussel was 22 mm in length. It is important that pink colored nacreous shell layers were ocasionally noted in some other bivalve species (e.g. *Calista chione, Venus verucosa, Pecten jascobaeus*), and in several mussels pink or brown pearls were found (Figure 2c). The named other bivalve species were found only at few different localities in very low number, and this result was not included in a systematic study. Southward along the shallow eastern Adriatic coast pink colored mussels have never been observed. In the research period of 1991 and 1992 a decrease in percentage of pink mussels and in their shell coloring intensity

was observed, and in 1993 pink colored mussels were not registered (Medaković et al., 1995a). The present results show a difference between the samples collected during short/long unexpected natural events - like anoxia, or by different human activities manipulating near harbors, city waste or sewage and chemical industry centers, indicating different levels of contamination or pollution (Degobbis, 1989; Hrs-Brenko et al., 1990).
Consequently, hard structure of marine organisms, i.e. biominerals should be found to provide useful data on determination of environmental status and also on estimation of pollution caused by anthropogenic influences as it was confirmed in several recent studies during whole east Adriatic coast (Hamer et al., 2008, 2010; Rončević et al., 2009, Kanduč et al., 2011).
Effects of environmental factors have to be considered in sampling strategies for monitoring programmes to prevent possible false interpretation of the results. In their research Hamer et al., (2008), examined salinity and temperature variations at investigated sites from northern to southern Adriatic coast (from Limski kanal to Dubrovnik). The results showed that mussel *Mytilus galloprovincialis* sensitively responded to subtle changes in the environmental conditions. Results of the isotopic analysis presented in this study support the hypothesis that the δ13C in mussel's shell might be used as an indicator of environmental salinity stress and the freshwater influx. The values of δ13C mussel was also reflected in lower *M.galloprovincialis* shell carbonates. Furthermore, the calculated temperatures for *M.galloprovincialis* shell growth in the investigated area range from 13.4 to 20.9 °C for calcite and from 16.6 to 23.1 °C for aragonite. According to the δ18O and δ13C values of shell layers, the investigated area could be separated into three locations: those with a bigger influence of freshwater, those with a smaller influence of freshwater and those with marine environments. Therefore such investigations might represent an additional tool for waste water management and environmental protection (Hamer et al., 2010). The results are confirmed by recent study of Kanduč et al., (2011). Oxygen and carbon isotopes were analysed for calcite and aragonite in separate shell layers, while major, minor and trace elements in the bulk shell were analysed to evaluate environmental conditions such as the temperature of carbonate deposition, freshwater influence and locations of anthropogenic pollution. The highest concentrations of manganese, barium, boron, arsenic, nickel and chromium were observed in shells from Omiš, Bačvice and Zablaće (Central Adriatic) and Sv. Ivan (South Adriatic), where chemical and heavy industries are located and where sewage is known to be discharged into coastal areas. The highest concentrations of zinc, lead and copper were measured in samples from Pula, Rijeka and Gruž, where there are also ports in addition to industry.
As a conclusion of this chapter and confirmation of Carriker et al., (1980 and 1991) findings that bivalve posses a remarkable capacity to incorporate other elements in their shells, the results of the metal content in shell samples of *Mytilus galloprovincialis* mussel will be presented. Using inductively coupled plasma atomic emission spectrometry (ICP-AES), Rončević et al., (2010) analysed bioaccumulation of metals in the shell layers of mussels sampled along eastern Adriatic coast. The efficiency of conversion of crude samples into solution by acid digestion in an open plate and in a microwave oven was examined by use of certified reference material of marine sediment and laboratory made standards of calcite and aragonite. Influence of high Ca content matrix on emission intensities of Al, Ba, Cd, Cu, Fe, Mg, Mn, Na, Ni, Pb, Sr and Zn was observed as depression of emission signal for most of the measured elements, ranging from 0.8% to 8%. Greater values were noted for Ba and Ni emission lines. Enhancement of signals was observed for Na and Mg lines. The

determination of As, Sb, Se and Sn was performed by HG/ICP-AES. A greater abundance of Sn was found in samples collected near the Al-processing industry centre. No detectable concentrations of As, Sb, and Se were found in the shell samples. Results of ICP-AES metal analysis showed that samples collected near harbours, city waste or sewage outlets, and chemical industry centres indicate a certain level of contamination. It is important that different mineralogical structures of the shell layers imply incorporation of metals with different ionic radii into a calcitic or aragonitic layer. These studies show that bivalve shell analysis provides useful data on determination of marine environment status.

3.1.2 Influence of antropogenic pollution on the crystal growth in *Ostrea edulis* oyster shell

Coastal resources are used and exploited for economic and social objectives: urbanization, industry, tourism, fishery, aquaculture, energy production and transportation. Such activities produce combined environmental impacts resulting in marine and freshwater contamination and/or pollution, loss of marine resources, endangered biodiversity, loss of public access to the coast, etc. The number of new chemicals that appear on the market is increasing each year. Most of these substances are insoluble and non-biodegradable and their effect on the environment is rarely or not understood completely. The effects and impact of compound tributyltin (TBT) released from antifouling paints on marine environment was taken as a case study.

Organotin pollution started in the mid-1960s, when tributyltin (TBT) was used in antifouling ship paints intensively. TBT is organo-metallic complex that leaches from the paint and enters the marine environment. TBT accumulates in the sediments, in particular, in areas with many ships like harbors and ports. This compound was responsible for the disruption of the endocrine system of marine mollusks, leading to the development of male sex characteristics in female marine snails and causing harmful effects on other marine organisms. The number of harmfull effects of TBT were recognized like: killing spat, causing growth deformities, causing shell thickening and layering and other malformations in oyster shells due to alterations in the calcium metabolism (Alzieu, 1991; Dyrinda, 1992). Letal and subletal effects have been found in other bivalves, mussels and cockles. One of the most singular effect of TBT is the masculinisation induced in female gastropods, termed imposex, occurs worldwide in over 150 species following exposure to low levels of TBT. Females exposed to TBT grow Accessory Sex Organs (ASO), including sperm duct, seminal vesicle, external sperm groove and penis, because TBT disrupts their immune and endocrine systems (Fent, 1996). TBT is widely distributed in the world. It has been found in phytoplankton, which lies at a base of most marine food webs, thus transfering contaminants to higher organisms, fish and also salmon held in net pens, dolphins, seals and sperm whales of the deep sea, and so enter the human food chain (Fisher & Reinfelder, 1995). It is toxic to humans.

An extremely low level of TBT in the seawater significantly affects shellfish larval growth, impairs the immune system of organisms and causes malformations in the shells of adult shellfish (Alzieu et al., 1982; Thain et al., 1989). Shell calcification anomalies caused by TBT, which consists of stacked 'chambers' filled with a gelatinous substance that gives the shell a ball-like shape, were observed in the oyster *Crassostrea gigas* sampled in the French Atlantic coast and in California (Alzieu et al., 1989; Stephenson, 1991). Similar deformation effects were seen in the shells of adult oysters *C. gigas* and European flat oyster *Ostrea edulis* sampled at several localities along the East Adriatic coast (Medaković, 1995). In our previous research, X-ray powder diffraction (XRD) was used to study calcification and changes in the biomineralization process of the bivalvia and other invertebrates influenced by environmental changes (Medaković et al., 1997, 2003). Although mollusk shells are considered to be a powerful tool for monitoring environmental conditions, alterations in the biomineralization processes influenced by TBT have been insufficiently investigated so far. In this chapter we follow the accepted idea that the elemental composition of each polymorph can vary within the organisms, although the calcified material is produced in similar environments.

In the work of Medaković et al., (2006), X-ray photoelectron spectroscopy (XPS) and scanning electron microscopy (SEM) equipped with X-ray energy dispersive microprobe (EDX) was used because these techniques provide qualitative and quantitative information about environmentally induced elements in oyster shell layers in a fast, sensitive and simple way. In the recent literature, numerous papers describe morphological changes in mollusk shells caused by TBT. No data are available on mineral components and minor elements incorporated in shell layers thus deformed, and this is the first result related to tin in the chambers of oyster *O. edulis*.

The shells of European flat oyster *Ostrea edulis* are composed mainly of calcite. Aragonite is present only in the small area of the inner shell layers on the imprints of the muscle attachment (Taylor et al., 1969). XRD patterns of the analysed samples gave the average mineral composition in the milled shell parts. Calcite is the dominant component in the outerprismatic shell layers of both samples (Table 1). Also, both samples indicated a small presence of halite (NaCl), which could be eliminated by additional washing in distilled water. Sample OE-RV1 contained traces of aragonite in this layer. Furthermore, SEM and XPS analyses confirmed that this shell part sample, OE-RV1, contained 0.1 to 0.3 at.% of silica, in contrast to OE-MS1, in which Si was present in amounts of 1.2 to 1.5 at.% (Table 1). It was shown that a longtime exposure of organisms to environmental conditions might result in the presence of foreign particles in the shell. Clays or other suspended particles could be 'captured' in the structure of the shells during periods of withdrawal of the mantle and fast growth (Carriker et al., 1991). In the outer prismatic layers of the oyster *O. edulis,* a small amount of silica mineral (quartz 1 to 2 mol%) was registered, and detailed studies showed that it was characteristic only for one locality in the northern Adriatic (Medaković, 1995).

The above inferences and our results indicate that the promoting mechanism of incorporation of silica in oyster shells can not only be related to the structural characteristic of the *Ostrea edulis* shell but probably even more to specific environmental factors. Factors for precipitation of carbonate polymers in the shells include a complex interaction between the organisms and environmental conditions, inorganic ions, dissolved organic materials, enzymatic activity, macromolecules, etc. Although the importance of organic tissue in the formation of shell polymorph is indispensable, only the presence of TBT in the environment induces tin in the oyster soft tissue. Moreover, accumulated tin in the digestive gland and in the gills of the oyster causes morphological changes in their shells, unusual chamber formation and suggested mechanisms of TBT action (Alzieu et al., 1986). TBT is known to inhibit oxidative phosphorylation and it has been suggested that this forms the basis of its action on the shell. Thain & Waldock (1986) confirmed in laboratory experiments that only TBT influences growth and causes shell abnormalities in oysters.

The results of the present research showed, for the first time, that the only malformed chambers of *Ostrea edulis* contained tin. The XPS spectra confirm that tin is present in sample OE-RV2 in amounts of about 0.1 at.%, and in the OE-MS2 in amounts ranging from ~ 0.1 to 0.2 at.% (Table 1). The Sn 3d XPS spectrum of the OE-MS2 sample clearly shows two characteristic Sn peaks that could be associated with the presence of tin oxide compounds. The quantitative SEM analysis confirms the results given by XPS, and shows a negligible amount of tin in these samples.

			Mineral components (M %)				Minor elements (at. %)	
Locality	Sample code	Shell layer	A	C	H	MC	Si	Sn
Stn 1	OE-MS1	Outer	-	dom	tr	-	1.2 - 1.5	-
	OE-MS2	Inner	-	min	-	dom	0.05 - 0.2	0.1 - 0.2
Stn 2	OE-RV1	Outer	tr	dom	tr	-	0.1 - 0.3	-
	OE-RV2	Inner	-	dom	-	-	0.0 - 0.1	0.1

Table 1. Mineral components and minor elements in different shell layers of the oyster *Ostrea edulis* expressed in molar percentages (M %), and in atomic percentages (at. %). A, aragonite; C, calcite; H, halite; MC, magnesium calcite; Si (silica); Sn (tin); dom, dominant component (mostly 100 %); min, minor component (about 2-3%); tr, traces (much less than 1%); - no component is present; Stn 1- shellfish farm, Stn 2 – marina, both in East Adriatic Sea, Croatia.

3.2 Benthic marine organisms and fouling communities

3.2.1 Calcareous algae

The corralinacean algae make an important contribution to sediment production and reef limestone formation (Lewin, 1962; Cabioch & Giraud 1986; James et al., 1988), producing over 50% of the calcareous material in coral reefs (Aharon, 1991). In shallow tropical waters, carbonate deposition of calcareous green and red algae can reach levels of 61% of bulk sediments (Wefer, 1980). Because of their habits and abundance, corralinacean algae play very important role in some assemblages (Peres & Picard, 1964).

Many papers have been published on diverse genera of calcareous algae, which attempt to explain still unknown mechanisms of biomineralization. Crystallization processes depend on a number of conditions, but the main ones which influence the type and structure of the biomineral are the following: supersaturation of medium with calcium ions, nature of critical nucleus, organic matrix, inhibitors of phase transitions, environmental temperature, salinity concentrations, pH range. McConnaughey (1989) stated that, regardless of the type of new forming minerals, the main conditions included: transport mechanisms, existence of semipermeable membranes for precipitation, divalent cations for removal of protons, presence of Ca^{2+} enzyme ATP-ase and passive electronuclear anion exchange. Furthermore, some authors have pointed to the significance of environmental stress and other endogenous factors.

In this chapter we present mineral composition of nine species of red calcareous algae (Corallinaceae, Rhodophyta) collected in the Adriatic Sea in 1987 and 1988, examined by X-ray powder diffraction (counter diffractometer, monochromatized Cu*K*α radiation). In addition, a comparison between the calcareous algae from the north Adriatic (Rovinj area) and the central Adriatic (Kornati Islands) with regard to genus, species and environmental factors was undertaken (Medaković et al., 1995).

The results of this study show that all analyzed samples contained magnesium calcite, which was dominant in all but two cases, where aragonite was the main phase (Table 2). The fact that magnesium calcite is the dominant component is in agreement with statements of numerous literature data (Linck, 1930; Babička, 1936; Vinogradov, 1953; Chave, 1954; Chave & Wheeler, 1965; Walker & Moss, 1984; James at al., 1988; Borowitzka, 1989; Aharon, 1991; Pueschel et al., 1992). Most of the authors found that the magnesium content in Corallinacea ranged from 2 to 30 molar % (M %). Diffraction lines of magnesium calcite in our study were broadened and shifted toward higher Bragg angles in relation to pure calcite. Supposing that in the calcite crystal lattice only magnesium replaces calcium, it follows that the fraction of magnesium in magnesium calcite, found from diffraction line shifts, would be 18 molar % (M%), for all analyzed species at all stations and sampled seasons. One may conclude that content of magnesium in magnesium calcite is rather constant in all species analysed, although the molar fraction of magnesium calcite varies from one sample to the other (Table 2). The broadening of the diffraction lines could be explained due to small crystallite sizes. Similar line broadening was also observed in X-ray diffraction patterns of the first larval shell, prodissoconch I, of oyster, *Ostrea edulis*, containing crystalline aragonite (Medaković et al., 1989). The broadening of aragnonite diffraction lines decreased with further development of larval stages, through prodissoconch II to the juvenile oyster veliconcha, before its settlement and metamorphosis to the adult oyster. In the present study the similar diffraction broadening observed in nine different algal species indicated similar crystallite size. This could indicate a similar age of algal plants sampled, but this is not possible because the sampled species of Corallinacea did not reproduce in the same seasons. In the present study all parts of a given alga were used, i.e. basal and middle parts and younger tips or "apices". King & Schramm, (1982), showed that in *Phymatolithon calcareum*, different parts of the tallus exhibit different calcification rates. The same authors observed that calcification of crustose corralines was slower than that in articulated species. Taking into consideration all these facts we may accept the hypothesis that the mineral composition of corralinacean algae varies with algal age (Vinogradov, 1953) and growth rate (Chave & Wheeler, 1965), but we cannot explain why algal species show similar crystallite sizes.

Unusual and unexplained results described in this study relate on high amount (75 to 80 M%) of mineral aragonite in the algae *Pseudolithophyllum expansum* collected at Kornati Islands in two independent sampling during 1987 and 1988. These two samples also contained magnesium calcite and a small fraction of calcite (5 to 10 M%). In other studied samples aragonite was detected in small fractions, up to 10 M%. The elemental analysis of corralinacean algae obtained by X-ray spectroscopy showed that the fraction of the metals Sr, Fe, Mn, Zn, Pb, Br, Cu and Rb was very small (15 to 2000 ppm). In most samples other expected minerals were detected in small fractions, such as sylvite (KCl, up to 2 M%), quartz (α-SiO2, up to 2 M%) and magnesite ($MgCO_3$, only in one sample, 1 M%). The results show that calcareous algae are able to deposit a mixture of magnesium calcite, calcite and aragonite, and even dolomite but such a large molar fraction of aragonite in the alga *P. expansum*, or in any other corallinacean algae, has not been noted in recent literature (Table 2).

Algae species	Sample station	Calcareous components				Other minerlas [a]		
		Magn. calcite $(Ca,Mg)CO_3$	Aragonite $CaCO_3$	Calcite $CaCO_3$	Dolomite $CaMg(CO_3)_2$	Sylvite KCl	Quartz α-SiO_2	Magnesite $MgCO_3$
Amphiroa cryptarthrodia	1	100	-	-	-	1	-	-
	7	100	tr	-	-	tr	-	-
	8	100	-	-	-	-	-	-
Amphiroa rigida	1	100	tr	-	-	-	-	-
	7,8	99	1	-	-	-	-	-
Jania longifurca	1	98	2	-	-	-	1	-
	7	100	tr	-	-	1	-	-
	8	99	-	-	-	1	-	-
	7,8	100	tr	-	-	-	-	-
Corallina officinalis	1	99	1	-	-	1	1	-
	1	99	tr	-	1	1	1	-
	7,8	100	-	-	-	-	-	-
Lithophyllum lichenoides	5	98	2	-	-	-	-	-
	7,8	98	2	-	-	-	-	-
Lithophyllum racemus	2	90	5	-	5	-	2	-
	4	100	-	-	-	-	-	-
	8	100	tr	-	-	-	-	-
Phymatolithon calcareum	5	97	3	-	-	-	-	-
Tenarea undulosa	7,8	100	-	-	-	-	-	-
	8	100	tr	-	-	-	-	-
Pseudolitho-phyllum expansum	1	97	3	-	-	2	-	-
	1	90	10	-	-	1	1	1
	3	97	3	-	-	-	-	-
	6	100	tr	-	-	-	-	-
	7,8	15	75	10	-	-	1	-
	7,8	15	80	5	tr	-	1	-

[a] In relation to: magnesium calcite + aragonite + calcite + dolomite = 100 %

Table 2. Phase composition (molar fraction, %) of calcareous algae from the Adriatic Sea. (*tr* trace component; - no data).

The other unusual biomineral, dolomite, was registered as a minor component (from traces up to 5 molar %) in calcareous algae *Coralina, Lithophyllum* and *Pseudolithophyllum sp.* (Table 2). The appearance of mineral dolomite in different living organisms in recent literature is well presented; however, all authors could not agree with the theory of the dolomite formation (Deelman, 2008). Detailed explanation of possible dolomite formation in the living organisms will be presented in the last chapter on unusual biomineralization in the freshwater and subterranean snails.

The hypotheses that algal anatomy and various organic constituents of the cell modify and regulate the effects of CO_2 (or HCO_3^-) on calcium carbonate nucleation and crystal growth (Borowitzka, 1989), and that under certain conditions the same organism can form different minerals from the same tissue (McConnaughey, 1989), confirm our statement that complexity of microclimatic and oceanographic characteristics may influence the diversity of algal mineral composition and that environmental effects were imprinted in the skeletal composition of the Corallinacea (Medaković et al., 1995). Due to numerous factors which could potentially influence biomineralization in the calcareous algae, more algal species should be monitored on a longer time scale with a higher sampling frequency in order to asses the influence of the environment on the mineral composition and phase fractions in the calcareous algae.

3.2.2 Long term influence on crystals growth of fouling communities or how marine calcareous organisms protect and preserve antique bronze Sculpture of Apoxyomenos during 25 centuries

Benthic organisms that colonize and attach themselves to the natural or artificial substrates in the sea are often called fouling organisms. The attachments differ in style and range in permanence from temporarily simple or multiple muscular attachments (e.g., limpets, octopus), or holdfasts by releasable byssus threads (mussels), by organic glues (some barnacle species), up to permanent whole life cementation (oysters), or boring by chemical or mechanical means (bivalve *Lithophaga sp.*). The adhesion between the organisms and surfaces is characterized by the extreme complexity and depends on numerous physical and biological features. One of the most important processes includes the interaction between the attached surface and biologically formed minerals. The factors that control these properties are, however, poorly understood and represent a crucial problem in understanding the biomineralization mechanisms of marine organisms. All types of fouling organisms are able to attach themselves to various surfaces, to erode and degrade the surfaces, and in the same time to incorporate some unusual ions from the attached surface and environment in their calcareous structure.

Chemical and physical properties of the sea water cause a severe corrosion of any submerged material. A complex combination of living creature activities named fouling organisms (communities) may erode, degrade and deteriorate natural or artificial substrates. The interaction of specific artificial surface with mineral composition of several groups of marine fouling organisms attached during long term period to the bronze antic sculpture of Apoxyomenos were studied by X-ray diffraction (XRD) and elemental analysis (ICP-AES). Whole empty bivalve shells, tubes of the serpulid polyhaete worms, calcareous algae and encrusted calcareous structures contained small parts of several groups of different marine organisms, were collected from the inside and outside surfaces of the antique bronze sculpture of Apoxyomenos found at the depth of 45 m on the Adriatic sea bottom close to Lošinj Island (Croatia) in the 1999.

Two kinds of effects and active interaction between fouling calcareous organisms and the artificial surface of the bronze structure were observed. The parts of the statue directly immersed in the sediment showed much more serious damages and decomposition of the material surfaces than external parts covered by mixture of fouling organisms, communities and hard calcareous structures. The results of X-ray diffraction showed that calcite and/or aragonite were dominant components in all analyzed samples. In some shell parts and calcareous structures both calcite and magnesium calcite were found. Small amount of mineral feldspar was registered in the shells of boring bivalve *Rocellaria dubia* and in the tube of serpulid polychaete worm *Hydroides elegans*. However, beside other components, mineral quartz in traces was found in several parts of the undetermined bivalve shells, in calcareous structure and calcareous algae *Pseudolithophyllum expansum* (Table 3, Figure 3).

O	Sample code	Sample description	Mineral components in molar fraction Calcite	Aragonite	Feldspar	Quartz
1	1As1	Shells from the sediment	traces	dom	-	-
2	1As2		-	dom	minor	-
3	1As3		traces	dom	traces	-
4	1As4	*Hydroides elegans*	dom 0.6 +Mg calcite	0.4	minor	traces
5	1As5	*Rocellaria dubia* +Mixture	dom 0.6	0.4	-	-
6	1As6	Shell parts	0.3 + Mg calcite	0.7	-	traces
7	1As7	Calcareous structure	dom +Mg calcite	minor	-	traces
8	1As9	Calcareous algae	dom 0.7	0.3	-	traces

Table 3. Mineral components in the analyzed samples. Calcite ($CaCO_3$); Magnesium calcite ($Mg_xCa_{1-x}CO_3$); Aragonite ($CaCO_3$); Feldspars (e.g. albite $NaAlSi_3O_8$ and microcline $KAlSi_3O_8$); Quartz (α-SiO_2); dom = dominant component; minor = minor component; tr = component in traces.

X-ray diffraction patterns showed that aragonite was the dominant component in the analyzed bivalve shells (samples 1As1, 1As2, and 1As3). Beside aragonite, the shells of *Abra alba* and *Rocellaria dubia* contained traces of calcite (less than 1 molar %, 1M %). Mineral feldspar (e.g. albite, $NaAlSi_3O_8$, and microcline, $KAlSi_3O_8$), as a minor component (~ 3 M%), was found in the shell of *Divaricella divaricata* and in *R. dubia* was present only in traces. The second group of samples (1As4 to 1As9) was characterized as mixtures of calcite, magnesium calcite and aragonite, with traces of quartz except for the sample 1As5 (Table 3). Moreover, mineral composition and phase fraction of thus attached bivalve shells, polyhaete tubes and calcareous algae have been changed and differ from the same organisms living on the natural surfaces (Table 4). Inductively coupled plasma atomic emission spectroscopy analysis showed all samples to have a high metal content, with the specific metals and their concentrations being consistent with the sculpture as the source rather than normal seawater. No metal-based crystalline phases were found by powder X-ray diffraction apart from the presence of iron and copper hydroxides in one sample that showed a distinctly green coloration. Accumulation of non-native metals in the calcareous structures of these organisms and the unusual phase compositions of the latter suggest metal-induced changes to the biological pathways controlling the transport of calcium and magnesium and incorporation of heavy metals in growth sites of inorganic matrix (Lyons et al., 2009).

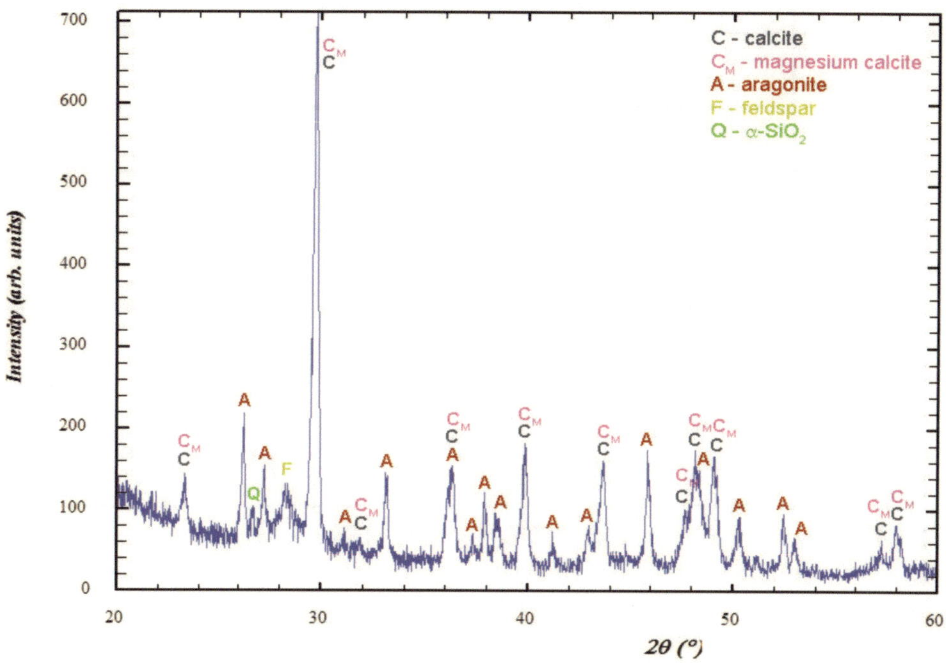

Fig. 3. Characteristic part of X-ray diffraction pattern of the tubes of the serpulid polyhaete worms *Hydroides elegans* (sample 1As4). Calcite (C) and magnesium calcite (C_M) are dominant (~ 60 M %), aragonite (A) is present in an amount of ~ 40 M %, feldspar is a minor phase (several M %), while quartz α-SiO_2 (Q) is present in traces.

The biomineral coating on the exposed part of the statue had accumulated over the centuries, reaching a thickness of 30 to 50 mm. The investigated samples removed from the outermost layer, indicated the gradual transport of metals from the statue surface to the most recently formed biomineral layer. The specific metal concentrations found in the minerals were higher than could be accounted for the bioaccumulation from seawater and were consistent with the composition of the ancient bronze, which was based on copper-tin alloys with lead and zinc added to improve the casting characteristics and durability of the sculptures. Evaluation of the mineral layers and fossilized organisms on the statue enable some presumptions and consideration that underwater fouling organisms and communities interacted with the statue as well as the effect of certain mineral deposits on the bronze sculpture slowing down its deterioration (Medaković, 2009).

	Attachment surface	Calcite (molar fraction)	Aragonite (molar fraction)
Hydroides elegans	natural	0.99	0.01
	bronze (sample 1As4)	0.60 ($C+C_M$)	0.40
Rocellaria dubia	natural	0.01	0.99
	bronze (sample 1As5)	0.60 ($C+C_M$)	0.40
Calcareous algae	natural-Istria	0.99 ($C+C_M$)	0.01
	natural-Kornati	0.10 (C) + 0.15 (C_M)	0.75
	bronze (sample 1As9)	0.70 ($C+C_M$)	0.3

Table 4. Comparison of phase analysis results of samples taken from different attachment surfaces. The tubes of the serpulid polychaete worm *H. elegans* attached to the outer sculpture surface (sample 1As4) contain 60 M% of $C+C_M$ and 40 M% of aragonite, while those found on the natural surface have 99 M% of calcite and 1 M% of aragonite. The shells of boring bivalves *Rocellaria dubia* sampled from natural surface contain 99 M% of A and 1 M% of calcite. That sample also contains traces of feldspar. Empty siphons of *R.dubia*, mixed with other calcareous structures contain calcite as the dominant phase, while the amount of A is 40 M% (sample 1As5). Similarly, the amounts of calcite and A in calcareous algae attached on the natural surfaces in the Adriatic Sea, Istrian region: 99 M% of $C+C_M$, and 1 M% of A; Kornati National park region: 10 M% of C, 15 M% of C_M, 75 M% of A; bronze sculpture surface: 70 M% of $C+C_M$, 30 M% of A (Medaković et al., 1995).

Studies of long-term biofouled manmade structure are limited; the finding an ancient sculpture immersed for two millennia in the sea provided a unique opportunity to test the long-term impact of a specific artificial substrate on biomineralizing organisms and the effects of biocorrosion (Medaković, 2009). The results showed that there was some metal sequestration by the marine organisms that were attached to the statue although this amount of metal, being significant to alter metabolic processes in the organisms, was relatively insignificant in view of the materials perspective. This give us a hint that we may be able to mitigate the negative effects of corrosion by using biogenic materials as passivating layers on metal surfaces.

While a lot more work has to be done on this topic, it has become an exciting area of research for the future to direct and controlled growth of these passivating layers directly on the metal surfaces, particulary by mimicking similar biological pathways that mineralizing organisms use. The advantage of growing these biolayers based on anchored enzymes, for example, is that the mechanical properties may be superior in many cases to those of the layers that can presently be deposited by physical or chemical deposition. Such efforts at reducing corrosion are also applicable to other surroundings such as land or subterranean areas, and not just marine environments. This study underlines the importance of long-term investigations when attempting to understand benthic community succession. Long-term studies allow detection of the effects of rare events and slow acting processes, revealing subtle but consistent trends and environmental changes.

3.3 Freshwater and subterranean snails

The present research is based on mineralogical and crystallographic characteristics of the gastropod shells that sometimes beside other taxonomical character could help in differentiating similar or overlapping species, or reflect specific environmental conditions of the locality. X-ray powder diffraction (XRD) was used to study the mineral composition and phase fractions of the shells of freshwater snails *Belgrandiella fontinalis* (F. Schmidt 1847) and *B. kuesteri* (Boeters 1970) collected from three springs in north-eastern Slovenia and subterranean snails *Zospeum alpestre* (Freyer 1855) and *Z. isselianum* (Pollonera 1886) sampled in four caves in Kamnik - Savinja Alps, Slovenia (Medaković et al., 1999, 2003; Slapnik & Medaković, 2007).

The gastropod shells contain an outer organic periostracum layer and an inner carbonate layer. Carbonates in the form of calcite, aragonite or vaterite could be distributed in two or more shell layers that differ in size, orientation and the way of crystals packing. The fractions of minerals depend on many factors, and are characteristics of a given family, genus, or even species. The recent research shows that crystal components in different organisms are influenced by biological conditions, and that crystal forms in specified tissues are adapted to their function. The mollusc shell morphology and mineralogy do not change gradually after organism's death, thus shells are equally important for basic biological research of the organisms and for monitoring environmental contamination (Addadi & Weiner, 1992; Albeck et al., 1993; Berman et al., 1993; Davidson et al., 1995).

Two independent measurements of as-found shells of both *Belgrandiella* species showed aragonite to be dominant in an outer inorganic layer (Table 5). Results of powdered snail shells showed the mineralogy of all shell layers. In the inner layers, aragonite was accompanied with calcite, quartz and dolomite as minor phases. The mineralogy differs according to locality and species indicating the influence of the environmental conditions on biomineralization processes in the inner inorganic layer (Table 5, Figure 4). The factors for precipitation of carbonate polymorphs in the shells include a complex interaction between organisms and environmental conditions, inorganic ions, dissolved organic materials, enzymatic activity, macromolecules, etc. Whereas the importance of organic tissue in formation of shell polymorphs is indispensable, several environmental factors that could influence different mixtures of minerals and the presence of dolomite in *Belgrandiella* shells will be described.

Species	Locality	Aragonite $CaCO_3$	Calcite $CaCO_3$	Quartz α-SiO_2	Dolomite $CaMg(CO_3)_2$	$SiCa_2$
Belgrandiella fontinalis	Stn 1	98	2	-	-	-
	Stn 2	78	2	-	20	-
Belgrandiella kuesteri	Stn 1	90	3	7	-	-
	Stn 2	80	-	-	20	-
	Stn 3	93	2	5	-	-
Zospeum alpestre	Stn 4	92	5	3	-	-
	Stn 5	60	37	3	-	-
Zospeum isselianum	Stn 6	85	10	-	-	5
	Stn 7	80	10	-	-	10

Table 5. The results of X-ray diffraction phase analysis of powdered snail shells (in molar %) for *Belgrandiella fontinalis, Belgrandiella kuesteri, Zospeum alpestre* and *Zospeum isselianum* inhabiting Kropa (Stn 1), Vidmar (Stn 2) and Markelc springs (Stn 3) and caves Kamniška jama (Stn 4), Jerohi 1 (Stn 5), Jama pod Mokrico (Stn 6) and Konečka zijalka (Stn 7). Both *Belgrandiella* species collected from Stn 2 contained 20 % of dolomite in the shells. *Zospeum isselianum* sampled at Stn 6 and Stn 7 have 5 to 10 % of unusual $SiCa_2$ phase.

Because of its crystallographic properties calcite in the shell layers provides a better temperature protection of the organism from the environment than aragonite, and for this reason most mollusks form their shells by both calcite and aragonite (Lowenstam, 1954). For example, subterranean gastropod *Zospeum alpestre* from the locality with a higher and more stable annual temperature contained in the inner shell layer 5 molar% of calcite, but the same species from a locality with a lower and oscillating annual temperature contained up to 37 molar% of calcite (Medaković et al., 1999) (Table 5). The measured hydrographic parameters at Stn 1 and Stn 3 were similar, resulting in similar calcite fractions (2–3 molar%) in the *Belgrandiella* shells (Table 5). Calcite in the *Belgrandiella* shells could be induced by the annual temperature variation, but it seems that quartz in *B. questeri* at Stn 1 and 3, and dolomite formations of both snails at Stn2, were promoted by other complex environmental activities.

The presence of mineral quartz in the adult mollusks could be environmentally induced by "capturing" small particles in the structure of the shell during periods of fast growth and intensive biomineralization (Carriker, 1992; Medaković, 1995; Medaković et al., 1999; Mutvei et al., 1996). During the sample preparation for XRD snail shells were occasionally washed with distilled water and by that procedure all possible impurities from inside and outside of the shells were removed (Medaković et al., 1997, 1999, 2003a). The fact that all analyzed species are characterized by a very slow annual growth (Slapnik, 1998), and therefore by slow biomineralization processes indicates that quartz in *B. questeri* and *Z. alpestre* shells is rather a structural characteristic of these species and not environmentally induced by specific conditions of the localities. A possible promoting mechanism of incorporation of quartz could be related to structural characteristic of the inner shell layer of *B. questeri* (it being a characteristic of this species) and/or specific hydrographic conditions at the given localities.

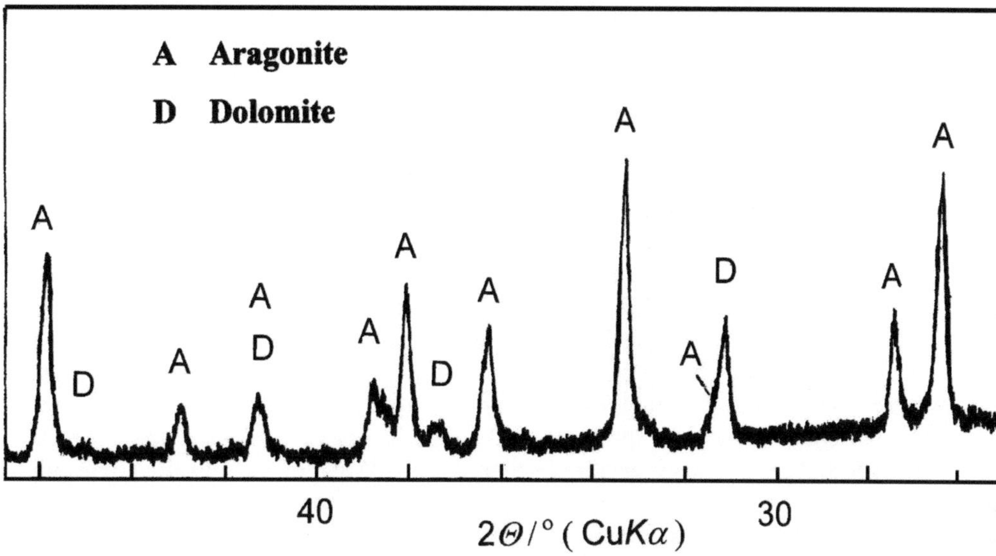

Fig. 4. *Belgrandiella kuesteri*. Characteristic part of X-ray diffraction patterns of freshwater snail shells sampled at Stn 2, Vidmar spring near Podkraj and Hrastnik, Slovenia. Diffraction pattern presents the mineral composition of powdered shells. The amounts of dominant aragonite (A) is 80 molar%, and dolomite (D) 20 molar%. *Belgrandiella fontinalis* sampled at the same station contained approximately 78 molar% of aragonite, 20 molar% of dolomite and 2 molar% of calcite. Both *B. kuesteri* and *B. fontinalis* were the first freshwater organisms found to precipitate dolomite. Mineral dolomite has been never before registered in the literature as a constituent of adult gastropod and even mollusks shells.

We postulate that higher concentrations of Mg ions in all springs result in formation of aragonite as a dominant phase in both *Belgrandiella* species. Obviously, the concentration of Mg ions in water could be a critical factor determining the shell mineralization, and maybe one of important factors that could lead to the dolomite formation in *Belgrandiella* shells from Stn 2. Unfortunately, relevant literature data about dolomite formation in the adult molluscs does not exist. Here we will discuss two possible factors for dolomite formation in *Belgrandiella.*

The first factor is related to magnesium ions concentration in the shell environment. Fritz et al., (1990) found that bivalves decrease high concentrations of certain elements in the tissue by secreting them into the extrapallial fluid as divalent cations. Falini et al., (1996) confirmed that double charged ions, particularly Mg^{2+}, present in $CaCO_3$ solution favours the formation of aragonite. The increase in the Mg concentration in the external solution is directly proportional to the increase of Mg concentration in aragonite of the shell, and at the same time high Mg concentrations in seawater inhibit the growth of calcite in the shells of *M. edulis* (Lorens & Bender, 1980). Lorens & Bender (1980) found that *M. edulis* may produce aragonite in the calcitic layer when Mg concentrations in the extrapallial fluid become very high, and that the mantle could mediate the transfer of Mg ions, thus, fractioning ions and producing carbonates that are not 'characteristic'. Contents of magnesium ions in the operculum and the shell layers of the snails of the genera *Neritina, Puperita* and *Astraea* can be directly correlated with differences in shell mineralogy (Adegoke, 1973).

An unusual mineral in the gastropods and even mollusks shell, dolomite $CaMg(CO_3)_2$, in amounts of 20 M% was present in both *Belgrandiella* species sampled from Stn 2. Another unusual and unexpected mineral phase, $SiCa_2$, was registered only in *Zospeum isselianum* shells. It is well established that mineralogy and composition of the calcified structure reflect endogenous and exogenous condition in the environment (Dodd, 1967; Kennedy et al., 1969; Carriker et al., 1991). Fritz et al., (1990) indicated that minerals of elemental composition similar to calcium carbonates could be mineralized by mollusks. Findings of a small amount of dolomite, beside other minerals in the embryos and larvae of the mussels *Mytilus edulis* (Medaković, 1995) in the embryos, early echinopluteus, and in spines, teeth and skeleton of adult Antarctic Sea urchins *Sterechinus neumayeri* (Medaković & Popović, 2000; Medaković et al., 2003) confirm the statement of the above authors, and indicate two possible formation mechanisms of dolomite in both *Belgrandiella* species, as well as $SiCa_2$ in *Zospeum isselianum*. We believe that dolomite in both *Belgrandiella* species is influenced by specific environmental characteristics of the locality at Stn 2. Fallini et al., (1996), and Fritz et al., (1990) showed that the presence of double charged ions (Mg^{2+} or Ba^{2+}) in the environment changes the "normal" biomineralization process and causes formation of unusual biominerals (e.g. barite in the freshwater bivalvia). The hardness of the water, ° dH measured at Stn 2, amounted 17.6 that was almost two times higher than at other two stations, Stn 1 and Stn 3. On the contrary, Stn 2 showed a much lower level of dissolved oxygen, 17.1 mg l^{-1} O_2, in comparison to 24.6 mg l^{-1} O_2, at Stn1 and 23.4 mg l^{-1} O_2 at Stn 3 (Medaković et al., 2003a; Slapnik, 1994). We believe that these hydrologic characteristics at Stn 2 and a complex interaction of other environmental conditions of the Stn 2 caused the formation of dolomite in both *Belgrandiella species*. Contrary to that, $SiCa_2$ phase in *Zospeunm isselianum* indicates possible selectivity in the mineral deposition that belongs to the individual shell structure (Dodd, 1967).

The appearance of mineral dolomite in different living organisms in recent literature is well presented; however, all authors could not agree with the theory of the dolomite formation (Deelman, 2008). In his book, in the chapter entitled *Organic or inorganic?*, Deelman (2008) refers in detail on mechanisms of natural low-temperature formation of dolomite by activity of sulphate or methane reducing and urea decomposing bacteria, like the components in two species of tape worms, in freshwater snails, in pearls, in bladder stones of the Dalmatian dog, as well as in human granulomas, tooth enamel, kidney and urinary stones. Our former research confirmed the formation of dolomite as a minor component (from traces up to 5 molar %) in calcareous algae *Coralina, Lithophyllum* and *Pseudolithophyllum sp.* from the Adriatic sea (Medaković et al., 1995). However, dolomite is registered as a starting biomineral between other unusual biominerals only in the embryos of the bivalve *Mytilus edulis*, and in the embryos, larvae and adults of the Antarctic sea urchins *Sterechinus neumayeri* (Medaković, 1995; Medaković et al., 2003).

A different approach could be used to explain the presence of dolomite in the organisms. It is well known that carbonate minerals are seldom pure but tend to incorporate other cations within their crystal lattice. The rhombohedral calcite tends to incorporate cations with smaller ionic radius, and orthorhombic aragonite large cations. Mg, as a small cation, substitutes Ca in calcite and this process can result equally in low Mg-calcite (Mg content < 1 molar %) or in high Mg-calcite (4-20 molar % of Mg). As the concentration of Mg ions in the sea water is almost 5 times higher than that of Ca ions, a number of marine organisms form their hard structures and skeletons of high Mg-calcite. Dolomite is formed when the amount of Ca ions is similar to that of Mg ions in the rhombohedral crystal lattice, and once formed dolomite is relatively a stable mineral. Mg ion is smaller than Ca ion; so the lattice spacings in pure dolomite are considerably smaller than those in calcite. The organisms are capable to selectively precipitate different minerals; their specific regulation can overcome the strict physical parameters of ambient ion concentrations. Because of thermodynamical *reasons,* organisms favour less energetically consuming processes. According to the Ostwald step rule related to the nucleation and growth of crystalline phases, the organisms form a sequence of metastable phases before the formation of the final stable phase (Navrotsky, 2004; Weiner, 2006). This could be an explanation of the dolomite deposition in *Belgrandiella* snails.

Another approach comprises influence of Mg/Ca seawater chemistry oscillation during geological periods on skeletal secretion of different marine organisms (algae, sponge, corals, bryozoans, molluscs) that are reflected on initial evolution on phosphatic to calcitic invertebrate biominerals (Kempe & Kazimiercak, 1994). Mg/Ca ratio of seawater appears to control also evolution of biomineralization and favoured evolution of calcitic skeleton in one geoperiod or high Mg-calcitic or aragonitic during another period (Stanly & Hardie, 1998). Yongding et al., (1977) observed, on the basis of marine sediments, that the mineral components in fossils show an evolution sequence: silica and dolomite—phosphate—organic matter—calcite—magnesium calcite and aragonite. Thomas et al. (2000) showed that evolution and strategy for constructing hard skeleton are imprinted in the organism RNA sequences and patterns of early embryonic development. Recurring features of parallel evolution of complex organic and inorganic skeletons are controlled by genes that were already established in ancestral organisms. An important aspect of the evolutionary history of biomineralization is the finding that from the first up to presently known mineralized organism *Cloudina* contained in its shell walls organic materials with only microdolomite crystals (Grant, 1990; Knoll, 2003).

Belgrandiella fontinalis and *B. kuesteri* are the first freshwater gastropods found to precipitate an unusual mineral, dolomite, in their shells. The mineral composition and fractions of the minerals in shells of both *Belgrandiella* species show that the analyzed species respond to stress, and that ecological factors of the locality influence the shell biomineralization. The findings have a taxonomic and ecological significance.

4. Conclusion

The final aim of the study will be also to describe ecological and morphological importance of the crystal formation in several „model" organisms, to express importance in understanding, and to distinguish terminology between unusual and usual biomineralization. Generally accepted statements have comprised that crystal components in different organisms are under biological control and that crystal forms and morphology in specified tissues are adapted to their function being a characteristic for family, genus and species.

As a support to this review our unpublished experimental results, related to the findings of "unusual" initial biomineral struvite, $(NH_4)MgPO_4 \cdot 6(H_2O)$, in the embryos of the Mediterranean mussels *Mytilus galloprovincialis*, brushite, $CaHPO_4 \cdot 2(H_2O)$, and small amount of feldspars in tropic bivalve species *Codakia orbicularis* and *Tivela mactroides*, and finally dolomite in embryos of the Nudibranch gastropod *Aplysia punctata*, lead us to the conclusion that in specific conditions contemporary marine invertebrates are capable of following ancient feature and models of the biominerals formation in the early ancestors.

Moreover, the presented experimental results clarify and explain well known biomineralization hypotheses:

- calcium carbonate and phosphate ions are abundant in the oceans and thus would prevail in biomineralization over much rarer materials (Lowenstam & Margulis, 1980; Lowenstam, 1981);
- the evolutionary pattern described as "exaptation" (Gould & Vrba, 1982; Kirschvink & Hagadorn, 2000) and some essential immunological similarities between macromolecules involved in the invertebrate and vertebrate biominerals arguing for such common ancestor;
- the powerful idea that molecular inhibitors evolved early as anti-calcification protectors in highly oversaturated Proterozoic oceans were later recruited for the physiological control of crystal formation (Marin et al., 1996, 2000);
- hypothesis of Westbroek & Marin (1998), and Marin et al., (2008) that origins of calcareous skeletons reflect multiple independent co-optations of molecular and physiological processes, which are widely shared by eukaryotic organisms.

5. Acknowledgements

The authors thank the Croatian Ministry of Science, Education and Sport (project No. 098-0982705-2727 and project No. 119-0982886-1009) for financial support of this work. The review is a partially supported by EU COST Action TD0903 ("Biomineralix", see http://www.biomineralix.eu/). Authors also thanks Dr. Frederic Marin from the University of Burgundy, Dijon, France, for helpfull discussion on the biomineralization of the marine and freshwater organisms.

6. References

Addadi, L., & Weiner, S. (1992). Control and design principles in biological mineralization. *Angewandte Chemie International Edition English,* 31, 153-169

Adegoke, O.S. (1973). Mineralogy and biogeochmistry of calcareous operculi and shells of some gastropods. *Malacologia,* 14, 39–46

Aharon, P. (1991). Recorders of reef environmental histories: stable isotopes in corals, giant clams and calcareous algae. *Coral Reefs,* 10, 71-90

Albeck, S., Aizenberg, J., Addadi, L., & Weiner, S. (1993). Interactions of various skeletal intracrystalline components with calcite crystals. *Journal of the American Chemical Society,* 115, 11691-11697

Alzieu, C. (1991). Environmental problems caused by TBT in France: assessment, regulations, prospects. *Marine Environmental Research,* 32, 7-17

Alzieu, C., Sanjuan, J., Deltreil, J.P., & Borel, M. (1986). Tin contamination in Arcachon Bay: effects on oyster shell anomalies. *Marine Pollution Bulletin* 17, 494-498

Alzieu, C., Sanjuan, J., Michel, P., Borel, M., & Dreno, J.P. (1989). Monitoring and assessment of butyltins in Atlantic coastal waters. *Marine Pollution Bulletin,* 20, 22-26

Babička, J. (1936). La tener de Padina pavonia de l' ile de Rab en manganese. *Sitzungsberichte der Königlichen Bohm. Gesellschaft Wissenschaften Trida II,* 5, 1-4

Bauerlein, E. (2000). Biominerals - an Introduction. In: *Biomineralization Progress in Biology, Molecular Biology and Application,* Bauerlein E (ed), pp 1-3, Willey -WCH, Weinheim

Baeuerlein, E. (2004). *Biomineralization. Progress in Biology, Molecular Biology and Application.* In: Baeuerlein E. (Ed.), pp. 1–14, Wiley-VCH, Weinheim

Berman, A., Hanson, J., Leiserowitz, L., Koetzle, T.F., Weiner, S. & Addadi, L. (1993). Biological control of crystal texture: a widespread strategy for adapting crystal properties to function. *Science NY* 359, 776-779

Borowitzka, M.A. (1989). Carbonate calcification in algae - initiation and control. In: *Biomineralization. Chemical and biochemical perspectives,* Mann, S., Webb, J., Williams, R.J.P. (eds), pp 63-95, VCH Verlagsgesellschaft, Weinheim

Brečević, Lj., Nothig-Laslo, V., Kralj, D., & Popović, S. (1996). Effect of divalent cations on the formation and structure of calcium carbonate polymorphs, *Journal of the Chemical Society, Faraday Transactions* 92, 1017-1022

Cabioch, J., & Giraud, G. (1986). Structurel aspects of biomineralization oin the coralline algae (calcified Rhodophyceae). In: Ledbeater BSC, Riding R (eds). *Biomineralization in lower plants and animals,* pp 141-156, Clarendon Press, Oxford

Carriker, M.R., & Palmer, R.E. (1979). Ultrastructural morphogenesis of prodissoconch and early dissoconch valves of the oyster *Crassostrea virginica. Proceedings of the National Shellfisheries Association* 69, 103-128

Carriker, M.R., Swann, C.P., Prezant, R.S., & Counts, IIIC.L. (1991). Chemical elements in the aragonitic and calcitic microstructural groups of shell of the oyster *Crassostrea virginica*: a proton probe study. *Marine Biology* 109, 287-297

Carriker, M.R., Palmer, R.E., Sick, L.V., & Johnson, C.C. (1980). Interactions of mineral elements in sea water and shell of oysters (*Crassostrea virginica,* Gmelin) cultured in controlled and natural systems. *Journal of Experimental Marine Biology and Ecology* 46, 279–296

Carriker, M.R. (1992). Prismatic shell formation in continuously isolated (*Mytilus edulis*) and periodically exposed (*Crassostrea virginica*) extrapallial spaces: explicable by the same concept? *American Malacological Bulletin* 9, 193–197

Chave, K.E. (1954). Aspects of the biogeochemistry of magnesium. I. Calcareous marine organisms. *Yournal of Geology* 62, 266-283

Chave, K.E., & Wheeler, B.D. Jr. (1965). Mineralogic changes during growth in the red algae, Clathomorphum compactum. *Science,* NY 147, 621

Davidson, E.H., Peterson, K.J., & Cameron, R.A. (1995). Origin of bilaterian body plans: evolution of developmental regulatory machanisms. *Science NY* 270, 1319-1325

Deelman, J.C. (2008). Low-temperature formation of dolomite and magnesite. Compact Disc Publications Geology series, Eindhoven, The Netherlands (http://www.jcdeelman.demon.nl/dolomite/bookprospectus.html)

Degobbis, D. (1989). Increased eutrophication of the northers Adriatic Sea. Second act. *Marine Pollution Bulletin,* 20(9), 452-457

Dodd, J.R. (1967). Magnesium and strontium in calcareous skeletons: a review. *Journal of Paleontology,* 41, 1313-1329

Dolenec, T., Medaković, D., & Lojen, S. (2000). The influence of marine anoxia on precipitation of *Mytilus galloprovincialis* shell carbonate in the coastal zone of the Rovinj Bay (Notrthern Adriatic). *Annales, Series Historia Naturalis,* 19(1), 55-60

Dyrinda, E.A., (1992). Incidence of abnormal shell thickening in the Pacific oyster *Crassostrea gigas* in Poole Harbour (UK), subsequent to the 1987 restrictions. *Marine Pollution Bulletin,* 24, 156-163

Fagan, J.F., & Ward, M.D. (1992). Building Molecular Crystals. *Scientific American,* Vol. 267, No. 1, (July 1992), pp 48-54.

Fallini, G., Albeck, S., Weiner, S., & Addadi, L. (1996). Control of aragonite or calcite polymorphism by mollusk shell macromolecules. *Science NY*, 271, 67-69

Fent, K. (1996). Ecotoxicology of organotin compounds. *Critical Reviews in Toxicology*, 26, 1-117

Fisher, N., & Reinfelder, J.R. (1995). The trophic transfer of metals in marine systems. In: Tessier, A, and Turner DR, (Eds.), *Metal speciation and bioavailability to aquatic systems,* pp. 661, John Wiley, Cheschester, UK

Fritz, L.W., Ragone, L.M., & Lutz, R.A. (1990). Biomineralization of barite in the shell of the freshwater Asiatic clam *Corbicula fluminea* (Mollusca: Bivalvia), *Limnology and Oceanography,* 35, 756-762

Fritz, L.W., Ferrence, G., & Jacobsen, T.R. (1992). Induction of barite mineralization in the Asiatic clam *Corbicula fluminea. Limnology and Oceanography,* 37, 442–448

Gould, S. J., & Vrba, S. (1982). Exaptation—a missing term in the science of form. *Paleobiology,* 8, 4-15

Grant, S.W.F. (1990). Shell structure and distribution of *Cloudina*, a potential index fossil for the terminal proterozoic, *American Journal of Science,* 290 A, 261-294

Hamer, B., Jakšić, Ž., Pavičić-Hamer, D., Perić, L., Medaković, D., Ivanković, D., Pavičić, J., Zilberberg, C., Schroder, H.C., Müller, W.E.G., Smodlaka, N., & Batel, R. (2008). Effect of hypoosmotic stress by low salinity acclimation of Mediterranean mussels *Mytilus galloprovincialis* on biological parameters used for pollution assessment, *Aquatic Toxicology,* 89(3), 137-151

Hamer, B., Medaković, D., Pavičić-Hamer, D., Jakšić, Ž., Štifanić, M., Nerlović, V., Travizi, A., Precali, R., & Kanduč, T. (2010). Estimation of Freshwater Influx Along Eastern Adriatic Coast as a Possible Source of Stress for Marine Organisms, *Acta Adriatica,* 51(2), 181-194

Hare, P.E. (1963). Amino acids in the proteins from aragonite and calcite in the shells of *Mytilus californianus. Science* NY, 139, 216-217

Hrs-Brenko, M., Medaković, D., & Zahtila, E. (1990). The Appearance of Pink Coloured Mussels (*Mytilus galloprovincialis,* Lamarck) on the Western Coast of the Istrian Peninsula, *Rapp.Comm.int.Mer Médit,* 32(1), 27-28

James, N.P., Wray, J.L., & Ginsburg, R.N. (1988). Calcification of encrusting aragonitic algae (*Peyssonneliaceae*): implications for the origin of late Paleozoic reefs and cements, *Journal of Sedimentary Petrology,* 58(2), 291-303

Jones, D.S. (1985). Growth increments and geochemical variations in the molluscan shell. In: Mollusks, Broadhead TW (ed). University of Tennesee, Knoxville, Tenn. Stud. Geol., 13: 72-87.

Kanazawa, T., & Sato, S. 2008, Environmental and physiological controls on shell microgrowth pattern of Ruditapes philippinarum (Bivalvia: Veneridae) from Japan. *Journal of Molluscan Studies,* 74, 89-95

Kanduč. T., Medaković, D., & Hamer, B. (2011). *Mytilus galloprovincialis* as a bioindicator of environmental conditions: the case of the Eastern Coast of the Adriatic Sea. *Isotopes in environmental and health studies,* 47(1), 1-20

Kempe, S., & Kazmierczak, J. (1994). The role of alkalinity in the evolution of ocean chemistry, organization of living systems, and biocalcification processes. *Bulletin de l'Institut Oceanographique.* Monaco 13, 61-117

Kennedy, W.J., Taylor, J.D., & Hall, A. (1969). Environmental and biological controls on bivalve shell mineralogy. *Biological reviews of the Cambridge Philosophical Society,* 44, 499-530

King R.J., & Schramm, W. (1982). Calcification in the maerl coralline alga, *Phymatolithon calcareum*: effects of salinity and temperature, *Marine Biology,* 70, 197-205

Kirschvink, J.L., & Hagadorn, J.W. (2000). A Grand Unified Theory of Biomineralization. In: *Biomineralization. Progress in Biology, Molecular Biology and Application.* Bäuerlein E (ed), pp.139-150, Wiley-VCH Verlag, Weinheim, Germany

Knoll, A.H. (2003). Biomineralization and Evolutionary History. *Reviews in Mineralogy and Geochemistry,* 54, 329-356

Lewin, J.C. (1962). Calcification. In: *Physiology and biochemistry of algae,* Lewin RA (ed), pp 457-465, Academic Press, New York

Linck, G. (1930). Der Strahlenkalk von Steinhein eine Cladophore. *Chemie der Erde,* 6, 72

Lorens, R.B., & Bender, M.L. (1980). The impact of solution chemistry on *Mytilus edulis* calcite and aragonite. *Geochimica et Cosmochimica Acta* 44, 1265–1278

Lowenstam, H.A. (1954). Environmental relations of modification compositions of certain carbonate secreting marine invertebrates. *PNAS,* 40, 39–48

Lowenstam, H.A. (1981). Minerals formed by organisms. *Science,* NY 211, 1126-1131

Lowenstam, H.A., & Margulis, L. (1980). Evolutionary prerequisites for early phanerozoic calcareous skeletons, *Biosystems,* 12, 27-41

Lyons, D.M., Medakovic, D., Skoko, Z., Popovic, S., Roncevic, S., Pitarevic - Svedruzic, L., & Karnis. I. (2009). Biomineralization on an Ancient Sculpture of the Apoxyomenos: Effects of a Metal-Rich Environment on Crystal Growts in Living Organisms. *Crystal Growth & Design,* 9(8), 3671–3675

Mann, S. (2001). *Biomineralization, Principles and Concepts in Bioinorganic Materials Chemistry.* Oxford University Press, Oxford.

Mann, S. (1991). Biomineralization: a novel approach to crystal engineering. *Endeavour* (New Ser), 15(3), 120-125

Marin, F., Luquet, G., Marie, B., & Medakovic, D. (2008). Molluscan Shell Proteins: Primary Structure, Origin, and Evolution. In: *Current Topics in Developmental Biology,* Schatten, P Gerald (ed.), pp. 209-276, Elsevier Science & Technology Books, Pittsburgh

Marin, F., Smith, M., Isa, Y., Muyzer, G., & Westbroek, P. (1996). Skeletal matrices, muci, and the origin of invertebrate calcification. *PNAS,* 93, 1554-1559.

Marin, F., Corstjens, P., de Gaulejac, B., de Vrind-de Jong, E.W., & Westbroek, P. (2000). Mucins and molluscan calcification. Molecular characterization of mucoperlin, a novel mucin-like protein from the nacreous shell layer of the fan mussel *Pinna nobilis* (Bivalvia, Pteriomorphia), *Journal of Biological Chemistry,* 275, 20667-20675

McConnaughey, T. (1989). Biomineralization mechanisms. In: *Origin, evolution, and modern aspects of biomineralization in plants and animals,* Crick, R.E., (ed), pp 57-73, Plenum Press, New York

Medaković, D. (1995). The calcification processes of larval, juvenile and adult shells of oysters (*Ostrea edulis*, Linnaeus) and mussels (*Mytilus galloprovincialis*, Lamarck). Ph.D.thesis, pp 1-216, University of Zagreb and R.Boskovic Institute, Zagreb, Croatia

Medakovic, D. (2009). 2.000-year-old statue sheds light on seawater corrosion. Information on corrosion control and prevention, *Materials performance*, 48(10), 12-14

Medaković, D., Hrs-Brenko, M., Popović, S., & Gržeta, B. (1989). X-ray diffraction study of the first larval shell of *Ostrea edulis*. *Marine Biology*, 101(2), 205-209

Medaković, D., Popović, S., Zavodnik, N., Gržeta, B., & Plazonić, M. (1995). X-ray diffraction study of mineral composition of calcareous algae (Corallinaceae, Rhodophyta). *Marine Biology*, 122(3), 479-485

Medaković, D., Hrs-Brenko, M., & Zahtila, E. (1995a). The incidence and distribution of the "Pink Mussels" *Mytilus galloprovincialis*, Lamarck in the Adriatic Sea. *International Workshop on Shell Disease in Marine Invertebrates: Environment-Hopst-Pathogen Interactions*, pp 17, Brest, France

Medaković, D., Popović, S., Gržeta, B., Plazonić, M., & Hrs-Brenko, M. (1997). X-ray diffraction study of calcification processes in embryos and larvae of the brooding oyster *Ostrea edulis*. *Marine Biology*, 129(4), 615-623

Medaković, D., Slapnik, R., Gržeta, B., & Popović, S. (1999). The shell mineralogy of subterranean snails *Zospeum alpestre* (Freyer 1855) and *Zospeum isselianum* (Pollonera 1886) (Mollusca: Gastropoda: Carychiidae). *Periodicum Biologorum*, 101(2), 143-149

Medaković, D., Slapnik, R., Popović, S., & Gržeta, B. (2003). Mineralogy of shells from two freshwater snails *Belgrandiella fontinalis* and *B.kuesteri*. *Comparative Biochemistry and Physiology* Part A 134; 123-129.

Medaković, D., Popović, S., & Manahan, D.T. (2003a). Biominerals in embryos, pluteus and adult Antarctic sea urchins *Sterechinus neumayseri*. In: *Antarctic Biology in a Global Context*, Huiskes, A.H.L., Gieskes, W.W.C., Rozema, J., Schorno, R.M.L., van der Vies, S.M., & Wolf, W.J. (Eds.), , pp 140-143, Backhuys Publishers, Leiden, The Netherlands

Medaković, D., Traverso, P., Bottino, C., & Popović, S. (2006). Shell layers of *Ostrea edulis* as an environmental indicator of TBT pollution: the contribution of surface techniques. *Surface and Interface Analysis*, 38 (4), 313-316

Mitchel, L., Fallick, A.E., & Curry, G.B. (1994). Stable carbon and oxygen isotope composition of molluscs shells from Britain and New Zealand. *Palaeogeography, Palaeoclimatology, Palaeoecology*, 111, 207-216

Mutvei, H., Dunca, E., Timm, H., & Slepukhina, T. (1996). Structure and growth rates of bivalve shells as indicators of environmental changes and pollution. In: Biomineralization 93. Proceedings of the Seventh International Biomineralization Symposium Monaco, Allemand, D., Cuif, J. P., (Eds.), numero special 14, 65-72, Bulletin de ÍInstitut Oceanographique, Monaco

Navrotsky, A. (2004). Energetic clues to pathways to biomineralization: Precursors, clusters, and nanoparticles. *PNAS*, 101, 12096-12101

Odum, H.T. (1951). Notes on the strontium content of sea water, celestite radiolaria, and strontianite snail shells. *Science* NY 114, 211–213

Peres, J.M., Picard, J. (1964). Noveau manuel de bionomie benthique de la Mediterranee. *Recl Trav Stn mar Endoume,* 31 (47), 5 - 137

Pueschel, C.M., Eichelberger, H.H., & Trick, H.N. (1992). Specialized calciferous cells in the marine alga *Rhodogorgon carriebowensis* and their implications for models of red algal calcification. *Protoplasma,* 166, 89-98

Rončević, S., Pitarević-Svedružić, L., Smetiško, J., & Medaković, D. (2010). ICP-AES Analysis of Metal Content in Shell of Mussel *Mytilus galloprovincialis* from Croatian Coastal Waters. *International Journal of Environmental Analytical Chemistry,* 90, 620-632

Schifano, G., & Censi, P. (1986). Oxygen and carbon isotope composition, magnesium and strontium contents of calcite from subtidal *Pattela coerula* shell. *Chemical Geology,* 58, 325-331

Slapnik, R., & Medaković, D. (2007). The shell mineralogy of some freshwater and subterranean snails (Gastropoda: Hydrobiidae and Carychidae). *Mollusca,* 25 (2), 125-129

Slapnik, R. (1998). Mobility of *Zospeum isselianum* Pollonera, 1886 (Gastropoda, Pulmonata, Carychiidae) in cave Jama pod Mokrico in Kamnik - Savinja Alps (Slovenia). *Natura Croatica,* 13, 115-135

Stanley, S.M., & Hardie, L.A. (1998). Secular oscillations in the carbonate mineralogy of reef-building and sediment-producing organisms driven by tectonically forced shifts in seawater chemistry. *Palaeogeography, Palaeoclimatology, Palaeoecology,* 144 (1-2), 3–19

Stenzel, H.B. (1962). Aragonite in the Resilium of Oysters. *Science,* NY 136 (3522), 1121-1122

Stenzel, H.B. (1963). Aragonite and calcite as constituents of adult oyster shells. *Science,* NY 142, 232-233

Stenzel, H.B. (1964). Oysters: composition of the larval shell. *Science,* NY 145, 155-156

Stephenson, M. (1991). A field bioassay approach to determining tributyltin toxicity to oysters in California. *Marine Environmental Research,* 32, 51-59

Taylor, J.D., Kennedy, W.J., & Hall, A. (1969). The shell structure and mineralogy of the Bivalvia. Introduction. Nuculacea-Trigonacea. *Bulletin of the British Museum (Natural History), (D:Zool)* 3 (Suppl*),* 1-125

Thain, J. E., & Waldock, M. J. (1986). The impact of tributyl tin (TBT) antifouling paints on molluscan fisheries. *Water Science* and *Technology,* 18, 193-202

Thain, J.E., Waldock, M.J., Michel, P., Borel, M., & Dreno, J.P. (1989). Monitoring and assessment of butyltins in Atlantic coastal waters. *Marine Pollution Bulletin, 20, pp. 22–26*

Thomas, R.D.K., Shearman, R.M., & Stewart, G.W. (2000). Evolutionary Exploitation of Design Options by the First Animals with Hard Skeletons Science, Vol. 288 no. 5469 pp. 1239-1242

Veis, A. (2008). Crystals and Life. In: *Metal Ions in Life Sciences,* Sigel, A., Sigel, H., & Sigel, R.K.O. (eds), Vol 4, 1-35, John Wiley& Sons Ltd

Vinogradov, A.P. (1953). Elementary composition of nonplanctonic marine algae, Chapter II. In: *The elementary chemical composition of marine organisms,* Parr, A.E. (ed), pp 17-129, Sears Foundation for Marine Research, New Haven

Walker. R., & Moss, B. (1984). Mode of attachment of six epilithic crustose Corallinaceae (Rhodophyta). *Phycologia,* 23(3), 321-329

Waller, T.R. (1981). Functional morphology and development of veliger larvae of the European oyster, *Ostrea edulis* Linne. *Smithsonian Contributions to Zoology,* 328, 1-70

Watabe, N. (1988). Shell structure. In: *The Mollusca. Form and function.* Wilbur, K.M. (ed),Vol. 11 (4), pp 69-104, Academic Press, New York

Wefer, G. (1980). Carbonate production of algae Halimeda, Penicillus and Padina. *Nature,* 285, 323-324

Weiner, S. (2006). Transient precursor strategy in mineral formation of bone. *Bone* 39, 431-433

Westbroek, P., & Marin, F. (1998). A marriage of bone and nacre. *Nature* 392, 861–862

Wilbur, K.M. (1964). Shell formation and regeneration. In: *Physiology of Mollusca,* Wilbur, K.M. & Yonge, C.M. (Eds.), pp. 243–282, Academic Press, New York

Wilbur, K.M., & Saleuddin, A.S.M. (1983). Shell formation. In: *The Mollusca, Physiology,* Saleuddin, A.S.M., & Wilbur, K.M. (eds), Vol. 4 (1), pp 235-287, Academic Press, New York

Yongding, D., Xieguang, J., Shengbai, Z., Hong, Y., Juying, L., & Kui, H. (1977). Classification and evolution of calcareous skeletal textures of fossil organisms. *Chinese Journal of Geology.* 12, 219-235

8

Biomimetic and Bioinspired Crystallization with Macromolecular Additives

Il Won Kim
Department of Chemical Engineering, Soongsil University, Seoul, South Korea

1. Introduction

Biominerals in nature have intrigued scientists and engineers alike because of their unusual structures and properties, and the biomineral examples of extraordinary properties are continually growing with further research efforts. Therefore, biomineralization has become one of the more promising inspirations for the new paradigm of the crystallization science and technology.

The controls found in the biomineralization have been investigated in detail to understand the mechanism, or at least the tools, of forming the intricate structures of the biominerals. The structural controls have been found in the hierarchical levels of the materials. Molecular packing is regulated to generate specific polymorphs of the crystals. Nano- and micro-domains are organized to form the bulk crystals that have unusual shapes and orientations. One of the well-known examples can be found in the abalone nacre. The metastable aragonite is the main constituent of the nacre, while calcite is the thermodynamically stable phase. [Belcher et al., 1996] Also, the nacre has layered structures of micro-plates, which in turn are the assemblies of nano-domains. [Li et al., 2004]

The exact mechanisms to steer the mineralization pathways to form such intricately arranged structures are still under various investigations to be fully unraveled. However, it is now well known that the biomacromolecules that possess the specific interactions with the mineral phases are of utmost importance in the mineralization controls.

This chapter discusses calcium-based biominerals to review the role of biomacromolecules in the mineralization. Biominerals based on calcium carbonate and calcium oxalate are the subject materials to explore. Also, some biomimetic approaches that partly replicate the biomineralization processes are discussed.

Biomimetic crystallization can be defined as the process that mimics the desired aspects observed in biomineralization. When the biomacromolecules implicated in the biomineralization have been directly used, identifying their roles in the mineral formation has been of the initial interests. Also, some fragments of the biomacromolecules have been investigated *in vitro* for the similar purpose. In addition, synthetic polymers and peptides have been applied to the crystallization that mimicked the special controls of the biomineralization.

2. Case studies of the biomineralization of calcium carbonate

Proteins implicated in the formation of mollusk shells (calcium carbonate) have been actively studied in their structures and functions. While the related publications have increased in their numbers, by examining diverse biological species, some of the early examples are represented by the studies of Pacific red abalone (*Haliotis rufescens*) and Japanese pearl oyster (*Pinctada fucata*). In both cases, the focus of many studies was on the formation of nacreous layers. The nacreous layers have aragonite, a metastable phase of calcium carbonate, as the inorganic constituent, and its formation mechanism over calcite, the stable phase of calcium carbonate under ambient conditions, has been investigated in terms of the selective control by biomacromolecules.

2.1 Biomacromolecules associated with *Haliotis rufescens*

AP7, AP8, and AP24 are the proteins identified from the abalone nacre (*H. rufescens*), and the names indicate the molecular weight (7, 8, and 24 kDa) of the aragonitic proteins (AP). AP7 and AP24 were isolated as EDTA-soluble proteins, and they appeared to form a complex through strong intermolecular interactions. [Michenfelder et al., 2003] AP8, on the other hand, was obtained as the soluble component in aqueous solutions of both acetic acid and ammonium sulfate. [Fu et al., 2005a] The compositions of these proteins suggested that the acidic amino acids were the vital part of the protein structures to interact with the calcium carbonate. The acidic amino acids appear beneficial in the protein interactions with calcium ions as well as the mineral surfaces of calcium carbonate.

Protein	Amino acid sequence or composition
AP7	DDNGNYGNGM ASVRTQGNTY DDLASLISYL TRHSFRRPFH ECALCYSITD PGERQRCIDM YCSYTN
AP8-α	Asx 36.5%, Thr 1.2%, Ser 4.7%, Glx 5.2%, Pro 2.1%, Gly 37.7%, Ala 5.2%, Try 0.4%, Phe 1.2%, Lys 2.8%, Arg 3.0%
AP8-β	Asx 34.5%, Thr 0.6%, Ser 2.8%, Glx 8.4%, Pro 1.9%, Gly 39.6%, Ala 3.0%, Lys 4.7%, Arg 4.3%
AP24	ADDDEDASSG LCNQYNQNVT TRPNNKPKMF LRKNINFEII SVHNIWRDPN TVYWCDFSLE EEDGIKHWRH YDFNATHWWV EKGCSGTFVV EECNTKDITN PGPRSTAGKS PMQGTLAAPK PVANWMSIMS RSRFDMGTWD KEGFNML

Note that one-letter and three-letter codes were used for amino acid sequences and compositions, respectively. Also, Asx = Asp or Asn, and Glx = Glu or Gln.

Table 1. Amino acid sequences of AP7, AP8, and AP24 proteins.

The secondary structure of AP7 (66 amino acids), among the AP proteins, was most thoroughly studied. [Collino et al., 2008; Kim et al., 2004a, 2006a; Wustman et al., 2004] Its N-terminal domain (AP7N, 30 amino acids) was structurally labile, while its C-terminal sequence (AP7C, 36 amino acids) possessed short α-helical regions. It was interesting that the *in vitro* crystallization revealed that the AP7 and AP7N demonstrated similar effects on the growth of calcium carbonate, while the AP7C did not display visible effects. The initial result on the secondary structures of AP8 similarly indicated combinations of unfolded and structured regions, while the local distributions of the various structures were not well defined. [Fu et al., 2005b] The N-terminal domain of AP24 also showed an unfolded structure comparable to AP7N, and their effects on the growth of calcium carbonate crystals were of the same nature. [Michenfelder et al., 2003; Wustman et al., 2004] These results indicate the importance of the labile structures in the intrinsically disordered proteins (IDP) that modulate the biomineralization of the abalone nacre.

In situ atomic force microscopy (AFM) has been one of the important tools to observe the *in vitro* biomimetic crystallization of calcium carbonate. For example, *in situ* AFM studies were performed for AP7N, AP8, and AP24N. [Fu et al., 2005b; Kim et al, 2006b] These studies revealed the surprising effects of growth acceleration that had not been observed before, and the underlying mechanism appeared to be through lowering the activation barriers for moving ions from the solution phase to the crystal growth sites [Elhadj et al., 2006; Kim et al, 2006b]. These results indicate the ability of the biomacromolecules to modulate the mineral/solution interfaces to control the kinetics of crystallization.

Peptide	Amino acid sequence
AP7N	DDNGNYGNGM ASVRTQGNTY DDLASLISYL
AP7C	TRHSFRRPFH ECALCYSITD PGERQRCIDM YCSYTN
AP24N	ADDDEDASSG LCNQYNQNVT TRPNNKPKMF

Note that one-letter codes were used for amino acid sequences.

Table 2. Amino acid sequences of AP7N, AP7C, and AP24N peptides.

Biomimetic crystallization to reproduce the nacre formation of *H. rufescens* was initially performed using the mixture of anionic proteins isolated from the aragonitic layer. The anionic proteins prompted the aragonite formation, irrespective of the presence or absence of the nucleating protein sheets also obtained from *H. rufescens*, which indicated that these proteins were the key components in the polymorphic control of the biomineralization. [Belcher et al., 1996] Recent study showed that the AP7 protein was able to form metastable aragonite as well as amorphous calcium carbonate. [Amos and Evans, 2009] It is noteworthy that AP7 also has a function of limiting the growth of calcite, which is the most stable polymorph of calcium carbonate, because suppressing the calcite formation should be helpful, if not required, to produce the metastable phases.

Lustrin A, a matrix protein of *H. rufescens* is also worth mentioning, although it does not appear directly related to the controlled biomineralization. It was identified as a protein (116 kDa, 1,428 amino acids) localized to the regions between aragonite layers in the nacre, and it contained alternating cysteine-rich and proline-rich domains. [Shen et al., 1997; Wustman et al., 2003a] The freshly cleaved nacre exhibited sawtooth-shaped force-extension curves during tensile pulling experiments using an atomic force microscope, and the Lustrin A protein is believed to act as a biomacromolecular adhesive to display the observed behavior and ultimately enhance the toughness of the nacreous layers of *H. rufescens*. [Smith et al., 1999] The study of some part of the cysteine-rich domains revealed a loop conformation that could be a critical factor in the elastomeric behavior of the Lustrin A. [Zhang et al., 2002] Interestingly, Lustrin A also possessed an acidic domain (D4) rich in aspartic acid with an unfolded structure that could act as the mineral binding region - another important aspect for its role as a biomacromolecular adhesive. [Wustman et al., 2003a, 2003b] Readers should note that the biomacromolecular adhesive is only one of several reasons responsible for the high toughness of the nacre. Nacre has lamellar architecture constructed with aragonite microplatelets, which in turn are made of polygonal nanograins, and the micro- and nano-structures are both important for its exceptional mechanical properties. [Li et al., 2004, 2006]

Protein or peptide	Amino acid sequence or composition
Lustrin A	Ser 16.39%, Pro 13.87%, Gly 13.38%, Gys 9.17%, Val 6.23%, Ala 5.95%, Thr 5.32%, Arg 4.90%, Leu 4.41%, Gln 3.92%, Asp 3.85%, Lys 2.24%, Glu 2.10%, Ile 2.03%, Tyr 1.96%, Phe 1.68%, Asn 1.26%, Trp 0.84%, His 0.35%, Met 0.14%
RKSY	YRGPIARPRS SRYLAKYLKQ GRSGKRLQKP
D4	GKGASYDTDA DSGSCNRSPG YLPG

Note that one-letter and three-letter codes were used for amino acid sequences and compositions, respectively.

Table 3. Amino acid composition and sequences of Lustrin A and its related peptides [Shen et al., 1997; Wustman et al., 2003a].

Fig. 1. Schematic of the N-terminal regions of Lustrin A that shows the alternating cysteine-rich and proline-rich domains; redrawn after Shen et al. (1997). C1 to C9 indicate cysteine-rich domains, and P1 to P8 specify proline-rich regions.

2.2 Biomacromolecules associated with *Pinctada fucata*

N16 (16 kDa) family and nacrein are the proteins originally identified from the nacreous shell layer and the pearl of the oyster (*P. fucata*). The N16 was a family of proteins associated with EDTA insoluble organic matrix, and it was separated by using a dilute alkali solution. It was also hydrophilic and acidic in its composition, and it showed an ability to regulate the crystallization of calcium carbonate *in vitro*. [Samata et al., 1999] The N- and C-terminal sequences of N16 (N16N and N16C, both 30 amino acids) were analyzed in their structural changes along with concentrations, pH, and the presence of calcium ions. [Amos et al. 2011; Kim et al., 2004b] They had a mixture of unfolded and β-stranded structures, and they appeared to form supramolecular assemblies regulated by pH and calcium ions. This is consistent with the preliminary analysis of the secondary structure of N16 and with the fact that N16 is a matrix protein. [Samata et al., 1999] Also found in the pearl oyster (*P. fucata*) was nacrein (60 kDa). It was isolated as the EDTA-soluble protein, and it appeared to have domains that regulated both calcium and carbonate components of the biomineral. The calcium-interacting domain had repeating structures of Gly-Xaa-Asn where Xaa was mainly Asp or Asn, and the carbonate-regulating regions showed homology with carbonic anhydrase. [Miyamoto et al., 1996] Especially, the carbonic anhydrase-like domains could play a critical role in conversion of CO_2 to HCO_3^-, ultimately supplying the CO_3^{2-} constituent of the calcium carbonate minerals. Nacrein was originally identified from the nacreous layer of *P. fucata*, but it now appeared responsible for the shell formation in general. [Takeuchi and Endo, 2006] The hypothesis of the general usage seems more plausible with the finding that the homologous proteins have been discovered in other related species. [Kono et al., 2000; Norizuki and Samata, 2008] Note that when oligosaccharides were found as linked to nacrein, alternative explanation for the Ca^{2+} uptake was suggested - the negatively charged sulfite and sialic acid in the N-glycan. [Takakura et al., 2008] More recently, an acidic matrix protein, termed Pif, was identified from the acid-insoluble material of *P. fucata*; it was an aragonite-specific protein found by a calcium carbonate binding assay. It is believed to have domains that can bind chitin and aragonite, and it appears to form protein complex with N16 to regulate the oriented aragonite formation in the nacreous layer. [Suzuki et al., 2009]

Protein or peptide	Amino acid sequence
N16-1	AYHKKCGRYS YCWIPYDIER DRYDNGDKKC CFCRYAWSPW QCNEEERYEW LRCGMRFYSL CCYTDDDNGN GNGNGNGNGL NYLKSLYGGY GNGNGEFWEE YIDERYDN
N16N	AYHKKCGRYS YCWIPYDIER DRYDNGDKKC
N16C	GLNYLKSLYG GYGNGNGEFW EEYIDERYDN

Note that one-letter codes were used for amino acid sequences. Also, N16-1 was the representative protein among N16 family.

Table 4. Amino acid sequence of N16 and its related peptides [Samata et al., 1999; Kim et al., 2004b].

Fig. 2. Schematic of the nacrein protein sequence; redrawn after Miyamoto et al. (1996). CA indicates domains homologous to carbonic anhydrase, and Xaa in Gly-Xaa-Asn is mainly Asp or Asn.

Protein/ Sequence	Amino acid sequence
Nacrein/ 61–109	GPHYWHTISK CF–IACGICQ RQSPINIVSY DAKFRQRLPK LKFKPHMEKL
CA/ 11–52	GPEHWH---K DFPIA--KGE RQSPVDIDTH TAK---YDPS LKPLSVSYDQ
Nacrein/ 379–416	YYTYEGSLTT PPCTESVLWV VQKCHVQV–S RRVL–HALRN
CA/ 189–218	YWTYPGSLTT PPLLECVTWI VLKEPISVSS EQVLKFRKLN

Note that one-letter codes were used for amino acid sequences.

Table 5. Some amino acid sequences of nacrein and carbonic anhydrase (CA) that show the homology between the proteins [Miyamoto et al., 1996].

Biomimetic crystallization to reproduce the nacre formation of *P. fucata* has been performed in the context of using the combinations of β-chitin, silk fibroin-like proteins, and acidic proteins to reconstruct the biological mineralization environments. [Falini et al., 1996] N16N showed binding affinity to β-chitin (but not to α-chitin), and induced aragonite formation

(occurrence rate, 89%). [Keene et al., 2010a] Note that N16N alone could induce aragonite growth, but the occurrence rate was much lower (27%). [Metzler et al., 2010] When silk fibroin hydrogels were introduced to the N16N–β-chitin complex, vaterite and amorphous calcium carbonate were obtained. [Keene et al., 2010b] Probably, the most compelling work in the mineralization mimicking the nacre formation of *P. fucata* was done with Pif protein. When Pif–N16 protein complex was added on the chitin-coated glass, highly oriented aragonite formed between chitin and glass. [Suzuki et al., 2009]
The proteins implicated in the crystallization of mollusk shells have been studied with a basic assumption of heterogeneous nucleation of inorganic phases on the organic frameworks. [Weiner and Addadi, 1997] One of the alternative, if not incompatible, hypotheses to this extracellular mechanism is the intracellular nucleation of the nascent crystals. [Mount et al, 2004] Accordingly, the mantle tissue of *P. fucata* was analyzed to find the genes involved in the control of the mineralization processes as opposed to those encoding the mineral-implicated proteins [Liu et al., 2007]. The highly expressed *P. fucata* mantle gene was found and designated as PFMG1. Its protein sequence showed homology with a known calcium-binding protein CBP-1 with two EF-hand regions, and, therefore, it was presumed to be involved in the signal transduction during nacre formation as well as the crystal nucleation. [Liu et al., 2007; Skelton et al., 1994] The secondary structure of the PFMG1 was studied for the terminal domains, and they possessed the so-called intrinsically disordered protein sequences (IDP) similar to AP7 and N16. [Amos et al., 2010] It is interesting that the terminal fragment peptides were able to generate aragonite, although the occurrence was not exclusive probably due to the lack of other components that usually participate in the biomineral generation. Clearly, much more multidisciplinary work is expected to be done until we fully understand the principles of mineralization in the biological species.

Protein or peptide	Amino acid sequence
PFMG1	MRFLTIIATV LVLGVVVCDA RKRWWRRATK KVELKPEIGR GGDWKVSGGI SISWRKKRSI EKQGKFNIEL TLNPCDLMSY DSNKDGVVTH EDIKHIFDNE KLAEVFFSEA DENDDEQIST SEFKDFKSRI NQCVKD
PFMG1-N	RKRWWRRATK KVELKPEIGR GGDWKVSGGI
PFMG1-C	FSEADENDDE QISTSEFKDF KSRINQCVKD

Note that one-letter codes were used for amino acid sequences.

Table 6. Amino acid sequence of PFMG1 and its related peptides [Liu et al., 2007; Amos et al., 2010].

2.3 Amorphous calcium carbonate

Amorphous phase has been found in some organisms as another form of calcium carbonate. [Lowenstam and Weiner, 1989] It is considered as the temporary storage of calcium and carbonate, and its usage as the structural component has been also discovered. [Aizenberg et al., 1996, 2002; Lowenstam and Weiner, 1989] The representative findings of using the amorphous calcium carbonate (ACC) as the skeletal building blocks were from the calcareous sponge *Clathrina* and the ascidian *Pyura pachydermatina*. [Aizenberg et al., 1996,

2002] The spicule of *Clathrina* was found to have layered structures of a thin calcitic sheath/a thick ACC/a core of calcite, and the presence of ACC was assumed to provide isotropic and less brittle characteristics that could not be achieved by crystalline calcite. [Aizenberg et al., 1996] It was also found that the proteins, implicated in biogenic ACC and calcite, were able to form stable ACC and calcite *in vitro*, respectively. Interestingly, the amino acid compositions of the proteins showed subtle differences between each other. ACC-associated proteins were rich in Glx (Glu or Gln), Ser, and Gly, and calcite-implicated proteins were rich in Asx (Asp or Asn). Other notable features of the ACC-associated biomacromolecules were the presence of polysaccharides, which were absent in the calcite proteins, and their ability to inhibit crystallization of calcium carbonate *in vitro* at an appropriate concentration range. [Aizenberg et al., 1996]

Spicule part	Amino acid composition
Amorphous layer	Asx 7.2%, Thr 4.3%, Ser 19.2%, Glx 17.7%, Pro 5.4%, Gly 19.9%, Ala 8.4%, Val 3.5%, Ile 1.7%, Leu 2.3%, Tyr 0.8%, Phe 1.0%, Lys 4.1%, His 3.2%, Arg 1.1%
Calcite core	Asx 28.5%, Thr 3.5%, Ser 8.8%, Glx 14.3%, Pro 4.2%, Gly 10.8%, Ala 10.5%, Cys 0.2%, Val 3.5%, Met 2.1%, Ile 2.0%, Leu 2.9%, Tyr 1.6%, Phe 2.3%, Lys 1.7%, His 1.1%, Arg 2.0%

Note that three-letter codes were used for amino acid compositions.

Table 7. Amino acid composition of proteins from *Clathrina* spicule [Aizenberg et al., 1996].

Similar finding was noted for the tunic spicules of the *Pyura pachydermatina*. [Aizenberg et al., 2002] Also noticeable in this study was the role of the Mg ions, which facilitated the ACC stabilization. This was initially presumed based on the larger Mg content in the ACC region (5.9%) than in the calcitic layer (1.7%). (Note that the Mg content of the ACC and calcite parts of the sponge *Clathrina* was 11.2% and 6.4%, respectively.) *In vitro* experiments also confirmed that the presence of Mg ions could reduce the critical amount of the ACC proteins required for the ACC stabilization. Analysis of the amino acid compositions of the phase-specific proteins from the *Pyura pachydermatina* showed a similar trend to that for *Clathrina*: a high content of Glx (Glu or Gln) for ACC and Asx (Asp or Asn) for calcite. [Aizenberg et al., 2002] Also note that a negatively charged glycoprotein involved in ACC formation was more recently identified from the crayfish *Cherax quadricarinatus* among crustaceans. [Shechter et al., 2008]

Spicule part	Amino acid composition
Amorphous region	Asx 7.1%, Glx 17.2%, Ser 11.2%, Thr 12.2%, Gly 12.6%, Ala 5.5%, Val 6.0%, Leu 5.3%, Ile 3.0%, Arg 8.9%, Lys 4.4%, Phe 3.3%, Tyr 0.4%, Cys 1.6%, Pro 1.2%
Calcite region	Asx 18.8%, Glx 8.0%, Ser 6.8%, Thr 4.6%, Gly 20.8%, Ala 6.5%, Val 4.6%, Leu 7.1%, Ile 1.2%, Arg 3.4%, Lys 4.0%, His 0.8%, Phe 1.2%, Tyr 2.2%, Cys 7.8%, Met 1.2%, Pro 1.0%

Note that three-letter codes were used for amino acid compositions.

Table 8. Amino acid composition of proteins from the spicule of *Pyura pachydermatina* [Aizenberg et al., 2002].

Biogenic ACC has been also found as a precursor to eventually form crystalline calcium carbonate. The larval shell of *Mercenaria mercenaria* showed ACC-to-aragonite transformation *in vivo*. [Weiss et al., 2002] The biomacromolecules extracted from the *Atrina rigida* demonstrated ACC-to-aragonite transition when applied *in vitro* along with β-chitin and silk scaffold. [Gotliv et al., 2003] Later Asp-rich protein family was identified from the prismatic layer of *Atrina rigida*, and named Asprich. [Gotliv et al., 2005] This protein appeared to induce ACC formation and inhibit crystallization of calcium carbonate to slowly fine-tune the growth of the prismatic layer. [Politi et al., 2007] It is interesting to note that this protein, that stabilizes ACC only transiently, is rich in Asp, while those found in longer stabilized ACC is rich in Glu (or Gln). Also, the transient ACC and stable ACC appeared different structurally (short range order) and compositionally (water molecules). [Raz et al., 2003] It would be interesting future work to clearly identify the correlations between the structures of the involved proteins and the the formation of ACC in different nature.

2.4 Additives to regulate the crystallization of calcium carbonate

Macromolecules, not specifically associated with calcium carbonate biominerals, have been investigated to explore their ability to interact with the inorganic phase and alter its crystallization pathways. Diverse synthetic and biological polymers have been studied to control the size, shape, and polymorphism of calcium carbonate. As in the previous sections dealing with biomineralization, only some representative examples of bioinspired crystallization are introduced in this section. Readers interested in more extensive examples may find the special issue (volume 108, issue 11; November 2008) of *Chemical Review* useful.

Fig. 3. SEM micrographs showing representative morphology of calcite, aragonite, and vaterite obtained with poly(vinyl alcohol). [Kim et al., 2005]

Poly(vinyl alcohol) was one of the simplest example that showed successful control in the crystallization of calcium carbonate. [Hosoda et al., 2003; Kim et al., 2005; Lakshminarayanan et al., 2003] Poly(vinyl alcohol) was able to adsorb extensively on calcium carbonate to inhibit its crystallization. The inhibition of the crystallization was effective in the order: calcite > aragonite > vaterite. Therefore, the polymorph of calcium carbonate obtained with poly(vinyl alcohol) could be adjusted by changing the amount of poly(vinyl alcohol) in the supersaturated solution. As the concentration of poly(vinyl alcohol) increased, (i) the size of the calcite decreased, (ii) calcite was inhibited to form aragonite almost exclusively, and (iii) both calcite and aragonite were inhibited to form vaterite. [Kim et al., 2005] A recent study using *in situ* atomic force microscopy on a single crystal of calcite also verified the inhibitory effects of poly(vinyl alcohol) to modulate the overgrowth of the calcite. [Kim et al., 2009]

Some biomacromolecules, not originally associated with biominerals, were also able to selectively crystallize aragonite. Bovine serum albumin (66 kDa) appeared to form aggregates via self-assembly that stabilized aragonite polymorph. Note that the critical concentration of the protein required for aragonite polymorph was much higher than that for aragonitic shell proteins. [Amos et al., 2010] Similar behavior was also seen with soy globulins, but it was unclear if the proteins were solely responsible for this result because the aragonite was prepared at slightly elevated temperature (50°C) that could induce some aragonite formation. [Liu et al., 2009]

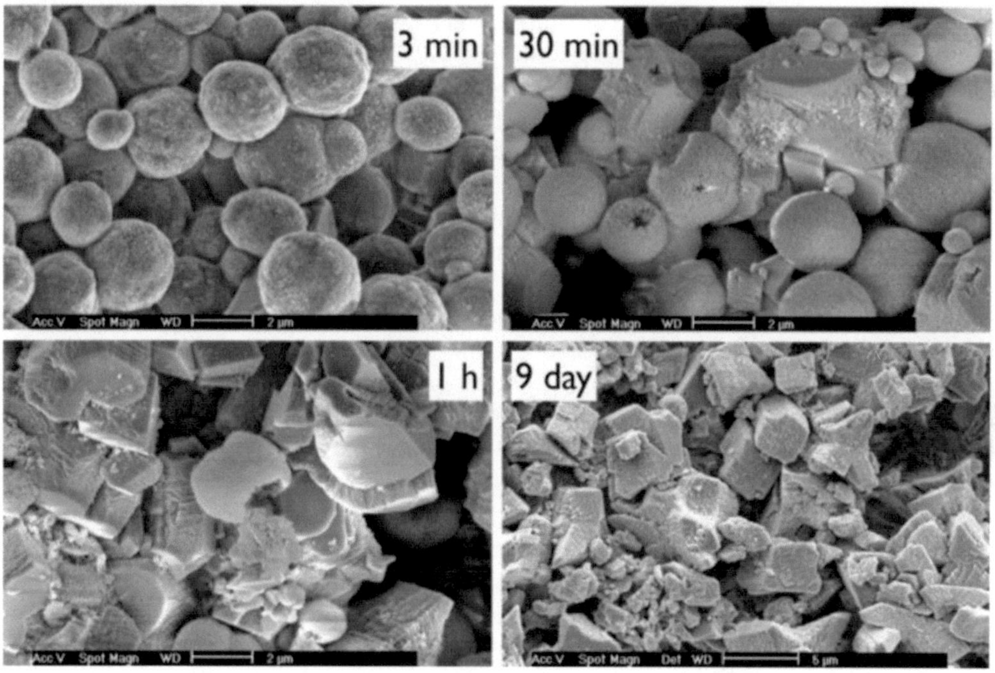

Fig. 4. SEM micrographs showing calcium carbonate crystallization with polyacrylamide. Initially formed vaterite transformed into mesocrystal-like calcite. [Kim, 2004]

Mesocrystal formation (probably through amorphous precursor particles) has been extensively studied and reviewed. [Cölfen and Antonietti, 2005] This non-classical crystallization pathway utilizes the building blocks of nanoparticles to generate superstructures of complex morphologies. Early examples of calcium carbonate mesocrystals utilized silica, gelatin, and polyacrylamide gels. [Dominquez-Bella and Garcia-Ruiz, 1986; Grassmann et al., 2002, 2003] More elegant and selective method was shown using poly(styrene sulfonate). Poly(styrene sulfonate) adsorbed selectively on highly polar (001) faces of calcite to generate porous mesocrystals with crystallographically well-ordered primary particles. [Wang et al., 2005] When polyacrylamide (0.50%) was used as an additive in a supersaturated solution of calcium carbonate (0.10 M), rather than as a gel matrix, mesocrystal-like calcite was also obtained. [Kim, 2004] Initially, spherical vaterite was obtained, and it transformed into calcite with time. As shown in Figure 4 (30 min), multiple vaterite particles were in contact with a calcite crystal probably acting as concurrent reservoirs for calcite growth, which appeared to contribute the moesocrystal-like features of calcite at the end. Also, the spherical shape of vaterite suggested that the initial phase could be amorphous calcium carbonate.

As stated in section 2.3, the amorphous calcium carbonate (ACC) is one of the important keys to accomplish the high level of controls observed in biomineralization. Some of the many examples, where simple ACC formation was successful, include those with ethanol, magnesium, surfactant, and phytic acid. [Lee et al., 2005; Loste et al., 2003; Li and Mann, 2002; Xu et al., 2005] When calcium carbonate precipitated in the ethanol solution of calcium chloride via the decomposition of ammonium carbonate to generate carbon dioxide, ACC formed with some included ammonia and carbon dioxide, which could be removed later through heating at 100°C. [Lee et al., 2005] When magnesium ion was present at high concentrations in the calcium carbonate solutions, ACC formed as a transient phase, and its stability was improved as the amount of magnesium ion increased. [Loste et al., 2003] When ACC nanoparticles were stabilized with a surfactant, its assembly process was manipulated to generate diverse forms of crystalline calcium carbonate. [Li and Mann, 2002] ACC formation with phytic acid presented an interesting scenario of making hollow ACC spheres. Anhydrous ACC nanoparticles initially formed, aggregated, and eventually transformed into hydrated ACC leaving only the exterior of the aggregated superstructures. [Xu et al., 2005]

ACC formation and/or stabilization via polymeric additives have been found in a number of publications. Some of the representative examples include those on utilizing polyaspartic acid, poly(propylene imine) dendrimers, and polyacrylate. [Dai et al., 2007; Gower and Odom, 2000; Donners et al., 2000; Jiang et al., 2009] Polyaspartic acid was employed to generate so-called a polymer-induced liquid-precursor (PILP) phase (a highly hydrated ACC), which could assemble and transform into various unusual morphology of calcium carbonate. [Dai et al., 2007; Gower and Odom, 2000] Poly(propylene imine) dendrimers (5th generation) that produced ACC particles were unique in that the kinetics of ACC-to-calcite transformation could be modulated with addition of surfactants, such as octadecylamine (stabilizing ACC) and cetyltrimethylammonium bromide (only calcite observed possibly because of ACC destabilization). [Donners et al., 2000] Polyacrylate, on the other hand, showed a simple example where the ACC formation was regulated in its size and stability by the molecular weight and amount of the polymer; it is interesting that metastable aragonite formed as the major phase when ACC was not stabilized. [Jiang et al., 2009]

ACC formation has been also found under spatial confinement. Well-defined ACC experiments that systematically studied the effects of confinement were performed with polycarbonate track-etch membranes and crossed half-cylinders. [Loste et al., 2004; Stephens et al., 2010] Cylindrical pores (diameters 0.2, 0.8, 3.0, and 10.0 μm) in the membranes were utilized to form confined calcium carbonate under transient ACC forming or non-forming conditions. ACC precursors were necessary as the initial phase to faithfully fill the pores (diameters $\leq$ 3.0 μm) and eventually form rod-shaped calcite, but 10.0-μm pores were too large to give such confinement effects. [Loste et al., 2004] The crossed half-cylinders were employed to probe the effects of spatial restriction on the stability of ACC. (The surface of the half-cylinder was self-assembled monolayers of mercaptohexadecanoic acid on Au.) The decreased gap between half-cylinders approaching the contact point demonstrated the kinetic stabilization of ACC due to confinement. [Stephens et al., 2010]

3. Case studies of the biomineralization of calcium oxalate

The examples in section 2 exhibited well-defined biominerals of calcium carbonate that were utilized in biological species for the structural and functional reasons. Similar examples of calcium oxalate biominerals could be found in various species. [Lowenstam and Weiner, 1989] However, the focus of this section will be limited to the pathological stones of calcium oxalate. Kidney stones are affecting about 10% of the populations, and their compositions include calcium phosphate and calcium oxalate. Calcium oxalate has been the focus of the majority of studies because it is the main constituent of the many kidney stones. The pathological biominerals of calcium oxalate differ from the previously discussed calcium carbonate in that their structures are relatively irregular with strong crystal aggregation to form kidney stones.

Calcium oxalate usually forms crystals as hydrates, and calcium oxalate monohydrate (COM) and calcium oxalate dihydrate (COD) are the most common forms. The research on calcium oxalate can be largely categorized into the studies of surface adhesion and growth modulation. In both cases, the atomic force microscopy (AFM) has played crucial roles in measuring adhesion forces at nanonewton scale and visualizing the growth of calcium oxalate at near molecular level.

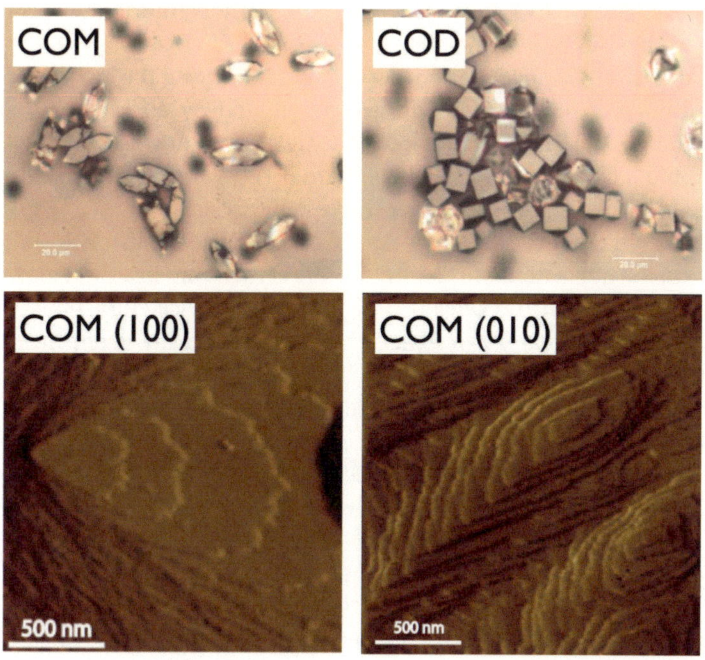

Fig. 5. Examples of COM and COD crystals (top optical microscope images) and the growth hillocks on COM (100) and (010) faces (bottom AFM images). [Kim, 2009]

3.1 Adhesion at calcium oxalate surfaces

The most common kidney stones form as COM aggregates, which makes the study of crystal adhesion extremely important in understanding the disease mechanism. Since the proteins implicated in the COM aggregation contain functional groups, such as carboxylates, amidinums, and alcohols, the specific interactions between the chemical groups and the surfaces of calcium oxalates have been investigated. [Sheng et al., 2003, 2005a, 2005b] Probably the most important finding in the adhesion study is the relative adhesion strength between COM and COD surfaces. Generally, COM showed stronger adhesion to the $Au{:}S(CH_2)_{10}COO^-$ and $Au{:}S(CH_2)_2NHC(NH_2^+)NH_2$ groups, which were prepared by reacting appropriate thiols on the gold-coated AFM tips. Especially, COM (100) and COD (101) faces showed the strongest and weakest adhesion, respectively. Since the COM (100) and COD (101) are most exhibited surfaces for the calcium oxalate hydrates, the results of adhesion measurements strongly corroborate the frequent observation of COM aggregate stones and the benign characteristic of COD. [Sheng et al., 2005b] Another interesting finding is the decreased adhesion force in the presence of citrate, a recognized inhibitor of COM formation and aggregation. [Sheng et al., 2005b]

3.2 Molecular modulation of calcium oxalate growth

COM growth was studied in detail to understand the role of the growth inhibitors, citrate and osteopontin. [Qiu et al., 2004] Also, synthetic Asp-rich peptides were attempted to mimic the effects of osteopontin. [Wang et al., 2006] *In situ* AFM was utilized in both studies to observe the effects of the additives on the specific surfaces of COM crystals. Citrate induced dramatic changes in the growth on COM (100) faces, and osteopontin showed strong inhibitory effects on COM (010) faces, clearly demonstrating the specific interactions between the additive molecules and the surfaces of COM crystals. [Qiu et al., 2004] When $(DDDS)_6DDD$ and $(DDDG)_6DDD$ peptides were present during the growth of COM crystals, the peptide with serine residue showed stronger inhibitory effect than that with glycine spacer. [Wang et al., 2006] The significance of the COM growth studies would increase when combined with the adhesion experiments (section 3.1) because COM crystals in the kidney stones are often found as aggregates rather than single crystals.

Before closing the section of pathological mineralization of calcium oxalate, two loosely related studies on the nucleation of calcium oxalate crystals are briefly reviewed. First, COM formation was promoted in the presence of brushite (or dicalcium phosphate dihydrate, $CaHPO_4 \cdot 2H_2O$). [Tang et al., 2006] Since the calcium oxalate stone formation is often associated with calcium phosphate, the role of the phosphate phase is of great interest to understand the initial stage of the stone formation. [Evan et al., 2003] When a supersaturated but metastable solution of calcium oxalate, which could represent the physiological situation in the urinary space, was observed for more than 2 days, no nucleation of COM was visible. However, when phosphate was present in the system, brushite formed as the initial phase, which in turn induced the COM formation. [Tang et al., 2006] This study suggested the crucial role of heterogeneous nucleation for COM formation. Second, COD formation induced by poly(acrylic acid) was via transformation of amorphous calcium oxalate. [Kim, 2010] When a supersaturated solution of calcium oxalate was prepared with poly(acrylic acid), COD formed instead of COM. [Sheng et al., 2005b] When the crystal formation was observed on self-assembled monolayers ($Au{:}S(CH_2)_{10}COO^-$ and $Au{:}S(CH_2)_{11}NH_3^+$), amorphous phase was detected. It is unclear at this stage whether the amorphous particles aggregate and transform into COD crystals in a similar fashion as the ACC-to-calcite transformation or they simply act as an initial phase that later dissolves to form more stable COD. However, it is interesting that similar behavior was not seen under COM-favorable conditions where the polymeric additive was absent. [Kim, 2010] This study suggested that when COM was prevented through the macromolecular inhibitors, COD could form via amorphous phase to effectively reduce the concentration of calcium oxalate in the originally supersaturated solution.

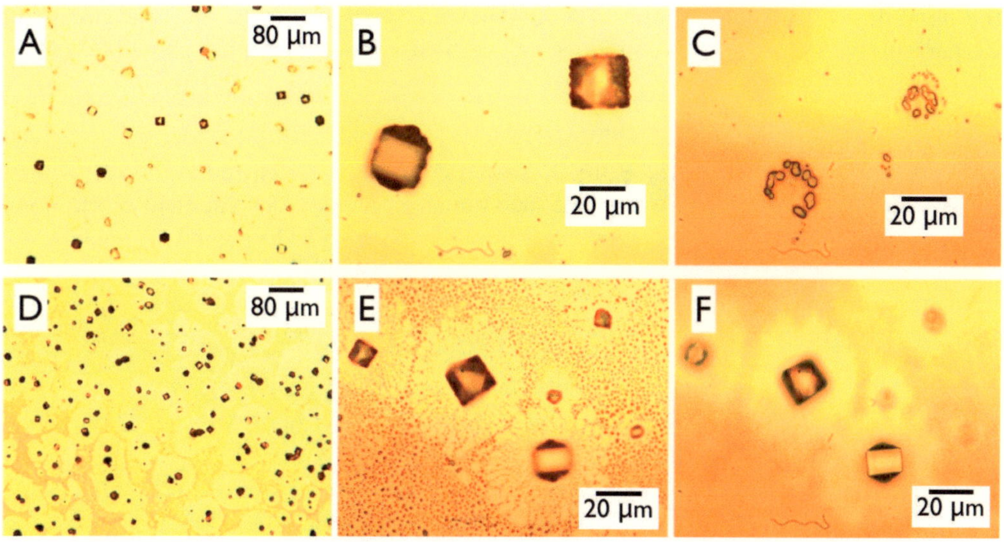

Fig. 6. COD formation on Au:S$(CH_2)_{10}COO^-$: (A) a large area view, (B) COD crystals with amorphous particles attached to them, (C) amorphous particles; COD formation on Au:S$(CH_2)_{11}NH_3^+$: (D) a large area view, (E) COD crystals surrounded by amorphous particles, (F) same image as (E) except that the focus was on COD. [Kim, 2010]

4. Summary

This chapter has dealt with biomineralization and biomimetic/bioinspired crystallization with special emphasis on macromolecular additives. Among the diverse types of biominerals, calcium carbonate and calcium oxalate have been discussed. Case studies of calcium carbonate biominerals have been focused on the shells of *H. rufescens* and *P. fucata*. Examples of calcium oxalate biominerals have been limited to the kidney stones. The shells of mollusks and the pathological stones of kidney are contrastingly different as the structures are highly well defined for the former and irregular for the latter. In dealing with the examples, it is clear that organic macromolecules play the central role in regulating the inorganic components.

Several unusual phenomena observed in the biomineralization of mollusk shells include selective formation of metastable aragonite polymorph and normally unstable amorphous calcium carbonate as well as the extraordinary control for the complex morphology. While the intricately controlled microenvironment *in vivo* is apparently important, the special functions of the biomacromolecules have been attributed to their amino acid compositions and secondary structures. Numerous examples exist to mimic the unusual phenomena *in vitro*, and some representative examples have been introduced. When the successful approaches are extended to other materials, the results of the aforementioned studies would be fully realized as the new paradigm of the crystallization engineering.

Calcium oxalate aggregates in kidney stones are mainly calcium oxalate monohydrate, and the dihydrate form is known as generally benign. This phenomenon is closely related to the strength of surface adhesion that is different between the monohydrate and dihydrate. Also, the molecules known as the inhibitors of stone formation appear to retard both adhesion and growth of the monohydrate. The understanding of the formation and inhibition mechanisms has opened new possibilities for the disease prevention and treatment.

5. Acknowledgment

This work was supported by the Korea Research Foundation Grant funded by the Korean Government (KRF-2008-331-D00117) and the Human Resources Development of the Korea Institute of Energy Technology Evaluation and Planning (KETEP) grant funded by the Ministry of Knowledge Economy, Republic of Korea (20104010100610).

6. References

Aizenberg J, Lambert G, Addadi L, and Weiner S (1996) Adv. Mater. 8, 222–226.
Aizenberg J, Lambert G, Weiner S, and Addadi L (2002) J. Am. Chem. Soc. 124, 32–39.
Amos FF and Evans JS (2009) Biochemistry 48, 1332–1339.
Amos FF, Destine E, Ponce CB, and Evans JS (2010) Cryst. Growth Des. 10, 4211–4216.
Amos FF, Ponce CB, and Evans JS (2011) Biomacromolecules 12, 1883–1890.
Belcher AM, Wu XH, Christensen RJ, Hansma PK, Stucky GD, and Morse DE (1996) Nature 381, 56–58.
Collino S, Kim IW, and Evans JS (2008) Biochemistry 47, 3745–3755.
Cölfen H and Antonietti M (2005) Angew. Chem. Int. Ed. 44, 5576–5591.
Dai L, Douglas EP, and Gower LB (2008) J. Non-Crystal. Solids 354, 1845–1854.
Dominguez Bella S and Garcia-Ruiz JM (1986) J. Cryst. Growth 79, 236–240.
Donners JJJM, Heywood BR, Meijer EW, Nolte RJM, Roman C, Schenning APHJ, and Sommerdijk NAJM (2000) Chem. Commun. 1937–1938.
Elhadj S, De Yoreo JJ, Hoyer JR, and Dove PM (2006) Proc. Natl. Acad. Sci. USA 103, 19237–19242.
Evan AP, Lingeman JE, Coe FL, Parks JH, Bledsoe SB, Shao Y, Sommer AJ, Paterson RF, Kuo RL, and Grynpas M (2003) J. Clin. Invest. 111, 607–616.
Falini G, Albeck S, Weiner S, and Addadi L (1996) Science 271, 67–69.
Fu G, Valiyaveettil S, Wopenka B, and Morse DE (2005a) Biomacromolecules 6, 1289–1298.
Fu G, Qiu SR, Christine AO, Morse DE, and De Yoreo JJ (2005b) Adv. Mater. 17, 2678–2683.
Gotliv B-A, Addadi L, and Weiner S (2003) ChemBioChem 4, 522–529.
Gotliv B-A, Kessler N, Sumerel JL, Morse DE, Tuross N, Addadi L, and Weiner S (2005) ChemBioChem 6, 304–314.
Gower LB and Odom DJ (2000) J. Cryst. Growth 210, 719–734.
Grassmann O, Müller G, and Löbmann P (2002) Chem. Mater. 14, 4530–4535.
Grassmann O, Neder RB, Putnis A, and Löbmann P (2003) Am. Mineral. 88, 647–652.
Hosoda N, Sugawara A, and Kato T (2003) Macromolecules 36, 6449–6452.

Jiang F, Yang Y, Huang L, Chen X, and Shao Z (2009) J. Appl. Polym. Sci. 114, 3686-3692.
Keene EC, Evans JS, and Estroff LA (2010a) Cryst. Growth Des. 10, 1383-1389.
Keene EC, Evans JS, and Estroff LA (2010b) Cryst. Growth Des. 10, 5169-5175.
Kim IW (2004) Ph.D. thesis, University of Michgan at Ann Arbor.
Kim IW (2009) J. Res. Inst. Frontier Tech. (Soongsil University) 39, 8-13.
Kim IW (2010) J. Res. Inst. Frontier Tech. (Soongsil University) 40, 1-6.
Kim IW, Morse DE, and Evans JS (2004a) Langmuir 20, 11664-11673.
Kim IW, DiMasi E, and Evans JS (2004b) Cryst. Growth Des. 4, 1113-1118.
Kim IW, Robertson RE, and Zand R (2005) Cryst. Growth Des. 5, 513-522.
Kim IW, Collino S, Morse DE, and Evans JS (2006a) Cryst. Growth Des. 6, 1078-1082.
Kim IW, Darragh MR, Orme C, and Evans JS (2006b) Cryst. Growth Des. 6, 5-10.
Kim R, Kim C, Lee S, Kim J, and Kim IW (2009) Cryst. Growth Des. 9, 4584-4587.
Kono M, Hayashi N, and Samata T (2000) Biochem. Biophys. Res. Comm. 269, 213-218.
Lakshminarayanan R, Valiyaveettil S, and Loy GL (2003) Cryst. Growth Des. 3, 953-958.
Lee HS, Ha TH, and Kim K (2005) Mater. Chem. Phys. 93, 376-382.
Li X, Chang W-C, Chao YJ, Wang R, and Chang M (2004) Nano Lett. 4, 613-617.
Li X, Xu Z-H, and Wang R (2006) Nano Lett. 6, 2301-2304.
Liu D, Tian H, Kumar R, and Zhang L (2009) Macromol. Rapid Commun. 30, 1498-1503.
Liu H-L, Liu S-F, Ge Y-J, Liu J, Wang X-Y, Xie L-P, Zhang R-Q, and Wang Z (2007) Biochemistry 46, 844-851.
Loste E, Park RJ, Warren J, and Meldrum FC (2004) Adv. Funct. Mater. 14, 1211-1220.
Loste E, Wilson RM, Seshadri R, and Meldrum FC (2003) J. Cryst. Growth 254, 206-218.
Lowenstam HA and Weiner S (1989) On Biomineralization; Oxford University Press: New York.
Mei L and Mann S (2002) Adv. Funct. Mater. 12, 773-779.
Metzler RA, Evans JS, Killian CE, Zhou D, Churchill TH, Appathurai NP, Coppersmith SN, and Gilbert PUPA (2010) J. Am. Chem. Soc. 132, 6329-6334.
Michenfelder M, Fu G, Lawrence C, Weaver JC, Wustman BA, Taranto L, Evans JS, and Morse DE (2003) Biopolymers 70, 522-533.
Miyamoto H, Miyashita T, Okushima M, Nakano S, Morita T, and Matsushiro A (1996) Proc. Natl. Acad. Sci. USA 93, 9657-9660.
Mount AS, Wheeler AP, Paradkar RP, and Snider D (2004) Science 304, 297-300.
Norizuki M and Samata T (2008) Mar. Biotechnol. 10, 234-241.
Politi Y, Mahamid J, Goldberg H, Weiner S, and Addadi L (2007) CrystEngComm 9, 1171-1177.
Qiu SR, Wierzbicki A, Orme CA, Cody AM, Hoyer JR, Nancollas GH, Zepeda S, and De Yoreo JJ (2004) Proc. Natl. Acad. Sci. USA 101, 1811-1815.
Raz S, Hamilton PC, Wilt FH, Weiner S, and Addadi L (2003) Adv. Funct. Mater. 13, 480-486.
Samata T, Hayashi N, Kono M, Hasegawa K, Horita C, and Akera S (1999) FEBS Letters 462, 225-229.
Schechter A, Glazer L, Cheled S, Mor E, Weil S, Berman A, Bentov S, Aflalo ED, Khalaila I, and Sagi A (2008) Proc. Natl. Acad. Sci. USA 105, 7129-7134.
Shen X, Belcher AM, Hansman PK, Stucky GD, and Morse DE (1997) J. Biol. Chem. 272, 32472-32481.
Sheng X, Ward MD, and Wesson JA (2003) J. Am. Chem. Soc. 125, 2854-2855.

Sheng X, Jung T, Wesson JA, and Ward MD (2005a) Proc. Natl. Acad. Sci. USA 102, 267–272.
Sheng X, Ward MD, and Wesson JA (2005b) J. Am. Soc. Nephrol. 16, 1904–1908.
Skelton NJ, Kordel J, Akke M, Forsen S, and Chazin WJ (1994) Nat. Struct. Biol. 1, 239–245. [YET TO FIND AND READ]
Smith BL, Schaffer TE, Viani M, Thompson JB, Frederick NA, Kindt J, Belcher A, Stucky GD, Morse DE, and Hansma PK (1999) Nature 399, 761–763.
Stephens CJ, Ladden SF, Meldrum FC, and Christenson HK (2010) Adv. Funct. Mater. 20, 2108–2115.
Suzuki M, Saruwatari K, Kogure T, Yamamoto Y, Nishimura T, Kato T, and Nagasawa H (2009) Science 325, 1388–1390.
Takakura D, Norizuki M, Ishikawa F, and Samata T (2008) Mar. Biotechnol. 10, 290–296.
Takeuchi T and Endo K (2006) Mar. Biotechnol. 8, 52–61.
Tang R, Nancollas GH, Giocondi JL, and Orme CA (2006) Kidney Int. 70, 71–78.
Wang L, Qiu SR, Zachowicz W, Guan X, De Yoreo JJ, Nancollas GH, and Hoyer JR (2006) Langmuir 22, 7279–7285.
Wang TX, Cölfen H, and Antonietti M (2005) J. Am. Chem. Soc. 127, 3246–3247.
Weiner S and Addadi L (1997) J. Mater. Chem. 7, 689–702.
Weiss IM, Tuross N, Addadi L, and Weiner S (2002) J Exp Zool 293, 478–491.
Wustman BA, Weaver JC, Morse DE, and Evans JS (2003a) Connect. Tissue Res. 44, 10–15.
Wustman BA, Weaver JC, Morse DE, and Evans JS (2003b) Langmuir 19, 9373–9381.
Wustman BA, Morse DE, and Evans JS (2004) Biopolymers 74, 363–376.
Xu A-W, Yu Qiu, Dong W-F, Antonietti M, and Cölfen H (2005) Adv. Mater. 17, 2217–2221.
Zhang B, Wustman BA, Morse D, and Evans JS (2002) Biopolymers 63, 358–369.

Part 4

Crystal Growth for Health Protection

9

Freezing Properties of Disaccharide Solutions: Inhibition of Hexagonal Ice Crystal Growth and Formation of Cubic Ice

Tsutomu Uchida[1], Satoshi Takeya[2],
Masafumi Nagayama[1] and Kazutoshi Gohara[1]
[1]Faculty of Engineering, Hokkaido University,
[2]Research Institute of Instrumentation Frontiers,
National Institute of Advanced Industrial Science and Technology (AIST),
Japan

1. Introduction

Numerous studies have been undertaken to preserve living bodies and cells from freezing by addition of cryoprotective agents including natural substances (e.g., sugars and proteins) and synthetic chemicals (e.g., glycerol and dimethyl sulfoxide). Trehalose and sucrose, which consist of fructose and glucose rings connected by a glycosidic bond, are naturally occurring disaccharide compounds found in cryoprotectants. Trehalose is primarily found in animals capable of enduring cold temperatures, whereas sucrose is typically found in plants (Crowe et al., 1988). Since these molecules are too large to permeate the biomembrane, the cryoprotective mechanism is considered to be different from those of cell-permeable substances, such as glycerol and dimethyl sulfoxide.

One of the considerable mechanisms of disaccharide molecules operating as a cryoprotective agent of living cells is that they protect the lipid bilayer by making the hydrogen bonding during extracellular ice formation. Based on the interactions between disaccharide molecules and the lipid bilayer, it has been suggested that the molecular mechanism underlying this cryoprotective effect is the hydrogen bonding of trehalose molecules to the bilayer head group (Sum et al., 2003). A simulation of the interaction of the lipid bilayer with trehalose has revealed that only marginal changes occur in the lipid bilayer.

Another considerable mechanism of disaccharide molecules of the cryoprotective effect for living cells is that they inhibit the growth of ice crystals in extracellular space. To reveal the inhibition mechanism of disaccharides on ice crystal growth, the interaction between the disaccharide and water molecules has been investigated both by macroscopic observations (for example, Sei et al., 2001; 2002; Sei & Gonda, 2004), where the melting point of disaccharide solutions has been determined, and by microscopic observations (for example, Wang & Tominaga, 1994; Kanno & Yamazaki, 2001; Akao et al., 2001; Branca et al., 1999a) where the interaction has been investigated via Raman or infra-red spectroscopy. Sei et al. (2001; 2002) found that the melting points of trehalose and sucrose solutions were lower than those expected by the molar freezing point depression. They explained this phenomenon by considering the ratio of the hydrated water around disaccharide molecules to that of free water, because these disaccharides interact with water strongly. Therefore, the cryoprotective mechanism of disaccharide is considered to be controlled by the hydration of disaccharide molecules in solution. Raman spectroscopic observations have revealed that disaccharides promote a destructive effect on the tetrahedral hydrogen-bond network of pure water (Branca et al., 1999a). The results of such studies have led to the hypothesis that disaccharides obstruct the crystallization process, thereby destroying the network of water in a manner similar to that occurring in ice. This hypothesis qualitatively coincides with the conclusions suggested by the ice single-crystal observations (Sei et al., 2002). However, the cyroprotective mechanisms of the living cells by these cell-impermeable protectants are still under consideration.

Based on the application of disaccharides to the sample preparation for the electron microscopic observation, it is believed that the quenching process allows forming amorphous phase, which inhibits the ice crystal growth and thus protects from damage the living cells or objective samples during freezing. However, the observations of molecular conditions in solution are very difficult due to the limitations of temporal and spatial resolutions. Uchida and coworkers have investigated the effects of cryoprotectant materials on the inhibition mechanism of ice growth. By using the freeze-fracture replica technique and a field-emission gun type transmission electron microscope (FEG-TEM), they investigated the nano-scale size molecular conditions in the disaccharide solutions (Uchida et al., 2007). To determine the materials observed in the replica images, they carried out the x-ray diffraction experiments on the similarly prepared disaccharide solution samples (Uchida & Takeya, 2010). They identified each phase observed in the TEM images, and they revealed the frozen state of both disaccharide molecules and water molecules. In addition, they discovered the formation of metastable phase of ice, ice Ic, and its anomalous stability at higher temperature conditions. In section 2, these morphological investigations of the quenched disaccharide solutions are reviewed. In order to investigate the structuring of solution prior to the freezing, which affects the growth process of ice crystal, the temperature and concentration dependences of viscosity of disaccharide solutions were required to be measured. Uchida et al. (2009) measured the viscosity of trehalose solutions by a dynamic light-scattering (DLS) spectroscopy (section 3). Since these investigations allowed us to discuss the inhibition process of disaccharide molecules for the ice crystal growth in the extracellular space, we discuss the role of disaccharide molecules serving as the cryoprotectant of living cells in the final section.

2. Microscopic observation of frozen disaccharide solutions

Uchida et al. (2007) aimed to observe the solute-solution interaction in the liquid phase using a FEG-TEM along with the freeze-fractured replica technique. This technique has usually been applied for biological investigations. They performed to show it to be very useful for investigating the microstructures and the dynamic features of solutes in the solution when a small droplet is quenched at liquid nitrogen temperature. However, the replica cannot determine the substances observed in TEM image. Thus they carried out the powder x-ray diffraction (PXRD) measurements for samples prepared under similar conditions (Uchida & Takeya, 2010).

They focused primarily on trehalose, which has been considered as one of the cryoprotective agents that reduces the freezing rate of ice (Sei & Gonda, 2004). Research-grade trehalose dihydrate crystals were donated by Hayashibara Biochemical Laboratories, Inc. To compare the properties observed in the various trehalose solutions, they conducted similar investigations using sucrose (of the highest quality; Sigma-Aldrich Japan, Inc.) and maltose (research grade, Hayashibara Biochemical Laboratories, Inc.). The purity of the distilled and deionized water used for the solutions was confirmed by the conductivity of approximately 5×10^{-6} S m^{-1}.

They prepared the trehalose solutions which mass concentrations are ranged from 3 wt% to 50 wt%. The concentration of the solutions was determined by measuring the masses of the disaccharides and pure water taking into consideration the hydrated water of crystallization. The accuracy of the concentration measurements was estimated ±0.1 wt% at room temperature. Then, these solutions (approximately 20 mm^3) were rapidly frozen in a liquid nitrogen atmosphere (approximately 80 K). To avoid the bubbling of the solution, the sample was kept just above the liquid nitrogen surface. Under these conditions, the freezing rate was approximately 3×10^2 K min^{-1}. The order of this value was almost similar (ranging within ±50 K min^{-1}) in all trehalose solutions, even though the solutions had a relatively wide concentration range. Then, the frozen sample was introduced into the freeze-fracture replica sample preparation system (JEOL, type JFD-9010). The sample was fractured under a vacuum (10^{-5} Torr) and at a low temperature (150–170 K). A replica film of this fractured surface was prepared by evaporating platinum and carbon on it. To observe the replica samples, we used FEG-TEM (JEOL, type JEM-2010) at an accelerating voltage of 200 kV. An imaging plate was used to record the observed images.

Figures 1a–1f show the TEM images of the freeze-fractured replica samples of trehalose solutions at various concentrations. These pictures showed three typical textures: (**S**) smooth surface, (**P**) fine particles (average diameter was 20-30 nm), and (**R**) remaining part of the boundary between the smooth-surface area adjacent to the fine particles (see Figure 1d). The area of **S** at lower concentrations was so large that their precise shapes and sizes could not be determined (see Figures 1a and b) from TEM images only. However, **S** area became sufficiently small at higher trehalose concentrations. As shown in Figures1c–1f, the **S** area appeared to have oval elongated shapes in a certain direction. Further increases in the trehalose concentration led to a reduction in the size of **S** area and exhibited the crystal facets. The intermediate space consisting of **P** area also becomes large simultaneously, thus the distance between each **S** area increased.

These phenomena indicated that the **S** area was considered to be the ice crystal, and the **P** area relates to the trehalose molecules. The decrease in size and the change in morphology of **S** area would be explained by both a reduction in the amount of H_2O molecules available to form ice crystals and a decrease in the growth rate of the ice crystals. The increase in the trehalose concentration may also affect the degree of supercooling of the solution, which would control the heat conduction formed in the ice crystal. Therefore, the growth rate was controlled either by the rate of supply of H_2O molecules from the surrounding solution to the ice surface or by the heat conduction from the reaction sites. If trehalose inhibited the ice crystal growth kinetically, we would find fine particles on the edge of a smooth area. However, the TEM images reveal that only a small amount of fine particles was located just along the boundary of the **S** area. Therefore, it appears that the main rate-controlling process is the rate of supply of H_2O molecules from the solution or the rate of the latent heat conduction.

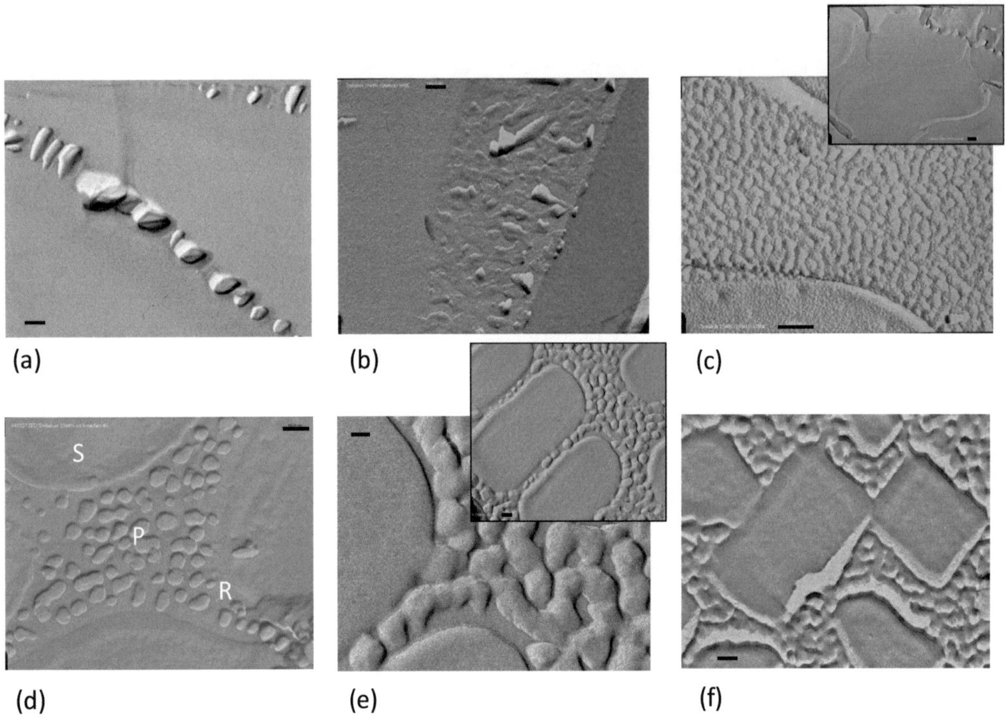

Fig. 1. (a) TEM image of an ice surface formed from trehalose solution (8.5 wt%). Scale bar: 100 nm. (b) from trehalose solution (15 wt%). Scale bar: 100 nm. (c) from trehalose solution (25 wt%). Scale bar: 100 nm. The inset was taken at a smaller magnification (scale bar: 500 nm). (d) from trehalose solution (35 wt%). Scale bar: 100 nm. (e) from trehalose solution (40 wt%). Scale bar: 100 nm. The inset picture was taken at a smaller magnification (scale bar: 200 nm). (f) from trehalose solution (50 wt%). Scale bar: 100 nm. (Reprinted from Uchida et al., 2007, with permission from Elsevier with slight arrangement.)

Figure 2 shows the distribution of ice crystals (**S** area) in terms of (a) size and (b) number concentration, depending on the trehalose concentration in the original solution. These results indicate that the size of each ice crystal is too large to be measured by TEM imaging at concentrations below 20 wt%, and very few ice crystals are present for TEM measurement. On the other hand, the size of the crystals is too small to be observed by an optical microscope at concentrations above 20 wt%. Then, TEM images revealed that the crystal size decreases exponentially to submicron orders of magnitude at 50 wt%, whereas the number density increases exponentially. Therefore, it appears that the growth rate of the ice crystals in the high-concentration trehalose solution is reduced either by a reduction in the phase-separation rate between free water and the condensed solution or by a large supply of nucleation sites for ice crystals.

Figure 3a and b show the variations in the average diameter (d_p) and the number density (N_p) of the fine particles (**P**), respectively, with respect to the trehalose concentration in the original solution. Figure 3a shows that the d_p value of each sample exhibited weak dependence on the trehalose concentration although there were several variations in terms of the size distributions. However, N_p values were found to increase exponentially with an increase in the trehalose concentration (see Figure 3b). Note here that the gradual saturation of N_p with larger trehalose concentrations is considered to primarily result from the existence of ice crystallites in the observed area. This is supported by the fact that N_p measured in the very rapid freezing method (open square) where no ice crystallites were observed is one order of magnitude larger than that measured in the quenched sample (solid circle) at the similar trehalose concentration (approximately 35 wt%). Therefore, it was observed that **P** was almost the same size and distributed in the intermediate space of **S** area. The area occupied by particles increased with the original trehalose concentration as follows from an increase in N_p, whereas the growth of ice crystals is inhibited. The total volume of fine particles (estimated as $N_p \times \pi d_p^3/6$) increases with the trehalose concentration. These results support the assumption that the object **P** was related to trehalose molecules.

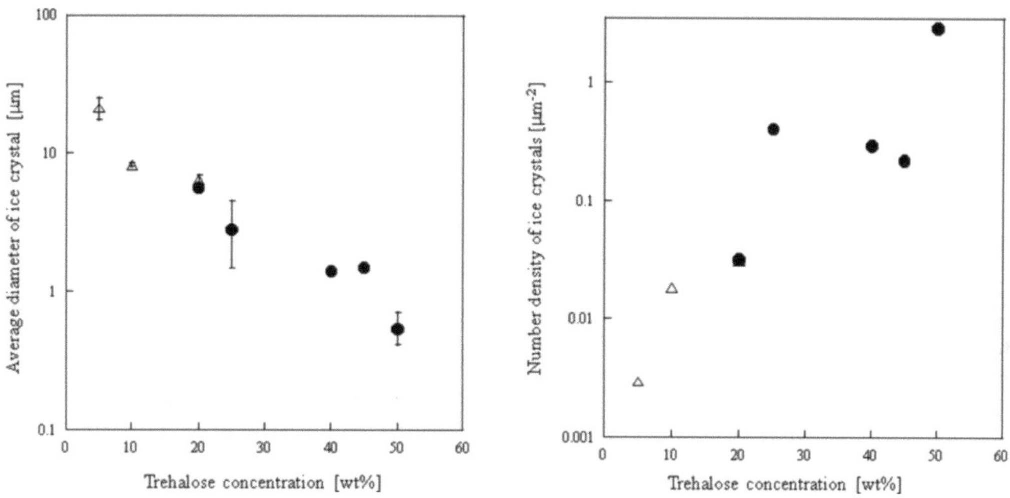

Fig. 2. (a) Dependence of the average diameter of ice crystals on trehalose concentration. Solid circles were observed via TEM and open triangles were via a polarized optical microscope (POM). (b) Dependence the number density of ice crystals on trehalose concentration. Solid circles were observed via TEM and open triangles were observed via POM. (Reprinted from Uchida et al., 2007, with permission from Elsevier.)

To determine each phase observed in the replica image, Uchida & Takeya (2010) carried out the PXRD method for the sample prepared similarity with the replica experiments. These frozen samples were finely powdered in a nitrogen atmosphere at a temperature below 100 K. The fine-powdered samples were top-loaded on a specimen holder made of Cu and mounted onto the powder x-ray diffractometer (Rigaku, type Ultima III). The PXRD measurements were done at 93 K in the $\theta/2\theta$ step scan mode with at a step 0.02 degree in the 2θ range from 6° to 43° using the CuKα radiation (40 kV, 40 mA).

The PXRD profiles for various concentrations of frozen trehalose solutions were shown in Figure 4. In the 23 wt% trehalose sample, typical diffraction pattern for hexagonal ice Ih (space group P_63/mmc) was observed. However, as the concentration of trehalose increased, background intensity in the 2θ range from 22° to 27° increased, which makes the line shapes broadening, and the intensity in these three peaks relatively decreased. Therefore it is considered that **S** area corresponds to the ice Ih crystallite. In addition, the relative peak intensity ratios of the ice Ih(110) and ice Ih(002) reflections changed since ice Ih(002) reflection overlaps with those of cubic ice Ic(111). These results suggest that the sample included some amount of ice Ic (space group *Fd3m*, Kuhs et al., 1987) in high concentration samples, and the volume ratio of ice Ic to Ih increased as the concentration of trehalose increased. This is the first evidence of ice Ic existence in the frozen sample of the disaccharide solutions. The ice Ic phase was observed in all the disaccharide solutions including both the sucrose and maltose solutions, especially at higher concentrations as well as in trehalose.

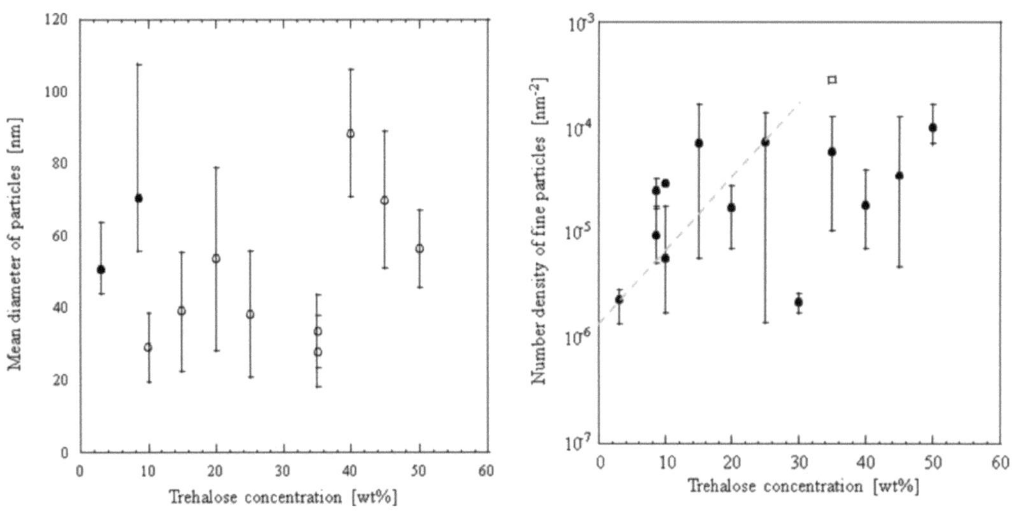

Fig. 3. (a) The average diameter of fine particles (d_p) on trehalose concentration. Open circles are estimated by fitting the log-normal distribution functions. The error bars show the standard deviations obtained by the curve fitting of each log-normal distribution. Solid circles are estimated by the arithmetic mean value because of the limited number of particles. The error bars for them denote the minimum and maximum diameter observed in the same sample. (b) The number density of fine particles (N_p) dependence on trehalose concentration. The open square denotes the average number density of the fine particles observed in the very rapid freezing method. The error bar shows the maximum variations of the measured N_p values in the same sample. Dashed line shows that N_p increases exponentially with the trehalose concentration. The gradual deviation from the dashed line at higher concentrations would be caused by the existence of ice crystallites in TEM images. (Reprinted from Uchida et al., 2007, with permission from Elsevier with slight arrangement.)

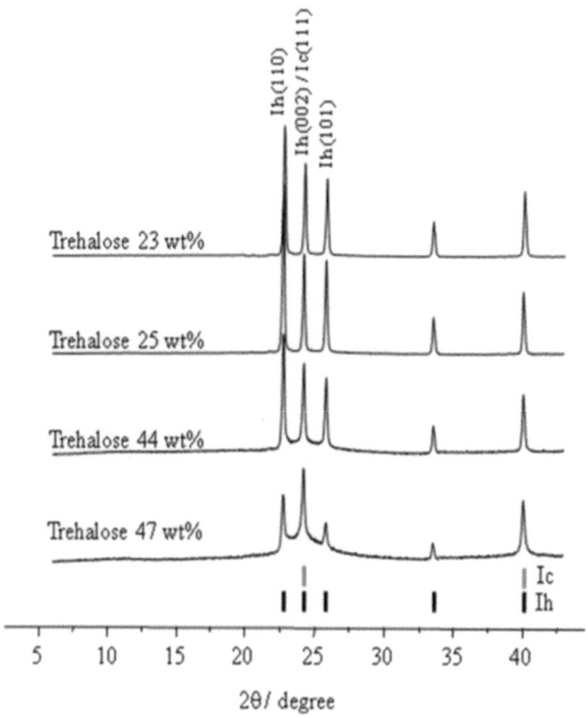

Fig. 4. PXRD patterns of rapidly frozen samples of various concentrations of trehalose solutions. The short bars indicate the positions of the diffraction patterns for both hexagonal ice Ih and cubic ice Ic. (Uchida & Takeya, 2010; reproduced by permission of the PCCP Owner Societies)

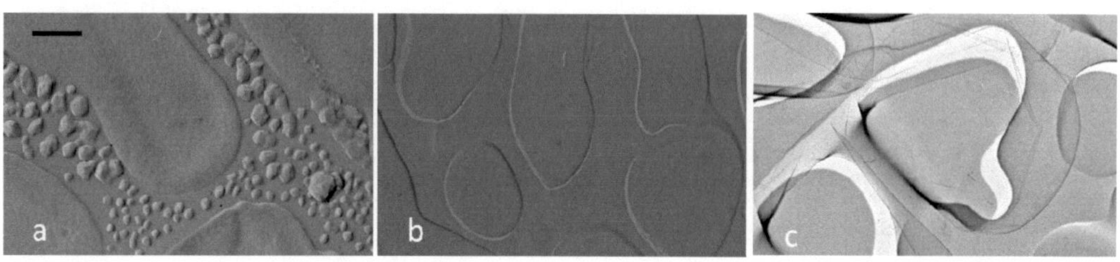

Fig. 5. TEM images of the ice surface formed from trehalose solution (35 wt%) fractured at temperatures (a) approximately 90 K, (b) approximately 180 K, and (c) approximately 250 K. Scale bar: 500 nm. (Reprinted from Uchida et al., 2007, with permission from Elsevier with slight arrangement.)

Uchida et al. (2007) investigated the temperature variation of **S** and **P** area distributions with the extent of sample annealing at higher temperatures. They prepared three frozen samples from the same 35 wt% trehalose solution. Then, sample (a) was fractured at a temperature of approximately 90 K to obtain the reference sample. Sample (b) was stored in a vacuum chamber and heated at a temperature of approximately 180 K. After approximately one hour the sample was fractured to form the replica. Sample (c) was heated to approximately 250 K, which is sufficiently higher than the vitrification temperature (T_g') of trehalose (approximately 238 K for 50 wt% trehalose solution; Branca et al., 2001). Figure 5 shows the comparison of the TEM images for these three samples taken at the same magnification. Figure 5a verified that the TEM image obtained for the sample (a) was the same as that shown in Figure 1d. However, in samples (b) and (c), they found that most of the fine particles disappeared, and spherical or elongated flat surface areas were observed (Figures 5b and 5c). The edge of the flat surface area in the sample (c) has a slightly larger contrast than that in sample (b).

To determine the material of each phase observed in Figure 5, and to investigate the stability of ice Ic, the temperature ramping PXRD measurements were carried out every 10 K from 93 K to 273 K for the frozen samples of 47wt% trehalose concentration solution quenched in liquid nitrogen (Uchida & Takeya, 2010). The initial profile of the 47 wt% trehalose solution at 93K (the farthest side profile of Figure 6a) was similar to those in Figure 4, which was closed up in the 2θ range from 22° to 27°. Three large peaks with a high background indicated that this sample included both ice Ih and ice Ic crystals. The PXRD profile did not change during the temperature ramping experiment up to 233 K, but the profile drastically changed at 243 K; the three large diffraction peaks became sharpened and the base line decreased down to the noise level. These results indicate that the ice Ic phase disappeared between 233 and 243 K whereas the ice Ih still remained even above 243 K. At 273 K, the whole ice Ih signals disappeared due to its melting. Instead, the trehalose dihydrate crystal was formed simultaneously.

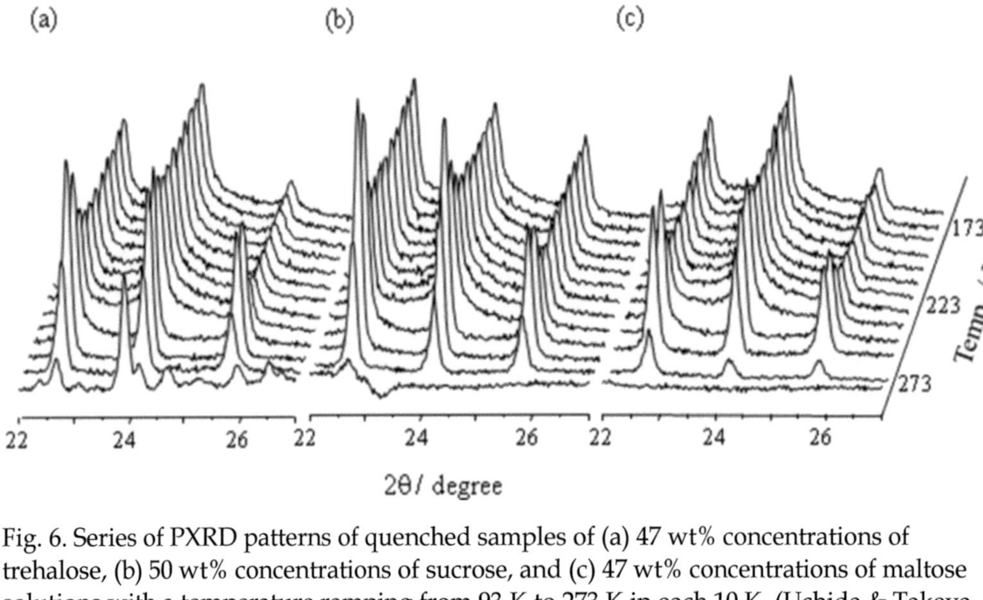

Fig. 6. Series of PXRD patterns of quenched samples of (a) 47 wt% concentrations of trehalose, (b) 50 wt% concentrations of sucrose, and (c) 47 wt% concentrations of maltose solutions with a temperature ramping from 93 K to 273 K in each 10 K. (Uchida & Takeya, 2010; reproduced by permission of the PCCP Owner Societies)

The temperature dependence of PXRD pattern was also observed in the 50 wt% sucrose (Figure 6b) and 47 wt% maltose (Figure 6c) solutions. The transition temperature from ice Ic to ice Ih in both samples was also observed between 233 and 243 K. These two samples showed, however, that no disaccharide crystal was formed at 273 K, that is, all solid phases were melted at the ice melting point. These results suggest that trehalose forms its dihydrate crystal at temperatures above the ice melting point, whereas sucrose and maltose do not form their crystals at the same temperature.

As a conclusion, three typical features observed on the replica via TEM were identified by PXRD measurements; the smooth surface **S** as ice Ih crystallite, fine particles **P** having relatively uniform sizes ranging from 20 to 30 nm in diameter as disaccharide precipitation, and remaining part **R** as glassy material including ice Ic. Dependence of disaccharide concentrations on both ice Ih crystallite size and number concentration of fine particles suggested that disaccharide molecules inhibit the ice crystal growth with increasing in their concentrations.

Figure 6 also shows that ice Ic formed in the quenched disaccharide solutions is stable under anomalously higher temperatures. In a pure water system, ice Ic can be obtained by the following several methods: (1) from the high-pressure phase of ice (ice III or VII) by a decrease of pressure during an increase in temperature (Hobbs, 1974), (2) from the temperature increase of amorphous ice condensed on a cold surface (below 110 K) (Dewell & Rinfret, 1960), or (3) the quenching of a water droplet at a rate of 10^4 K s^{-1} (Bruggeller & Mayer, 1980; Dubochet & McDowall, 1981; Mayer & Hallbrucker, 1987; Johari, 2005). The ice Ic phase is metastable and it transforms to the stable ice Ih phase when it is warmed up above approximately 143 K and the reverse transformation is never observed.

However, the crystallographic observations of the frozen samples of the glycerol solution (Vigier et al., 1987; Berejnov et al., 2006), glucose solution (Kajiwara et al., 2008; Thanatuksora et al., 2008), and ionic salt solutions (Murray et al., 2005) have revealed that ice Ic was also formed in these quenching samples. Vigier et al. (1987) estimated the average diameter of ice Ic in the frozen glycerol solutions to be 5 to 15 nm. The ice Ic phase formed by quenching of the aqueous solutions mentioned above transformed to ice Ih above 200 K, which was obviously higher than that observed in the pure water system. This process is considered to occur naturally, that is, small droplets of water in the upper atmosphere may often freeze directly into ice Ic (Kobayashi et al., 1976; Whalley, 1981; Whalley, 1983; Takahashi & Kobayashi, 1983; Whalley & McLaurin, 1984). Mayer & Hallbrucker (1987) demonstrated that ice Ic formed in the quenched aqueous aerosol droplets and that it was stable up to approximately 200 K. Other studies had revealed that the ice Ic forms in confined space (Takamuku et al., 1997; Steytler et al., 1983; Christenson, 2001; Morishige & Iwasaki, 2003; Murray & Bertram, 2006). Takamuku et al. (1997) reported that ice Ic was formed from water existed in a porous silica of 3 nm and 10 nm in diameter. It was found that the ice Ic formed in the confined space was also stable at temperatures higher than that in the pure water system.

Taking into account these findings, the formation of ice Ic in disaccharide solutions is expected to occur in a confined meso-pore space formed during the quenching process. In general, the full-width of the half maximum intensity of the x-ray diffraction peaks expands (Jenkins & Snyder, 1996) when the crystallite size is several hundred nanometres or less. However, the diffraction peaks of ice Ih in the present study were not broadened even in the highest concentration sample. This suggest that the crystallite size of ice Ih would be ranged from micron to sub-micron order even if small fraction of amorphous ice appears in the presence of trehalose under condition of rapid quenching. Therefore, it is considered that the formation sites of ice Ih and ice Ic were different, and the crystallite size of ice Ih would be ranged from micron to sub-micron order. This is consistent with the results obtained by the TEM image observations. Thus the ice Ic with less than a hundred nanometre in size was formed at the grain boundary between the ice Ih crystallites, and the disaccharide precipitated fine particles would provide the confined meso-pore space in the grain boundary.

The ice Ic would co-exist with the disaccharide precipitated particles, which may correspond to the giant disaccharide clusters (Gonda & Sei, 2005; Lerbret et al., 2005; Uchida et al., 2007; 2009). Moreover, the TEM images on the replica annealed at 150 K indicated that the fine particles disappeared whereas no profile changes in the PXRD were observed. If the ice Ic phase was formed on the fine particles, the ice Ic should be changed into the stable state at that moment. Thus we consider that the ice Ic would be formed between the disaccharide clusters by freezing of free water molecules, which locate around disaccharide molecules but are independent of the disaccharide clusters or glass phase.

The temperature ramping PXRD measurements on the rapid frozen sample of the highest concentration solutions indicated that the ice Ic phase was stable up to 233 K, which is much higher than that expected from the pure water system (approximately 143 K (Dewell & Rinfret, 1960)). This anomalous stability is similar to those reported in other solution systems (including glycerol (Vigier et al., 1987; Berejnov et al., 2006), glucose (Kajiwara et al., 2008; Thanatuksora et al., 2008), salts (Murray et al., 2005) or aqueous aerosol (Mayer & Hallbrucker, 1987)), or in the confined space systems (in porous silica (Takamuku et al., 1997) or in emulsion (Murray & Bertram, 2006)) as mentioned previously. Then the ice Ic phase is considered to be stabilized by foreign molecules. In the disaccharide solution system, the present study suggests that ice Ic exists between the disaccharide clusters. Then the water molecules constructing the ice Ic would slightly interact with disaccharide molecules even though they are not bound to the disaccharide molecules directly. The transition temperature from ice Ic to ice Ih was around 240 K which coincides well with the vitrification temperature (T_g') of the disaccharide solution; T_g' was approximately 238 K, 234 K and 233 K for the 50 wt% concentration of the trehalose, sucrose and maltose solutions, respectively (Branca et al., 2001). This is also the case for T_g'=230 K of glucose (Kajiwara et al., 2008; Thanatuksora et al., 2008) and for the eutectic point of 67 wt% glycerol at 230 K (Vigier et al., 1987; Berejnov et al., 2006). These coincidences allow us to consider that the transition of ice Ic to ice Ih is difficult to occur spontaneously in frozen solution systems because the foreign molecules might stabilize the system, but is induced by the high mobility of the foreign molecules. Therefore the transition of ice Ic to ice Ih observed in the present study may occur with the devitrification of the disaccharide molecules.

3. Inhibition process of ice crystal growth based on the viscosity measurement of disaccharide solutions via DLS measurement

To investigate the inhibition process of ice crystal growth by disaccharide molecules in the solution, we needed to know the dynamic properties of both water and disaccharide molecules. By comparison with the microscopic observations mentioned in the previous section, the macroscopic properties of the disaccharide solution, such as viscosity, would provide useful information. Uchida et al. (2009) measured the temperature and concentration dependence of viscosity of disaccharide solutions, especially for trehalose. They employed the dynamic light-scattering (DLS) spectroscopy method (Uchida et al., 2003) to measure the viscosity of aqueous solutions under wide-ranged conditions of temperature and concentration.

The brief explanations of the viscosity measurements are given here. Approximately 20 mm^3 of the solution was put in the sample cell for the DLS system. The temperature of the cell was controlled between 268 and 343 K with the accuracy of ±0.2 K by circulating silicone oil through a cooler. These temperatures are chosen to make the ratio between the glass-transition temperature T_g and the experimental temperature T, T_g/T, ranging from 0.4 to 0.6. The monochromatic light from an Ar-ion laser (20mW, λ=488 nm) was focused on the center of the sample cell. The light scattered by the uniform-sized latex particles in the solution was collected by an objective lens, and the time-dependent intensity of the mixing of scattered light through two pinholes was measured with the photomultiplier of the DLS spectrophotometer (Photal, DLS-7000). Since the Stokes-Einstein law was obeyed for the disaccharide solutions under the experimental conditions, the autocorrelation function $q_e(\tau)$ of the temporal fluctuations is computed by the photocurrent output and the translational diffusion coefficient of the solution D_τ from

$$q_e(\tau) = \exp(-D_\tau q^2 \tau) \quad (1)$$

where τ is the relative time and q is the scattering vector. The apparent particle diameter d, the viscosity η and the diffusion coefficient D_τ are related by the Stokes-Einstein law:

$$D_\tau = k_B T / A \eta d \quad (2)$$

where T is the absolute temperature, k_B is Boltzmann's constant, and A is a constant that depends on the type of friction between a material with d and its surrounding liquid (Corti et al., 2008). Substituting D_τ obtained by the DLS measurement and the latex particle size d = 204 nm, we obtained the viscosity of the solution around the small amount of light scattering materials (mass fraction of latex particles to be below 5×10^{-6}), which equals the absolute viscosity of the solution. Concerning the uncertainties of measurements and variation in the refractive indices, the uncertainty of viscosity measured by the DLS method was estimated to be ± 6%.

Figure 7 indicates η_{treha} (dots) with different trehalose concentrations C (wt%) measured at room temperature (300.3 ± 2.9 K). This figure indicates that η_{treha} increases with C. The data obtained by the DSL method was fitted by following equations (with a correlation of 0.99):

$$\eta_{treha} = 2.08\times10^{-4}C^3 - 8.68\times10^{-3}C^2 + 0.141C + 0.467 \quad (3)$$

which is shown in figure 7 as a dotted line. The cubic η-C relationship was proposed by Elias & Elias (1999). Recently, Longinotti & Corti (2008) summarized that the specific viscosity (η/η_w, where η_w is the viscosity of pure water at the same T (PROPATH, 1990)) related exponentially to molar concentration c (mol/kg). The obtained data were also checked by fitting the relation between η/η_w and c and the following fitting equation was obtained:

$$\ln (\eta_{treha}/\eta_w) = 0.900c - 0.0283 \quad (4)$$

with a correlation of 0.99 (data not shown). Equation (4) indicates the agreement of the data measured by the DLS method to the previous data. These results show that the relation between η and C is fitted by both models within the experimental conditions.

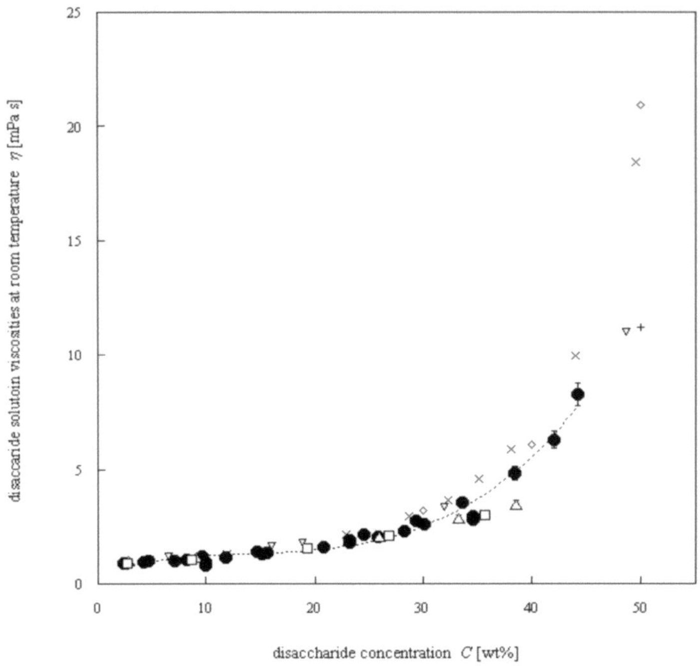

Fig. 7. Viscosities (η) of disaccharides vs. concentration (C) at room temperature. The dots indicate the relative viscosities of trehalose (η_{treha}), the open triangles indicate the relative viscosities of sucrose, and the open squares indicate the relative viscosities of maltose. The error bars for all the trehalose solution data indicate the maximum variations in η_{treha} observed in the DLS measurements, which are almost the same as the measurement uncertainties. The dotted line indicates the fitting curve for η_{treha} in Eq. (3). This figure also shows the η_{treha} near room temperature obtained by various conventional methods: open diamonds (Matsuoka et al., 2002), open inverted triangles (Magazù et al., 1999), ×-marks (Elias & Elias, 1999), and +-marks (Rampp et al., 2000). (Reprinted from Uchida et al., 2009, with permission from Elsevier.)

In addition, the data of $\eta_{\rm treha}$ were comparable with values in the literature for the bulk $\eta_{\rm treha}$ obtained by several conventional methods. In Figure 7, several reported viscosities measured using the conventional techniques are presented simultaneously. Most of the bulk viscosities of disaccharide solutions were measured by mechanical viscometers: the Ubbelohde viscometer (Magazù et al., 1999; Elieas & Elias, 1999; Branca et al., 2001), the Marion-Krieger Sisco capillary viscometer (Matsuoka et al., 2002), or the rotational viscometer (Nagasawa et al., 2003). The selected literature data plotted in Figure 7 were measured in a temperature range of 293.2-313.2 K.

Viscosities of other disaccharide solutions, sucrose ($\eta_{\rm suc}$) and maltose ($\eta_{\rm mal}$), were found to be similar to that of $\eta_{\rm treha}$ at a given concentration within the experimental uncertainties (shown in Figure 7). These three disaccharide solutions had very similar viscosity dependences of the viscosity-concentration relationships at room temperature, particularly in dilute solutions (less than 30 wt%).

On the other hand, the viscosity of a trehalose solution increases with decreasing temperature under all concentration conditions. Figure 8 depicts an Arrhenius plot of the viscosity of trehalose solutions (all solid marks, + and ×). As a reference, the data for pure water is denoted with a thick dashed line. This figure also indicates that $\eta_{\rm treha}$ follows an Arrhenius temperature dependence ($\eta_{\rm treha} = \eta_0 \exp[Q_{\rm treha}/k_B T]$, where η_0 is constant and $Q_{\rm treha}$ is the temperature-dependence parameter for trehalose solution) under the experimental conditions. The temperature dependence fits the Arrhenius curve even below the melting point for solutions with any concentration. Therefore, below the melting point, the viscosity of a supercooled trehalose solution can be predicted by extrapolating the temperature dependence of the viscosity in the liquid state, at least within the experimental validity (i.e, the temperature is not below $T_g/0.6$ (Corti et al., 2008)).

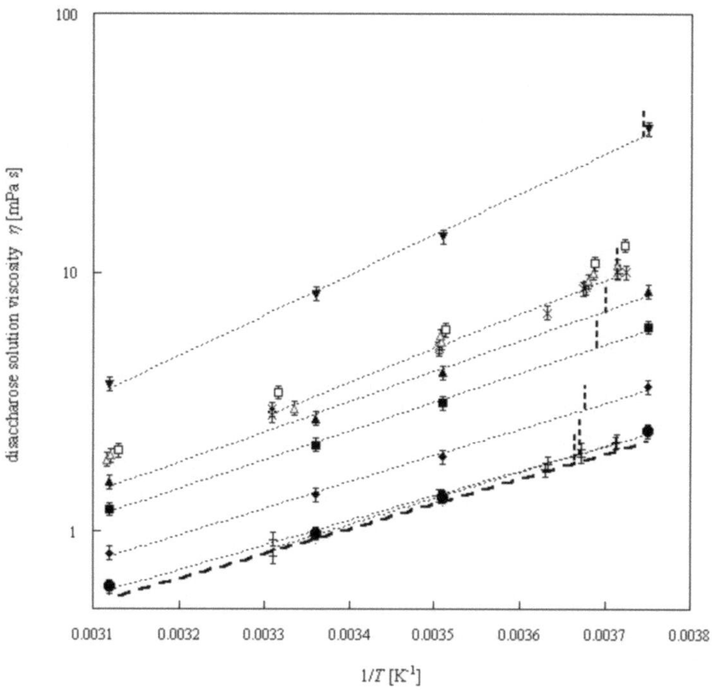

Fig. 8. Temperature dependence of viscosities for various concentrations of disaccharide solutions: η_{treha}, η_{suc}, and η_{mal}. Dots represent η_{treha} at a concentration of 4.75 wt%, crosses represent η_{treha} at a concentration of 9.93 wt%, solid diamonds represent η_{treha} at a concentration of 14.7 wt%, solid squares represent η_{treha} at a concentration of 24.5 wt%, solid triangles represent η_{treha} at a concentration of 29.4 wt%, ×-marks represent η_{treha} at a concentration of 34.7 wt%, and solid inverted-triangles, represent η_{treha} at a concentration of 44.3 wt%. The thin dotted line denotes the Arrhenius curve fit for each concentration. The thick dashed line indicates the temperature dependence of the viscosity of pure water, η_W (PROPATH, 1990). The thick short dotted vertical line on each curve is the melting point of the solution (Gonda & Sei, 2005). Open squares denote 38.5 wt% concentrations of sucrose solutions (η_{suc}), and triangles denote 35.8 wt% concentrations of maltose solutions (η_{mal}). (Reprinted from Uchida et al., 2009, with permission from Elsevier.)

The variation of the temperature-dependent parameter of trehalose solutions ($Q_{\rm treha}$) with the solute concentration C indicates interesting properties: $Q_{\rm treha}$ for dilute solutions with concentrations of less than 10 wt% is almost concentration-independent, which is the same as that of pure water ($Q_w = 17.3 \pm 1.4$ kJ mol^{-1}). This indicates that the dynamic properties of dilute solutions are similar to those of pure water. This interesting result is very important when considering the cryoprotective effect of trehalose used in biological and engineering applications because trehalose is usually applied in dilute conditions of less than 10 wt%.

However, Q_{treha} increases linearly with an increase in C above 10 wt%. This result indicates that trehalose solutions have two types of structures in the liquid phase: a dilute solution (C<10 wt%) and a condensed solution (C>10 wt%). Since the values of Q_{treha} in the condensed solutions increased linearly with C, the macroscopic properties of solutions were assumed to be continuously changing with trehalose concentration.

In the dilute trehalose solutions, the major matrix of the solutions is pure water (or free water) and the trehalose molecules exist as minor solute materials. Since the viscosity in these solutions is nearly the same as pure water viscosity, it is concluded that the trehalose molecules make only a small contribution to the macroscopic solution properties, or to the hydrogen-bonding network of the free-water molecules, even though the trehalose molecules tend to break the hydrogen-bond network of water (Branca et al., 1999a; Branca et al., 1999b; Branca et al., 1999c; Magazù et al., 2004; Nagasawa et al., 2003).

In the condensed trehalose solutions, most of free-water molecules are bound to trehalose molecules because of their high hydration abilities (the hydration number of trehalose is estimated to be 11 ± 4 (Lerbret et al., 2005)). The dynamic structure of the solution is then changed from a hydrogen-bond network of free water to a network structure related to hydrated-trehalose molecules with increasing in the trehalose concentration, which contributes to the viscosity of the solution. Thus, Uchida et al. (2009) employed the parameter W as the free-water ratio of the solution, W = (free water)/(total water), to consider the change in the network structure of the solution by the following equation:

$$W = (1 - C/C_s)/(1 - C/100) \quad (5)$$

where C_s is the saturation concentration of the trehalose solution. Based on the trehalose hydration number mentioned above, C_s is estimated to be 63 ± 7 wt%. Using this equation, the dilute condition corresponds to W exceeding approximately 93%.

As expected, Q_{treha} increases linearly with a decrease in W for W < 90% (Figure 9). Since Q_{treha} is an energy term, the increase in Q_{treha} for the condensed solutions is a reflection of the higher energy barriers in the translational motion of bound molecules, or an increase in the activation energy of the self-diffusion coefficients of both water and trehalose molecules. The self-diffusion coefficients of either solute or solution molecules in disaccharide solutions were previously measured by NMR spectroscopy (Ekdawi-Sever et al., 2003; Rampp et al., 2000). From these reported values, the activation energies for both molecules were found to increase gradually with a decrease in W. Therefore, the energy barrier for the translational motion of the matrix is caused mainly by the lack of free water.

A solute–solution model of low-W (or high-C) solutions has been proposed by molecular dynamics (MD) simulation (Lerbret et al., 2005). The simulation study suggested that the hydrated clusters of disaccharide molecules began to interact, forming compartment-like hydrogen-bonding networks. These clusters were consistent with the giant hydrated-saccharide clusters (Gonda & Sei, 2005; Lerbret et al., 2005), which also corresponded to the fine particles found in TEM images (Uchida et al., 2007). Therefore, a network structure with a high-energy barrier for the translational motion of the matrix was assumed to be built with these giant hydrated disaccharide clusters. Since Q_{treha} changes continuously, the phase change of the network structure from a free-water matrix to a disaccharide cluster matrix would progress partially in the solution.

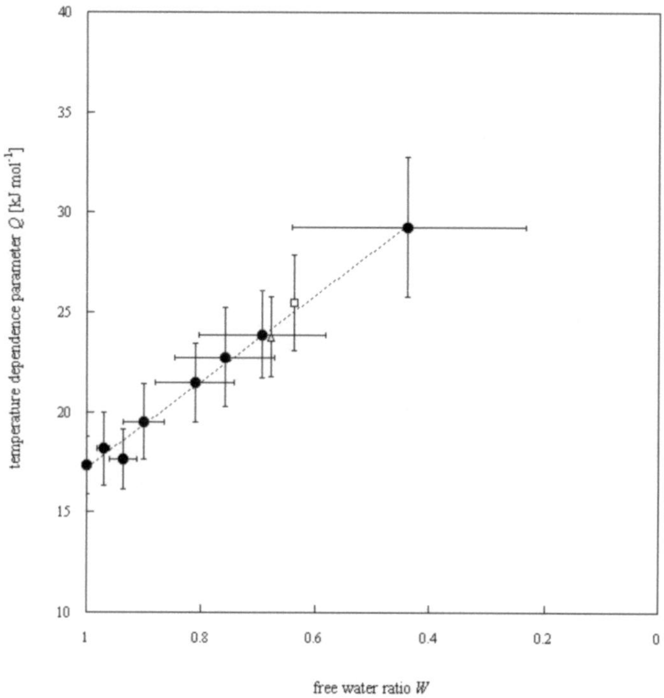

Fig. 9. Temperature dependence parameter of trehalose solutions (Q_{treha}) (dots) vs. the free-water ratio (W) estimated by Eq. (5). The Q_{suc} estimated from the experimental data presented in Fig. 8 is denoted by open squares, and Q_{mal} estimated from the experimental data presented in Fig. 8 is denoted by open triangles. Error bars for the W estimation were caused by the uncertainty of the hydration number (Lerbret et al., 2005), and those for Q were from the standard deviations of the Q estimate in Fig. 8. (Reprinted from Uchida et al., 2009, with permission from Elsevier.)

4. Frozen mechanism of trehalose solution and its cryopreservation effect for living cells

Above the melting point, the volume ratio of free water W decreases with an increase in C. The remaining region is the trehalose-binding (or hydrated) water which forms trehalose-rich clusters and the clusters tend to bind each other to construct the macroscopic cluster-binding network in high C solutions. We assumed that this network would correspond to a compartment-like structure in the solution (Lerbret et al., 2005), which gradually develops with an increase in C. Since this network develops all over the solution, its bulk viscosity increases.

When a small amount of solution was quenched at the liquid nitrogen temperature, the disaccharide molecules forming the giant clusters would construct the fine particles observed in the TEM images. On the other hand, the free water molecules free from the disaccharide molecules form ice Ih crystallites. The crystallite size of the ice Ih is larger than the sub-micron order. Then ice Ih crystallite is determined as the smooth surface area in the TEM images. The intermediate region of these phases is constructed by water molecules which are located in the compartment-like structure comprised by the giant clusters. These water molecules form the metastable ice Ic crystal because the compartment-like structure is considered to be the meso-pore space stabilizing ice Ic (Fig. 10). When the disaccharide concentration is high, a large amount of the compartment-like structures would be formed between the giant clusters. Then the total amount of ice Ic was larger in higher disaccharide concentration samples. In addition, the compartment-like structure would divide the space for free water, a number of nucleation sites of ice Ih also increases although the volume of free-water used for the ice Ih formation decreases. Thus the crystal growth of ice Ih is inhibited in the higher-concentration disaccharide solutions. Since these ice Ic crystals were anomalously stable up to 233 K, the compartment-like structure would be maintained until the disaccharide molecules change their molecular network at their vitrification temperatures.

On the other hand, when the trehalose solution is frozen slowly, the dihydrate trehalose crystal is formed prior to ice formation. At that moment, approximately 80% of water molecules bounded to a trehalose molecule is released. Thus the formation of dihydrated trehalose crystal increases the ratio of free water *W* and reduces the cluster-binding network in the solution. Therefore, the slow freezing process would lose the effect of cryopreservation effect of trehalose.

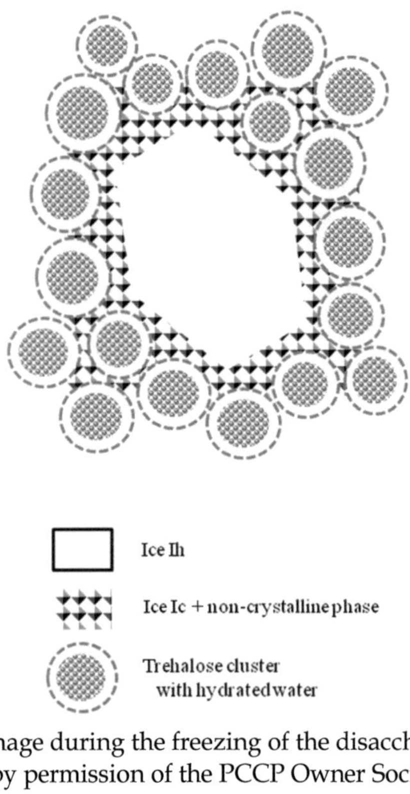

Fig. 10. Phase separation image during the freezing of the disaccharide solutions. (Uchida & Takeya, 2010; reproduced by permission of the PCCP Owner Societies)

Since the size of disaccharide molecules are too large to permeate the biomembrane, they cannot contribute to the inhibition of the intracellular freezing by reducing the melting point due to the molar freezing point depression effect. Thus the cryopreservation mechanism of disaccharide is different from the cell-permeable cryoprotectants, such as glycerol or of DMSO. Some cryopreservation tests for the primary cells having weak tolerance for cryopreservations demonstrate the difference between these cyroprotectants. Motomura et al. (2007) reported the provability of cryopreservation of the primary cells of neonatal rat cortical neurons with 10% DMSO but not with trehalose. Later, we carried out further experiments and verified the viability of the neurons related to the concentrations of DMSO (solid circles) and trehalose (open squares) at wider range (Fig. 11). On the other hand, Miyamura et al. (2010) reported the viability of cryopreserved neonatal rat cardiac myocytes depending on the DMSO concentration. We expanded their researches to the cryopreservation with trehalose. Figure 12 shows the viability of cryopreserved cardiac myocytes depending on the DMSO (solid circles) and trehalose (open squares) concentration. Figures 11 and 12 indicate that both neurons and cardiac myocytes can be cryopreserved with DMSO, which has the optimum concentration at approximately 10%. However, both primary cells cannot be cryopreserved by trehalose at any concentrations under their experimental conditions. Since these cryopreservation tests were carried out by relatively slow freezing process (approximately -10 K min^{-1} in the cooling rate), it is considered that trehalose cannot act as the effective cryoprotectant at those cooling rates. This is consistent with the prediction of the slow freezing process of disaccharide solutions mentioned previously. Concerning the fact that the cryoprotectant DMSO loses its cryoprotective effect when the neuron was quenched in liquid nitrogen (Uchida et al., 2007), it is considered that cryoprotectants have optimized cooling rates in addition to concentrations which relate to the frozen mechanism of the solution.

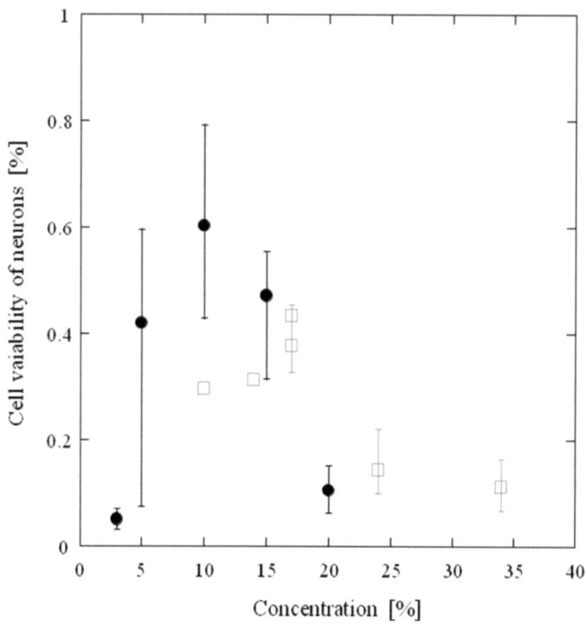

Fig. 11. Cell viability of rat cortical neurons cryopreserved by DMSO (solid circles) and trehalose (open squares) at various concentrations.

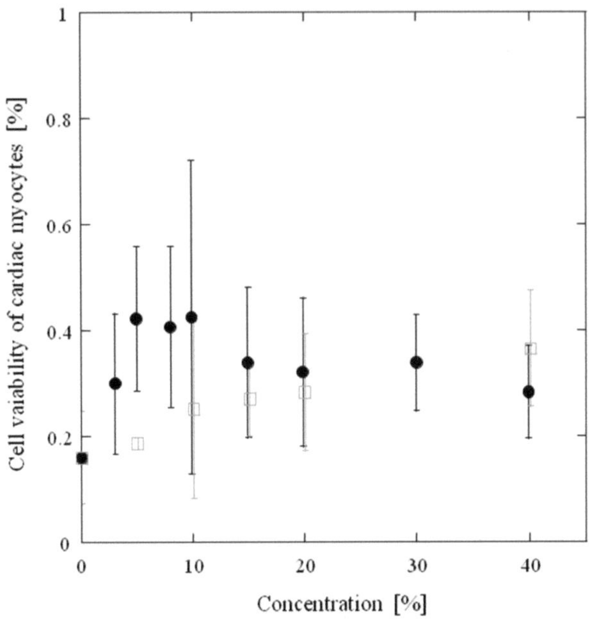

Fig. 12. Cell viability of rat cardiac myocytes cryopreserved by DMSO (solid circles) and trehalose (open squares) at various concentrations. (DMSO data are modified after Miyamura et al., 2010)

Therefore, we conclude that the cell-permeable cryoprotectants act well in the slow freezing process whereas the cell-impermeable cryoprotectants are feasible in the quench process. Since DMSO, as cell-permeable cryoprotectants, reduces the melting point of cytosol to protect the intracellular freezing, they require the habituation time to penetrate into cytosol. On the other hand, trehalose, as a cell-impermeable cryoprotectant, cannot act well in the slow freezing process since it crystallizes at temperatures above the melting point. Thus the optimization of trehalose would be revealed in the quenching process. At that condition, the formation of ice Ic would be a useful marker showing the growth inhibition of ice Ih crystallites.

5. Conclusion

Disaccharides are expected to be the natural cryoprotective agents for living bodies and cells. However, since they are too large to permeate the biomembrane, the cryopreservation mechanism has not been revealed. This review shows several experimental approaches to investigation of the freezing process of disaccharide solutions, especially trehalose solution.
The combination of the transmission electron microscopic observations of freeze-fractured replica films with the powder x-ray diffraction (PXRD) measurements of the quenched trehalose solutions revealed that the crystal growth of ice Ih crystallites in the quenched sample is inhibited by the formation of the giant hydrated-trehalose clusters which construct the compartment-like structures binding with each other. This is confirmed by the formation and anomalous stability of ice Ic. The stability of ice Ic is extended as high as approximately 243 K because the meso-pore space stabilizing ice Ic is maintained until the trehalose molecules change their hydrogen-bonded network at the vitrification temperature. This

condensed phase is confirmed to exist in liquid phase above the melting point by measuring the viscosity of the bulk liquid with dynamic light scattering method and by PXRD measurements with temperature ramping. However, when the trehalose solution is cooled slowly, the dihydrate trehalose crystal is formed prior to the ice formation. Thus trehalose would lose the cryopreservation effect in the slow freezing method.
Therefore, the disaccharide molecules are considered to act as the cell-impermeable cryoprotectants when the cell-dispersed solution is rapidly frozen. These predictions are qualitatively verified by several cryopreservation tests for the primary cells having weak tolerance for cryopreservation, such as the rat cortical neurons and rat cardiac myocytes.

6. Acknowledgment

Trehalose and maltose were donated by Hayashibara Biochemical Labs., Inc. The freeze-fracture replica observations by TEM were technically supported by Mr. A. Okutomi (JEOL), Dr. Shibayama, and Dr. N. Sakaguchi (Hokkaido Univ.). The DLS measurements were technically supported by Dr. H. Kaga (AIST). The cryopreservation experiments were carried out by Mr. J. Motomura and Mr. K. Miyamura (Hokkaido Univ.). Authors acknowledge for the reuse permission of the published artworks to Elsevier and the Royal Society of Chemistry. Part of this work was supported financially by the Grant-in-Aid for Scientific Research from the Japan Society for the Promotion of Science.

7. References

Akao, K., Okubo, Y., Masago, H., Asakawa, N., Inoue, Y., Sakurai, M. (2001). FTIR study of the properties of anhydrous form II of trehalose (in Japanese with English abstract). *Cryobio. Cryotech.*, Vol. 47, pp. 23–26

Berejnov, V., Husseini, N. S., Alsaied, O. A., Thorne, R. E. (2006). Effects of cryoprotectant concentration and cooling rate on vitrification of aqueous solutions. *J. Appl. Cryst.*, Vol. 39, pp. 244-251

Branca, C., Magazù, S., Maisano, G., Migliardo, P. (1999a). Anomalous cryoprotective effectiveness of trehalose: Raman scattering evidences. *J. Chem. Phys.*, Vol. 111, pp. 281–287

Branca, C., Magazù, S., Maisano, G., Migliardo, P., Villari, V., Sokolov, A. P. (1999b). The fragile character and structure-breaker role of α,α−trehalose: viscosity and Raman scattering findings. *J. Phys.: Condense. Matter*, Vol. 11, pp. 3823-3832

Branca, C., Magazù, S., Maisano, G., Migliardo, P. (1999c). α,α−Trehalose -water solutions. 3. Vibrational dynamics studies by inelastic light scattering. *J. Phys. Chem. B*, Vol. 103, pp. 1347-1353

Branca, C., Magazù, S., Maisano, G., Migliardo, P., Migliardo, F., Romeo, G. (2001). α,α−Trehalose/water solutions. 5. Hydration and viscosity in dilute and semidilute disaccharide solutions. *J. Phys. Chem. B*, Vol. 105, pp. 10140-10145

Bruggeller, P., Mayer, E. (1980). Complete vitrification in pure liquid water and dilute aqueous-solutions. *Nature*, Vol. 288, pp. 569-571

Christenson, H. K. (2001). Confinement effects on freezing and melting. *J. Phys.: Condens. Matter*, Vol. 13, No. 11, R95-R133

Corti, H. R., Frank, G. A., Marconi, M. C. (2008). Diffusion-viscosity decoupling in supercooled aqueous trehalose solutions. *J. Phys. Chem. B*, Vol. 112, pp. 12899-12906

Crowe, J. H., Crowe, L. M., Carpenter, J. F., Rudolph, A. S., Wistrom, C. A., Spargo, B. J., Anchrodoguy, T. J. (1988). Interactions of sugars with membranes. *Biochim. Biophys. Acta.*, Vol. 947, pp. 367–384

Dewell, L. G., Rinfret, A. P. (1960). Low-temperature forms of ice as studied by x-ray diffraction. *Nature*, Vol. 188, pp. 1144-1148

Dubochet, J., McDowall, A. W. (1981). Vitrification of pure water for electron-microscopy. *J. Microsc.*, Vol. 124, RP3-RP4

Ekdawi-Sever, N., de Pablo, J. J., Feick, E., von Meerwall, E. (2003). Diffusion of sucrose and α,α–Trehalose in aqueous solutions. *J. Phys. Chem. A.*, Vol. 107, pp. 936-943

Elieas, M. E., Elias, A. M. (1999). Trehalose+water fragile system: properties and glass transition. *J. Molecular Liquids*, Vol. 83, pp. 303-310

Gonda, T., Sei, K. (2005). The inhibitory growth mechanism of saccharides on the growth of ice crystals from aqueous solutions. *Progress in Crystal Growth and Characterization of Materials*, Vol. 51, pp. 70-80

Hobbs, P. V. (1974). *Ice Physics*, Clarendon. Oxford, Oxford

Jenkins, R., Snyder, R. L. (1996). Introduction to X-ray powder diffractometry. *Chemical analysis* (ed. Winefordner J. D.), John Wiley & Sons, Inc., New York, pp. 89-91

Johari, G. P. (2005). Water's size-dependent freezing to cubic ice. *J. Chem. Phys.*, Vol. 122, pp. 194504-1-5

Kajiwara, K., Thanatuksom, P., Murase, N., Franks, F. (2008). Cubic ice can be formed directly in the water phase of vitrified aqueous solutions. *CryoLett.*, Vol. 29, No. 1, 29-34

Kanno, H., Yamazaki, Y. (2001). Raman study of aqueous solutions of disaccharides including trehalose (in Japanese with English abstract). *Cryobio. Cryotech.*, Vol. 47, pp. 76–79

Kobayashi, T., Furukawa, Y., Takahashi, T., Uyeda, H. (1976). Cubic structure models at the junctions inpolycrystalline snow crystals. *J. Cryst. Growth*, Vol. 35, pp. 262-268

Kuhs, W. F., Bliss, D. V., Finney, J. L. (1987). High-resolution neutron powder diffraction study of ice Ic. *J. Physique*, Vol. 48, No. C-1, 631-636

Lerbret, A., Bordat, P., Affouard, F., Descamps, M., Migliardo, F. (2005). How homogeneous are the trehalose, maltose, and sucrose water solutions? An insight from molecular dynamics simulations. *J. Phys. Chem. B*, Vol. 109, pp. 11046-11057

Longinotti, M. P., Corti, H. R. (2008). Viscosity of concentrated sucrose and trehalose aqueous solutions including the supercooled regime. *J. Phys. Chem. Ref. Data*, Vol. 37, No. 3, pp. 1503-1515

Magazù, S., Maisano, G., Migliardo, P., Tettamanti, E., Villari, V. (1999). Transport phenomena and anomalous glass-forming behaviour in α,α−trehalose aqueous solutions. *Molecular Physics*, Vol. 96, No. 3, pp. 381-387

Magazù, S., Maisano, G., Migliardo, P., Mondelli, C. (2004). α,α–trehalose /water solutions. VII: an elastic inhoherent neutron scattering study on fragility. *J. Phys. Chem. B*, Vol. 108, pp. 13580-13585

Matsuoka, T., Okada, T., Murai, K., Koda, S., Nomura, H. (2002). Dynamics and hydration of trehalose and maltose in concentrated solutions. *J. Molecular Liquids*, Vol. 98–99, pp. 317-327

Mayer, E., Hallbrucker, A. (1987). Cubic ice from liquid water. *Nature*, Vol. 325, pp. 601-602

Miyamura, K., Nagayama, M., Gohara, K., Taira, T., Shimizu, K., Sakai, M., Uchida, T. (2010). Evaluation of viability of cryopreserved rat cardiac myocytes and effects of dimethyl sulfoxide concentration on cryopreservation (in Japanese with English abstract). *Cryobio. Cryotech.*, Vol. 56, No. 2, pp. 111–117

Morishige, K., Iwasaki, H. (2003). X-ray study of freezing and melting of water confined within SBA-15. *Langmuir*, Vol. 19, No. 7, pp. 2808-2811

Motomura, J., Uchida, T., Nagayama, M., Gohara, K., Taira, T., Shimizu, K., Sakai, M. (2007). Effects of additives and cooling rates on cryo-preservation process of rat cortical cells. *Physics and Chemistry of Ice* (Ed. Kuhs, W. F.), Royal Society of Chemistry, London, pp. 409-416

Murray, B. J., Knopf, D. A., Bertram, A. K. (2005). The formation of cubic ice under conditions relevant to Earth's atmosphere. *Nature*, Vol. 434, pp. 202-205

Murray, B. J., Bertram, A. K. (2006). Formation and stability of cubic ice in water droplets. *Phys. Chem. Chem. Phys.*, Vol. 8, pp. 186-192

Nagasawa, Y., Nakagawa, Y., Kenmochi, J., Okada, T. (2003). Microscopic viscosity of aqueous solution of saccharides: A study by ultrafast pump-probe spectroscopy. *Cryobio. Cryotech.*, Vo. 49, No. 2, pp. 87-95

PROPATH - a Program Package for Thermophysical Properties of Fluids, Ver. 7.1. (1990). Corona Publishing, Tokyo

Rampp, M., Buttersack, C., Lüdemann, H. -D. (2000). c, T-dependence of the viscosity and the self-diffusion coefficients in some aqueous carbohydrate solutions. *Carbohyd. Res.*, Vol. 328, pp. 561-572

Sei, K., Gonda, T., Arima, Y. (2001). Freezing of the solution of trehalose and water (in Japanese with English abstract). *Cryobio. Cryotech.*, Vol. 47, pp. 9–12

Sei, K., Gonda, T., Arima, Y. (2002). Growth rate and morphology of ice crystals growing in a solution of trehalose and water. *J. Crystal Growth*, Vol. 240, pp. 218–229

Sei, K., Gonda, T. (2004). Melting points of ice crystals growing in sugar solutions (in Japanese with English abstract). *Cryobio. Cryotech.*, Vol. 50, pp. 93–95

Steytler, D. C., Dore, J. C., Wright, C. J. (1983). Neutron-diffraction study of cubic ice nucleation in a porous silica network. *J. Phys. Chem.*, Vol. 87, No. 14, pp. 2458-2459

Sum, A. K., Faller, R., de Pablo, J. J. (2003). Molecular simulation study of phospholipids bilayers and insights of the interactions with disaccharides. *Biophys. J.* Vol. 85, pp. 2830–2844

Takahashi, T., Kobayashi, T. (1983). The role of the cubic structure in freezing of a supercooled water droplet on an ice substrate. *J. Cryst. Growth*, Vol. 64, pp. 593-603

Takamuku, T., Yamagami, M., Wakita, H., Masuda, Y., Yamaguchi, T. (1997). Thermal property, structure, and dynamics of superlooled water in porous silica by calorimetry, neutron diffraction, and NMR relaxation. *J. Phys. Chem. B*, Vol. 101, pp. 5730-5739

Thanatuksora, P., Kajiwara, K., Murase, N., Franks, F. (2008). Freeze-thaw behaviour of aqueous glucose solutions – the crystallisation of cubic ice. *Phys. Chem. Chem. Phys.*, Vol. 10, pp. 5452-5458

Uchida, T., Ohmura, R., Nagao, J., Takeya, S., Ebinuma, T., Narita, H. (2003). Viscosity of aqueous CO_2 solutions measured by dynamic light scattering. *J. Chem. Eng. Data*, Vol. 48, No. 5, 1225-1229

Uchida, T., Nagayama, M., Shibayama, T., Gohara, K. (2007). Morphological investigations of disaccharide molecules for growth inhibition of ice crystals. *J. Crystal Growth,* Vol. 299, No. 1, pp. 125-135

Uchida, T., Nagayama, M., Gohara, K. (2009). Inhibition process of ice crystal growth in trehalose solutions: Interpretation by viscosity measurements using dynamic light scattering method. *J. Crystal Growth,* Vol. 311, No. 23-24, pp. 4747-4752

Uchida, T., Takeya, S. (2010). Powder X-ray diffraction observations of ice crystals formed from disaccharide solutions. *Phys. Chem. Chem. Phys.,* Vol. 12, No. 45, pp. 15034-15039

Vigier, G., Thollet, G., Vassoille, R. (1987). Cubic and hexagonal ice formation in water-glycerol mixture (50% w/w). *J. Cryst. Growth,* Vol. 84, pp. 309-315

Wang, Y., Tominaga, Y. (1994). Dynamical structure of water in aqueous solutions of D-glucose and D-galactose by low-frequency Raman scattering. *J. Chem. Phys.* Vol. 100, pp. 2407–2412

Whalley, E. (1981). Scheiners halo - evidence for ice Ic in the atmosphere. *Science,* Vol. 211, 389-390

Whalley, E. (1983). Cubic ice in nature. *J. Phys. Chem.,* Vol. 87, No. 21, pp. 4174-4179

Whalley, E., McLaurin, G. E. (1984). Reraction halos in the solar-system 1. Halos from cubic-crystals that may occur in atmospheres in the solar-system. *J. Opt Soc. Am.,* Vol. A1, No. 12, pp. 1166-1170

10

Crystal Growth of Pharmaceuticals from Melt

J.S. Redinha[1] and A.J. Lopes Jesus[2]
[1,2]University of Coimbra / Department of Chemistry,
[2]University of Coimbra / Faculty of Pharmacy,
Portugal

1. Introduction

The preparation of crystalline powders and single crystals is an activity of utmost importance in many fields of modern industry. Single crystals are the basis of many daily necessities such as electric heaters, strain gauges, laser controllers and piezoelectric devices (Ropp 2003). In the pharmaceutical field, the search for compounds that have suitable properties to be used in drug formulation is nowadays a big challenge. In an increasingly demanding industry, science is asked to prepare crystalline forms of well defined structural patterns, with the size and shape that is required for a certain purpose. Despite recent advances in molecular modelling, it is not yet possible to accurately predict the molecular structures from our knowledge of the chemical compositions (Gdanitz 1997), which means that the study of crystal growth is still done on an experiment basis.

Crystallization is a conventional technique used to prepare solid forms. It can be carried out from solutions by decreasing temperature, solvent evaporation or by the diffusion of a much poorer solvent in solution. Alternatively, the crystalline forms can be obtained from the melt upon cooling. Unlike crystallization from solution, this does not require recovery, storage and disposal of liquids, what this means reduced environmental impact, low cost and time saving industrial processes. It therefore accomplishes the desired "green chemistry requirements". Since the pioneering work carried out in the early 1960's (Sekiguchi and Obi 1961), melt-extrusion has become an important drug delivery technology (Mollan 2003), and is currently being applied in the manufacture of a variety of dosage forms and formulations (Chokshi and Zia 2004, Breitenbach 2002, Crowley, et al. 2007, Repka, et al. 2007, Repka, et al. 2008).

Different crystals of a certain organic compound may be obtained by crystallization. These forms, called polymorphs, differ from one another by the packing and/or conformation of the molecules in a crystal lattice (Hilfiker 2006, Bernstein 2002, Threlfall 1995, Brittain 2009). Thus, polymorphs are solid forms constituted by molecules indistinguishable in gas or liquid state but exhibiting different structures in solid phase. When this difference lies just in the conformation, they are designated as conformational polymorphs (Bernstein 2002). Since the discovery that different polymorphs may have different bioavailabilities as drugs (Aguiar, et al. 1967), polymorphism has been a matter deserving great attention by pharmaceutical technology over the last half-century (Brittain 2009, Hilfiker 2006). For a given compound, only the one having the lowest Gibbs energy form is thermodynamically stable. However, in some cases, as consequence of the height of the energy barrier separating some polymorphs and the most stable one, they remain as metastable forms for a period of time long enough to be used for practical purposes as if they were stable forms (Brittain 2009).

This chapter is focused on the study of phase transitions occurring during the melt-cooling of active pharmaceutical ingredients or excipients and in the establishment of experimental conditions to obtain different kinds of solid forms. Another aim is to obtain information on the relationship between molecular and crystal structures. The compounds selected for this study were erythritol (*meso*-1,2,3,4-tetrahydroxybutane), pindolol [1-(1H-indol-4-yloxy)-3-((1-methylethyl)amino)-2-propanol], atelolol [4-(2-hydroxy-3-((1-methylethyl) amino)propoxy) -benzeneacetamide] and betaxolol hydrochloride [1-(4-(2-cyclopropylmethoxy) ethyl) phenoxy)-3-((1-methylethyl)amino)-2-propanol hydrochloride], whose structures are shown in Figure 1. Melt-growing has been frequently applied to prepare polymers (E.Donth 1992); however, very little attention has been given to compounds of smaller or medium molecular weight such as those treated in this chapter.

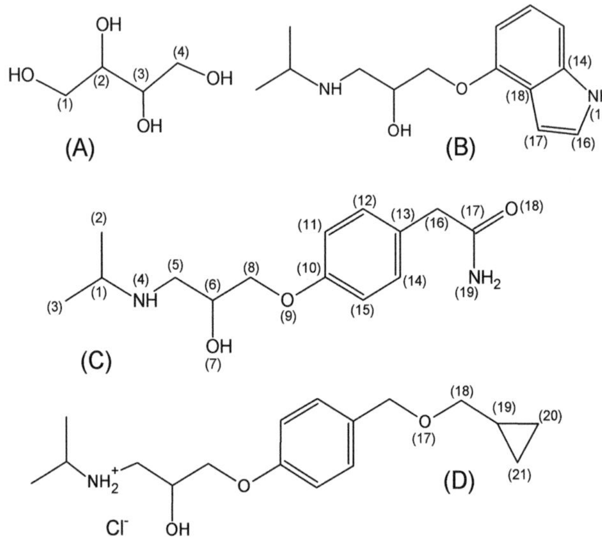

Fig. 1. Molecular structures and atom numbering of the compounds: (A) erythritol, (B) pindolol, (C) atenolol and (D) betaxolol hydrochloride. Only carbon, oxygen and nitrogen atoms are labelled. The 2-isopropylaminoethanol moiety in betaxolol and pindolol is numbered as in atenolol. In erythritol the oxygen atoms are numbered according to the carbon atom at which they are bonded.

Erythritol is widely used as a low caloric sugar. In addition, due to its sweet taste, low toxicity (Munro, et al. 1998), and high compatibility with drugs (Zhou, et al. 2000), it is an ideal excipient to be used in pharmaceutical formulations (Endo, et al. 2005, Gonnissen, et al. 2007, Ohmori, et al. 2004). The remaining compounds are used in medical practise in the treatment of hypertension and cardiovascular diseases. They belong to a group of medicines known as beta-blockers or beta-adrenergic compounds. Since betaxolol has a low solubility in water, the hydrochloride form is the most used in practise. A common feature of beta-adrenergic is the presence of the 2-isopropylaminoethanol group which enables the drug to be recognized by the characteristic centres of the biological receptors (Coulson 1994, Silverman 2004). A second moiety, varying from one compound to the other, gives some specificity to the drug. This set of compounds was selected to study the effect of various molecular features, particularly the molecular size and flexibility on the type of structures formed by crystallization.

The best quality compounds commercially available were used for this study. Attending to the effect that trace impurities may have in the crystallization process, purity of the compounds deserved great attention. Even though the compounds were labelled or certified as high purity substances, this was checked by HPLC or GC methods. A purity degree higher than 99% was found for all the original compounds. The instrumental methods used to produce the results necessary for this research program were differential scanning calorimetry (DSC), polarized light thermal microscopy (PLTM), infrared spectroscopy (IR) and X-ray or neutron diffraction.

2. Outlines of solid formation in super-cooled liquids

Before proceeding with the analysis of the results obtained for the systems under consideration, it may be useful to present an overview concerning the theoretical aspects of the formation of a solid phase from a super-cooled liquid.
When a liquid is cooled at temperatures below its freezing point, super-cooled liquid, the following phase transitions can be observed: a) precipitation of a crystalline solid; b) formation of a disordered solid (glass) which may crystallize afterwards upon heating; c) liquid-liquid separation followed by solidification of the components.

2.1 Nucleation and crystal growth

The crystallization of a homogeneous liquid phase is initiated by formation of a molecular cluster. The probability (P_n) of formation of a cluster containing n molecules more rapidly than the time derivative of the order parameter fluctuation is given by (Landau and Lifshitz 1969):

$$P_n \propto \exp\left(-\frac{\Delta G_n}{kT}\right) \tag{1}$$

where ΔG_n represents the Gibbs energy variation of the cluster formation and k the Boltzmann constant. According to the classical nucleation theory (Kelton 1991, Myerson 1999) this Gibbs energy variation has a negative and a positive contribution: the first results from the molecular aggregation and the second from the creation of the liquid/solid interface. Assuming a spherical cluster with radius r, the following expression accounting for the Gibbs energy variation can be written:

$$\Delta G_n = \frac{4}{3}\pi r^3 \Delta g + 4\pi r^2 \gamma \tag{2}$$

Δg is the Gibbs energy variation per volume unit and γ is the interfacial energy per unit area. Since $\Delta g < 0$ and $\gamma > 0$, the variation of ΔG_n with r passes through a maximum corresponding to the value of r^* given by the condition $dG_n/dr = 0$. The maximum of G_n (G_n^*) is obtained for the radius of the nucleus r^*, given by:

$$r^* = \frac{2\gamma}{|\Delta g|} \tag{3}$$

$$\Delta G_n^* = \frac{16\pi\gamma^3}{3(\Delta g)^2} \tag{4}$$

For $r < r^*$ the aggregates will dissolve into monomers because $\Delta G_n > 0$, while for $r > r^*$, $\Delta G_n < 0$ and the aggregates grow to form a crystalline phase. The molecular structure corresponding to r^* is called nucleus. The nucleation process of the new solid phase has an energy barrier height given by (4) which decreases as the super-cooling increases. In fact, the driving force for the occurrence of molecular aggregation is given approximately by $\Delta\mu=\Delta S_m \Delta T$, wherein $\Delta\mu$ is the variation of chemical potential, ΔS_m is the fusion entropy and ΔT (T_m-T, T_m is the fusion temperature) represents the super-cooling. Hence, Δg becomes more negative as ΔT increases, i.e., the energy barrier decreases as super-cooling increases. The nucleus, once formed, grows by deposition of new solid material on it. The crystal size of the material obtained by crystallization depends on the balance between nucleation and growth rates.

Thermodynamically, the temperature decrease of a super-cooled liquid favours crystal formation since the driving force of nucleation increases. However, from the kinetic point of view, crystallization becomes more difficult to occur due to viscous retardation. The maximum overall rate of crystallization is reached at a temperature for which the positive contribution given by the super-cooling equalizes the negative one arising from molecular motion (Hancock and Zografi 1997).

The nucleation just described arises by fluctuations in a homogeneous liquid phase free from any solid particles including dust and having no contact with a container. The nucleation in the presence of a solid phase (heterogeneous nucleation) has lower activation energy than the homogeneous one.

2.2 Glass transition

The metastable limit of a super-cooled liquid is sometimes not nucleation. A liquid may require such high super-cooling to crystallize that before this temperature is reached the molecules almost completely loose their translational motion. A solid is obtained by a sudden viscosity increase. This solid is amorphous because it has the short-range organization of the liquid at the temperature it was formed (glass transition temperature, T_g), but does not have the long-range order of the crystal. The glass transition is characterized by a more or less pronounced change in the derivatives of the thermodynamic properties by temperature variation (Wong and Angell 1976). For example, a sudden increase in volume and heat capacity with temperature increasing is observed. When melt-cooling is followed by DSC, the heat flow rate increases sharply as temperature decreases. Another property of a glass phase to bear in mind as background for the present work is the variation of T_g with the cooling rate. As the cooling rate increases less time is left for the molecules to adapt to the liquid structure and so higher T_g values are observed. Most glasses crystallize upon heating. At a temperature near T_g the glass devitrifies and crystallizes at a higher temperature. The DSC curve will show an increase in heat flow rate and a relaxation peak in the glass transition region, followed by an exothermic peak corresponding to crystallization.

2.3 Liquid phase separation

Another phenomenon that can occur in a super-cooled liquid before any of the transformations described above take place is liquid-liquid phase separation. This happens frequently on cooling viscous liquid mixtures (Turnbull and Bagley 1975), particularly on mixtures of polymeric materials (E.Donth 1992). As this phenomenon is observed in systems under consideration a brief note on this matter is left here.

For two miscible liquids, the variation of the Gibbs energy of mixing as a function of composition is a convex downward curve as displayed in Figure 2(a). However, for some systems a convex upward curve within a certain composition range can be observed, Figure 2(b). In this curve the points B and B′ correspond to the equilibrium phase composition or binodal composition (E.Donth 1992, Kelton 1991). The local limit of stability is given by the inflection points S and S′, whose composition is called spinodal. The mixture is unstable between S and S′ and metastable between BS and B′S′. The projection of binodal and spinodal in the T-ρ plane is shown in Figure 2(c). The maximum C is common to both curves and is called critical point. Systems with a critical point exhibit a miscible gap at temperatures below T_C. Inside the binodal nucleation of critical size particle of a certain composition and its growth *via* diffusion takes place. Inside the spinodal growth of a new phase goes *via* decomposition in which amplitude of concentration fluctuations is crucial. Liquid phase separation has also been observed for one component systems providing a long-range fluctuation alters an order-parameter of the liquid such as, for example, the density (Klein, et al. 1989, Kiselev 1999).

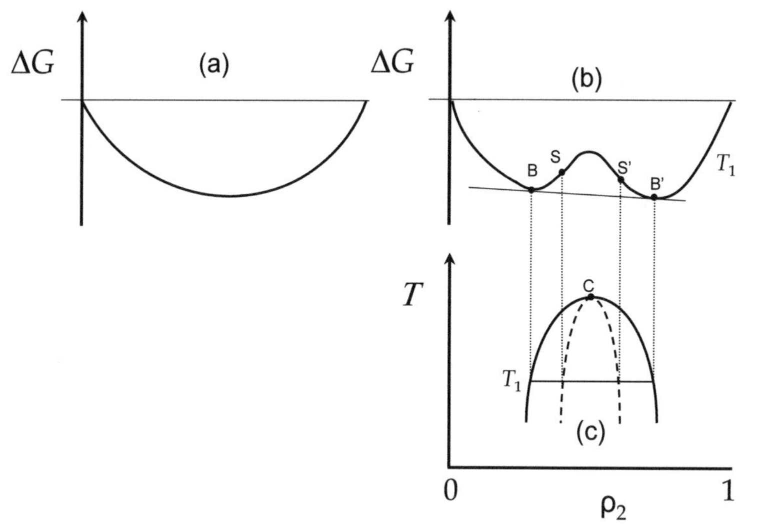

Fig. 2. Spinodal decomposition of viscous liquids: (a) ΔG versus density for two miscible liquids; (b) ΔG versus density for two partially immiscible liquid mixtures; (c) Temperature-density diagram for two partially immiscible liquid mixtures.

3. Thermodynamic information taken from DSC data

Despite being an old technique, DSC is still widely used to study the structural modifications occurring in a compound as it is cooled or heated. In the present case, it is used to investigate the phase transitions occurring in the melt-cooling process. Crystal or glass formation is observed and the phase transitions are characterized by their thermodynamic properties (temperature, enthalpy and heat capacity). Some of the structural aspects of the frozen solid can be evidenced upon heating. Hence, the following general procedure was used: samples encapsulated into aluminium pans were melted and the temperature was raised to approximately 20°C above the respective melting points.

The melt was then cooled down to approximately -160ºC at different scanning rates. The thermal cycle was completed by performing a heating run at a certain scanning rate until the melting point is reached. Apparatus and technical details of the DSC experiments can be found in previous works published by the authors (Lopes Jesus, et al., Canotilho, et al. 2010, Nunes, et al. 2004). The DSC results obtained for all compounds are summarized in Tables 1 and 2.

Compound	Cooling run	Heating run
Erythritol	Low cooling rates: crystallization $T_{cryst} \approx$ 14-16ºC or $\approx$ 40ºC $\Delta H_{cryst} \approx$ -20 kJ mol^{-1} Higher cooling rates: glass formation Tg = -44.4 ± 0.8 ºC ; $\Delta Cp \approx$ 0.1 J mol^{-1}K^{-1}	Crystallization from glass: $T_{cryst} \approx$ -20 to 20ºC Fusion: *Stable polymorph* T_{fus} = 117.9 ± 0.4 ºC ΔH_{fus} = 39.4 ± 0.9 kJ mol^{-1} *Metastable polymorph* T_{fus} = 104 ± 1 ºC $\Delta H_{fus} \approx$ 34 kJ mol^{-1}
Atenolol	Crystallization at any cooling rate: T_{cryst} = 147.1 ± 1.2 ºC ΔH_{cryst} =-36.3 ± 0.7 kJ mol^{-1}	Fusion: T_{fus} = 152.4 ± 0.5 ºC ΔH_{fus} = 36.5 ± 1.1 kJ mol^{-1}
Pindolol	Crystallization at higher temperature: T_{cryst} = 143 ± 7 ºC ΔH_{cryst} =-50± 3 kJ mol^{-1} Crystallization at lower temperature: T_{cryst} = 75.3 ± 1.1 ºC ΔH_{cryst} =-30.3 ± 1.4 kJ mol^{-1}	Crystallinity improvement of the solid formed at 75.3ºC T_{cryst} = 110 - 140ºC Fusion: T_{fus} = 169.7 ± 0.3 ºC ΔH_{fus} = 58 ± 2 kJ mol^{-1}
Betaxolol hydrochloride	Glass transition $Tg \approx$ -14ºC ; $\Delta Cp \approx$ 200 kJ mol^{-1}K^{-1} $Tg \approx$ -142ºC ; $\Delta Cp \approx$ 30 J mol^{-1}K^{-1}	Crystallization from glass: T_{cryst} = 64.4 ± 0.9 ºC ΔH_{cryst} =-23.6± 0.7 kJ mol^{-1} Fusion: T_{fus} = 113.8 ± 0.2 ºC ΔH_{fus} = 32.9 ± 1.2 kJ mol^{-1}

Table 1. Summary of the DSC data.

To obtain information concerning the thermal stability of the substances, a preliminary study was carried out in order to establish the best working conditions. It was found that atenolol and pindolol require special attention in sample preparation. During fusion, the amide group of atenolol is slowly converted into imine, which by thermal degradation gives rise to smaller molecular fragments. The degradation is substantially reduced in the absence of oxygen. Thermal oxidation of dialkylamines was observed by some authors (Dagnac, et al. 1997). To avoid degradation, the samples should be prepared in an inert atmosphere and scanning rates higher than 10ºC/min should be used in order to reduce the exposure time of the sample to high temperatures. In such conditions, no significant oxidation occurs (Esteves de Castro and Redinha 2002). In liquid state, the indol group of pindolol is oxidized to 3-hydroxy-indol which in turn is converted into indigo (Stenlake 1979). In fact, just a few percent of degradation of pindolol is found when the sample is encapsulated in a free air atmosphere and the melt left for one hour at 160°C. However, no degradation was detected when the samples were prepared under the conditions described for atenolol. The remaining compounds are thermally stable in the temperature interval needed for the experiments and no special precautions were taken in sampling.

A typical set of DSC curves are presented in Figures 3 and 4 for erythritol. When the compound is cooled at rates below 10°C/min just a narrow exothermic peak corresponding to crystallization is generally observed at temperatures between 16 and 14°C, or in a few cases at *ca.* 40°C. At 10°C/min, a wider crystallization peak is found at *ca.* 0°C/min, sometimes accompanied by a glass transition at approximately -44°C. For higher cooling rates (≥20°C/min), only glass transition is observed. The DSC patterns show the competition between nucleation and glass transition. At low cooling rates (≤5°C/min) only crystallization takes places, whereas for cooling rates higher than 20°C/min the solidification gives glass. For medium values both transitions occur.

The frozen solid upon heating can melt at two different temperatures, 104 or 118°C (most common), with enthalpies of fusion of 34 and 39 kJ mol^{-1}, respectively. In some cases, both peaks are observed simultaneously, curves (c) and (d) of Figure 4. This behaviour indicates that erythritol grown from melt may exhibit two polymorphs with the contribution of each one varying from 0 to 100%. According to the heat of fusion rule (Burger and Ramberger 1979) erythritol is a monotropic system, the higher melting point being the stable form. The solids forms obtained at medium or high cooling rates show, upon heating, devitrification followed by crystallization before fusion.

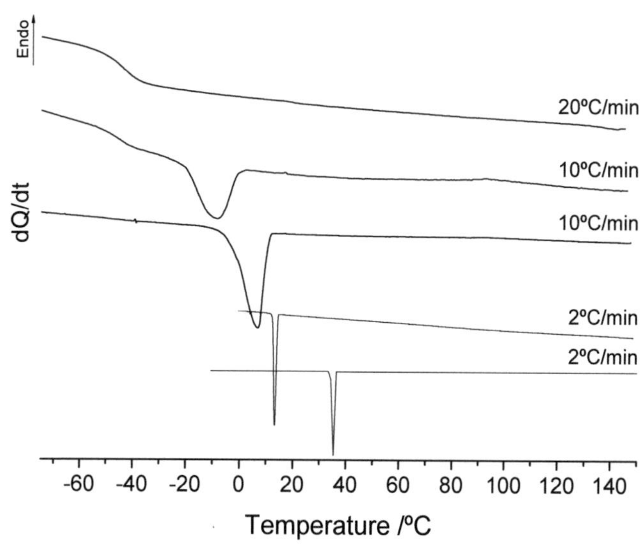

Fig. 3. DSC cooling curves of molten erythritol at different scanning rates. Reprinted from International Journal of Pharmaceutics, Vol 388, A.J. Lopes Jesus et. al. / Erythritol: Crystal growth from the melt, Page 130, Copyright (2010), with permission from Elsevier.

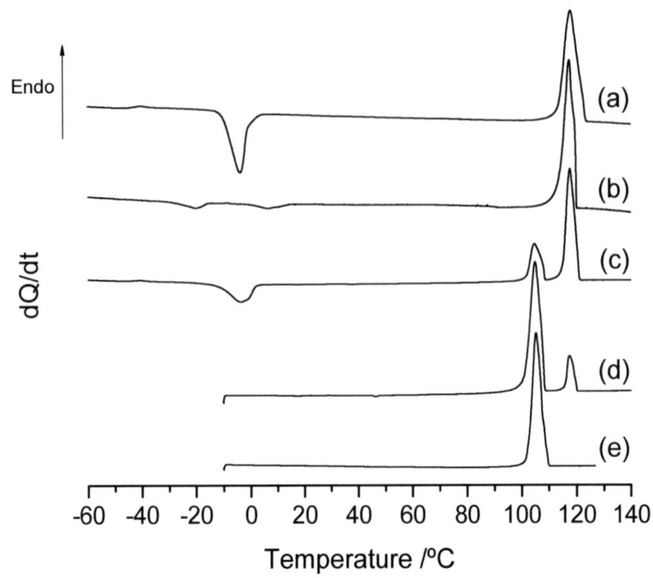

Fig. 4. DSC heating curves (10ºC/min) of solid erythritol grown from melt at 20 ºC/min (a), 10 ºC/min (b) and (c), 2 ºC/min (d) and 0.5 ºC/min (e). Reprinted from International Journal of Pharmaceutics, Vol 388, A.J. Lopes Jesus et. al. / erythritol: crystal growth from the melt, Page 131, Copyright (2010), with permission from Elsevier.

The shape of the fusion curves corresponding to the most stable form reveal that they are uncommonly asymmetric and have a large base width. This complex curve was decomposed by curve-fitting analysis using a parametric modified Gaussian function developed by Leitão *et al.* (Leităo, et al. 2002). A typical curve-fitting pattern is shown in Figure 5. A good fit was achieved using two or three individual curves, which have been characterized by their peak maximum temperatures (T_{max}). Using a *K*-means cluster analysis the values obtained for T_{max} were then grouped into different clusters. For 45 samples, three clusters were considered. The criterion used to establish the number of clusters (*K*) was that no significant decrease in the variance within each cluster was observed by increasing *K*. The following values of T_{max} were obtained: 118.5 ± 0.5 (0.18), 119.9 ± 0.4 (0.42) and 122.1 ± 0.7 (0.40). The figure in parenthesis corresponds to the probability of occurrence.

Let us now consider the thermal behaviour of the three beta-adrenergic compounds. The cooling DSC curves of molten atenolol show only a crystallization peak (~147ºC) at any scanning rate up to 100ºC/min. Therefore, when the melt-grown solid is re-heated only a fusion curve is denoted at about 152ºC. Another relevant feature evidenced by the cooling-heating cycle is the low super-cooling (5ºC) of this compound. Such a low super-cooling is not common in most organic compounds for which values as high as 100ºC can be observed. Crystallization of pindolol occurs predominantly at 143±7ºC or 75±1ºC, with percentages of occurrence of 28 and 65%, respectively. The values of the enthalpy of crystallization are -50±3 for the first case and -30±3 kJ mol^{-1} for the second one. Sometimes, crystallization at the higher temperature is by-passed and it occurs at the much lower temperature. Samples solidified under these conditions show upon heating a slow crystallization process between 110 and 140ºC followed by another one just preceding fusion (170ºC).

Unlike the two beta-adrenergic just considered, betaxolol hydrochloride does not exhibit any crystallization peak for cooling rates down to 1° C/min; only glass transitions are observed at approximately -14 and -142°C. In the subsequent heating process devitrification and recrystallization at ~65°C occur, followed by fusion at 114°C.
Following an identical procedure as that described for erythritol, the DSC fusion curves of the solid beta-adrenergic compounds after being submitted to a cooling/heating cycle were also decomposed by curve-fitting. Apparently, for all studied compounds, the solid phases obtained by crystallization from melt are mixtures of three or four unique structures (Table 2). A discussion on the origin and characterization of these structures will be given later in this chapter.

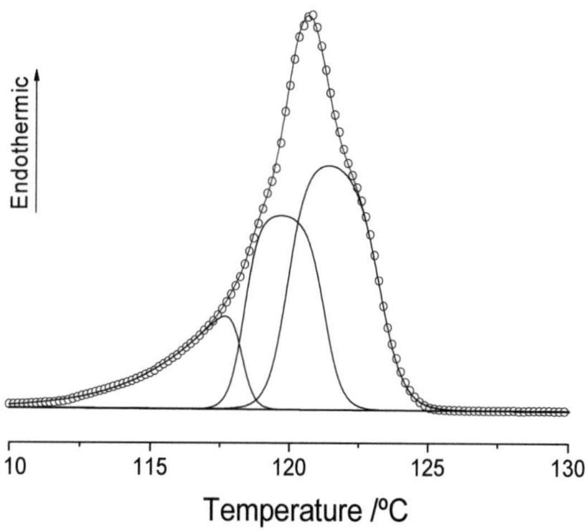

Fig. 5. Peak-fitting analysis of the fusion curve of the most stable form of erythritol. Values of T_{max} for the decomposed peaks are given in Table 2. Reprinted from International Journal of Pharmaceutics, Vol 388, A.J. Lopes Jesus et. al. / Erythritol: Crystal growth from the melt, Page 131, Copyright (2010), with permission from Elsevier.

Compound	Fusion temperatures (°C) of the individual components of melt-grown solids after the cooling/heating cycle[a]
Erythritol	118.5 ± 0.5 (17.8) ; 119.9 ± 0.4 (42.2) ; 122.1 ± 0.7 (40.0)
Atenolol	153.0 ± 0.3 (33.3) ; 154.0 ± 0.4 (38.9) ; 155.7 ± 0.3 (27.8)
Pindolol	169.7 ± 0.2 (13.4) ; 170.2 ± 0.1 (36.1) ; 170.6 ± 0.1 (33.0) ; 171.1 ± 0.2 (17.5)
Betaxolol hydrochloride	111.3 ± 0.2 ; 112.7 ± 0.4 ; 114.2 ± 0.3 ; 115.3 ± 0.4

Table 2. Results of the curve-fitting analysis of the fusion curves for the solids crystallized from melt. [a] The figures in parenthesis correspond to the probability of occurrence.

4. Phase transitions observed by PLTM

The cooling/heating cycles previously described were observed with a polarized light microscope. This method allows a direct observation of the texture of anisotropic solids, phase transitions, existence of polymorphism, etc (Kuhnert-Brandstatter 1982). By combining PLTM with DSC, valuable information on the mechanism of phase transitions can be obtained as will be seen later in this chapter (Wiedemann and Bayer 1985).

The equipment used consists of a cell holder hot stage platform (Linkam DSC 600) coupled to a Leika DMRB microscope to which a video camera was adapted. The equipment has also a photometric absorption device which helps to follow the phase transformations taking place inside the cell. Small solid aggregates of the substance to be studied are dispersed through the bottom of the cell. The sample is then melted giving rise to small liquid drops which are then submitted to cooling/heating cycles similar to those performed in the DSC study. A few drops were observed at the same time which is an advantage since different thermal events may occur in different droplets or aggregates. Recording of the images in a computer allows a detailed study and also following the crystallization front and determining its velocity. This is valuable information regarding the mechanism of the phase transitions and of the new structures being formed.

4.1 Erythritol

Two main types of melt cooling are observed for the liquid erythritol, which are illustrated in Figures 6 and 7. For low cooling rates (Figure 6), nucleation is initiated inside a liquid drop at ~11°C and a grained texture grows towards the surface. Fractures transversely oriented relatively to the growth direction are observed as the fresh solid phase is being deposited. The crystallization is extended to the whole liquid drop. On the subsequent heating a solid-solid transformation is generally observed at ~65°C: the grained texture is replaced by a phase boundary texture through a diffusional process as illustrated in Figure 6(C). In this Figure the phase boundaries are indicated by arrows as eye guides. At around 120°C, fusion takes place giving rise to a liquid with numerous air bubbles resulting from the solid fractures, Figure 6(D).

A different melt-cooling behaviour can be observed in experiments performed at 10°C/min, which are described by the following typical example and illustrated in Figure 7: At 5°C, crystallization is initiated on the surface of the liquid drop, progressing more rapidly along the surface than towards the interior of the liquid. At -70°C, a crystalline shell with a grained texture encloses a glass phase resulting from vitrification of the innermost liquid. The relative amount of ordered/disordered material varies from one experiment to another. On heating, the solid particle exhibits two transformations: slow crystallization of the glass into a fine grained structure which occurs between -8 and 2°C, followed by a solid-solid transition between 15 and 43°C, from which a sintered particle texture results, Figure 7(C). The final solid melts at 120°C.

In both behaviours just described the solid-solid transition corresponds to the transformation of the metastable form into the most stable one. Since no heat is involved as proved by DSC (see Figure 4), this process is fundamentally entropically driven. It is to be noted that in some experiments this transformation is not observed and the metastable phase remains unaltered until fusion (104°C).

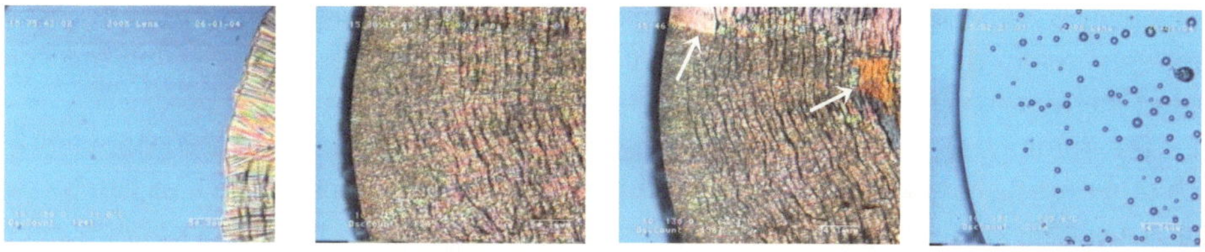

(A) (B) (C) (D)

Fig. 6. Hot-stage polarized micrographs showing the transformations occurring in erythritol as it is cooled from the melt at 2°C/min and re-heated at 10°C/min. Upon cooling: (A) T=11°, (B) T=-70°C. Upon heating: (C) T=62°C, (D) T=125°C.

(A) (B) (C) (D)

Fig. 7. Micrographs illustrating alternative transformations of erythritol during the cooling/heating at 10°C/min. Upon cooling: (A) T=5°C, (B) T=-70°C. Upon heating: (C) 20°C, (D) T=43°C.

In conclusion, erythritol cooled from the melt originates a crystalline metastable phase or a mixture of this phase with glassy material. On heating, three situations can occur: (a) all metastable form is converted into the stable one and just one fusion peak is observed at 118°C; (b) the transformation is by-passed and the fusion of the metastable occurs 14°C below; (c) only part of the metastable phase is transformed into the stable form and two fusion peaks are observed with areas depending on the relative amounts of both forms presented in the solid.

4.2 Atenolol

The crystallization of liquid atenolol is a very fast process. As super-cooling reaches a value of ~5°C a very fast liquid/solid transformation takes place. Nucleation arises inside the liquid drops and crystallization spreads almost instantaneously all over the liquid.

4.3 Pindolol

In pindolol, the beginning of the solid formation at 143°C is accompanied by liquid phase decomposition. As shown in Figure 8, the apparent homogeneous liquid divides into two domains just as the solid appears. The crystallization goes on in the first domain until it becomes exhausted. The second liquid domain (β) under observation crystallizes at about 70°C. The two crystallization temperatures are in agreement with the DSC data. The textures of the solids originated by the spinodal decomposition are quite different. That obtained at higher temperature is plate-like whilst the other has a central part of fine grains surrounded by a dendrite-like texture. Keeping in mind the close structures of pindolol noted in the analysis of the fusion curves, we can conclude that the differences between these textures are linked with the temperatures at which they are formed rather than with structural differences. In the example just referred, the liquid phases resulting from spinodal decomposition are kept far apart and therefore each one has its own crystallization temperature.

Besides the domains separated from each another by longer distances, micro-domains closer to each other arise sometimes from the liquid phase decomposition. In fact, as can be seen in Figure 9(A), the texture of the solid phase formed at higher temperature is not homogenous, highlighting the short interruptions occurring during crystallization, as can be confirmed by the plot of the velocity against time included in the Figure. In some cases, liquid-liquid separation is by-passed and the crystallization occurs at 70-75°C exhibiting a texture of fine particles as shown in Figure 9(B).

From the results obtained from the PLTM study of pindolol crystallization the following conclusions can be drawn: crystal nucleation at higher temperature is accompanied by liquid-liquid decomposition; since it is difficult to see which phenomenon occurs first, one can conclude that these transitions take place near the critical point where the metastable region is very narrow; lower super-cooling crystallization is by-passed if decomposition does not occur or *vice-versa*: at a lower temperature the probability of nucleation is high because the metastable region between the binodal and spinodal curve is wider. As shown in the plots depicted in Figure 9, the crystallization front is an oscillating curve because the heating release of the new phase and its sequent removal from the medium disturbs the steady state conditions giving rise to super-cooling fluctuations (Jackson 1975).

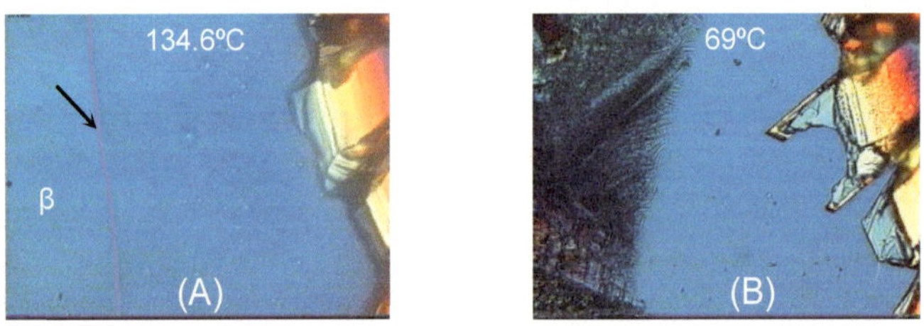

Fig. 8. Micrographs registered during cooling of pindolol from melt. (A): As crystallization is initiated, the β phase is separated out; the boundary of β is indicated in the picture by an arrow as eye guide. (B): crystallization of both domains.

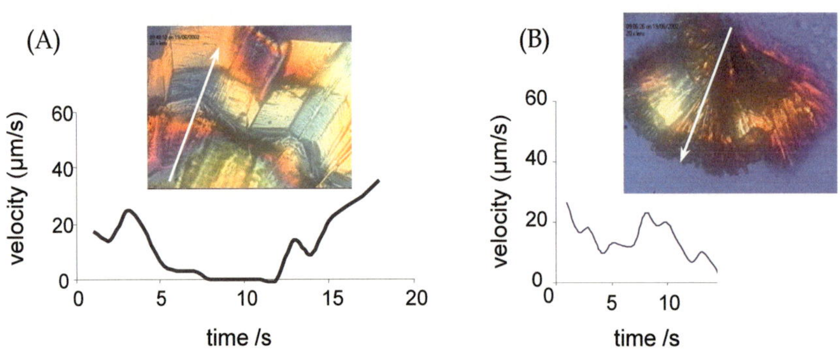

Fig. 9. (A): Texture and velocity of crystallization front for the solid grown at higher temperature. The arrow indicates the direction of front progress determinations. (B): Texture and velocity of crystallization not accompanied by liquid phase separation.

4.4 Betaxolol hydrochloride

The PLTM technique proves that betaxolol hydrochloride is manifestly the best glass former among the compounds under consideration. No crystallization is observed on cooling. Such transition is only observed on heating when the glass is frozen at a very low temperature, say -160ºC. On heating, the liquid resulting from devitrification crystallizes at 50-60ºC with slow formation of spherulites, see Figure 10. Alternatively, crystallization can occur under isothermic conditions at temperatures between 25 and 40ºC by annealing for approximately one hour. During crystallization, liquid is rejected from the solid phase and is accumulated in black-spots, which melt at about 108ºC.

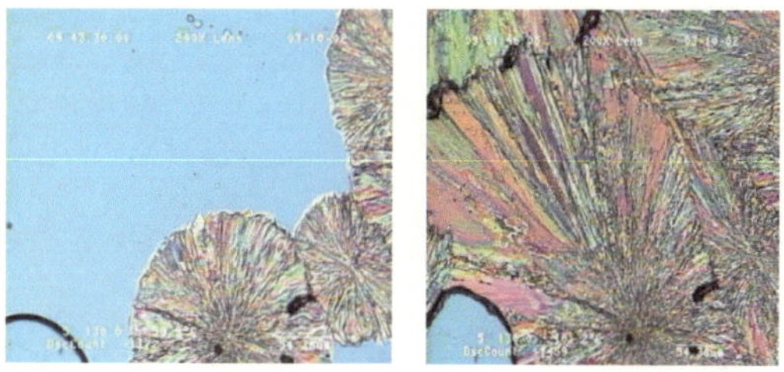

Fig. 10. Spherulites formed upon heating the melt-grown betaxolol hydrochloride.

5. Characterization of the hydrogen bond network by infrared spectroscopy

The molecules we are considering have various functional groups containing atoms able to participate in H-bonds as electron donors or acceptors. As this interaction plays a decisive role on the final structure of the solids as well as on crystallization, the knowledge of the H-bond network is indispensable to understand the melt-cooling process. We must have in mind that H-bonding is the main interaction in molecular recognition (Redinha and Lopes Jesus 2011). A hydrogen bond established between a Y atom carrying one or more electron lone-pairs and a hydrogen atom bonded to another electronegative atom (X), is generally represented as X-H···Y. As its formation weakens the X-H bond, the band maximum corresponding to its stretching vibration is displaced to lower frequencies. Another spectral manifestation of this interaction is the increase of the bandwidth and of its integrated intensity. Various methods can be used to obtain the reference spectrum of the molecule free from intermolecular hydrogen bonds (Jesus and Redinha 2009, Jesus, et al. 2003, Jesus, et al. 2006). Here, it has been calculated by applying quantum chemical methods to the monomer with the conformation exhibited in the crystalline structure.

Spectra of the compounds dispersed in a KBr pellet were registered from the melt down to -170ºC at intervals of about 5ºC during the cooling/heating cycle. The spectral region we are interested in is that between 4000 and 3000 cm^{-1}, corresponding to the OH and NH stretching vibrations. These are the vibration modes mostly affected by hydrogen bonding. The interval between 3000 and 2500 cm^{-1} is occupied by CH stretching vibrations. Due to the strong overlap of the bands in the spectral region we are considering, a peak-fit analysis is required. Good-fitts were achieved using Lorentzian functions. In this discussion we will give details of the procedure followed for erythritol while for the remaining compounds only a summary of the final results is presented.

The spectra of erythritol on cooling can be seen in Figure 11(A). The OH absorption of liquid erythritol shows a broad and symmetric band centred at ~3400cm^{-1}. As cooling is continued the absorption band becomes asymmetric and for lower temperatures two overlapping bands are clearly distinguished. At the end of the cooling run two band components at 3346 cm^{-1} and 3243 cm^{-1} were determined as illustrated in Figure 11(A). Further spectral changes are observed on heating from -170°C until melt is achieved, which can be followed through the band profile, Figure 11(B). Near the end of the heating process the spectrum of the solid exhibits three band components located at 3414, 3256 and 3154 cm^{-1}. A conclusion driven at once from the comparison of the spectra taken at -170 and 115°C is that the frozen solid does not have a perfect crystalline structure and it should be heated to get to this stage. Indeed, its crystallinity has been confirmed by X-ray diffraction.

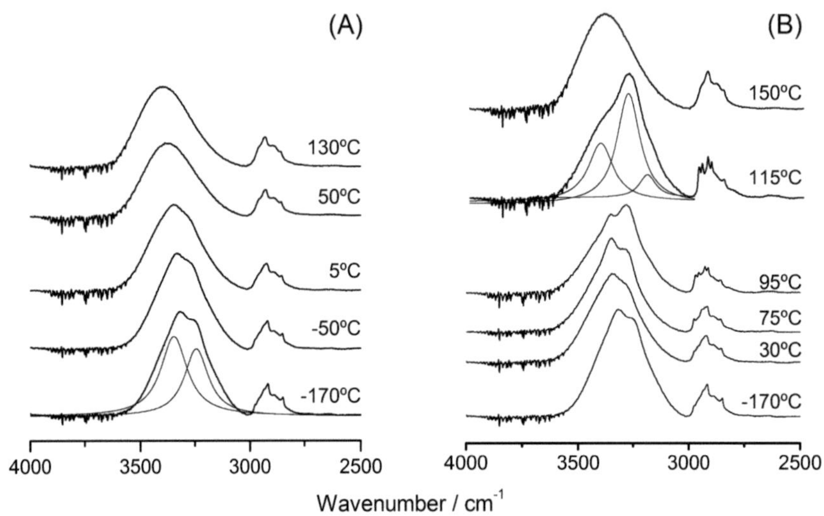

Fig. 11. Spectra of erythritol at different temperatures in the 4000-2500 cm^{-1} region during a cooling (A)/heating (B) cycle.

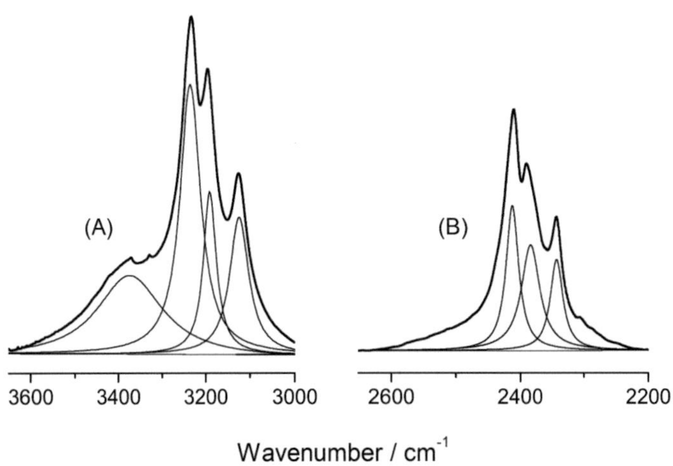

Fig. 12. Experimental spectra (bold lines) of erythritol at -258°C in the OH (A) and OD (B) stretching regions and resolved bands (thin lines) obtained by curve-fitting.

For an accurate band identification, a spectrum of solid erythritol crystallized from water was recorded at very low temperature (-258°C). The spectral resolution improved by lowering the temperature allows for the identification of four overlapped bands in the region under consideration as shown in Figure 12(A). Since by deuteration the band located at 3373 cm^{-1} has no counterpart in the OD region, Figure 12(B), it is not a fundamental OH stretching band but is rather a Fermi resonance. All the other absorption bands correspond to different OH stretching vibrations. It should be noted that the spectrum at this temperature differentiates the band located at 3234 cm^{-1} from that at 3190 cm^{-1}, which doesn't happen in the spectrum at 115°C as shown in Figure 11(B).
Having identified three absorption bands as OH stretching vibrations, a step forward to their complete assignment can be made by correlating the spectral with the crystallographic data: higher values of the frequency shift correspond naturally to stronger H-bonds. Fortunately, structural data is available not only for erythritol but also for all the other compounds abridged in this study. The structural parameters of the H-bonds and the assignment of the NH and OH stretching vibrations can be found in Tables 3 and 4, respectively.

Hydrogen bond	H···Y / Å	X-H···Y / °
Erythritol		
$O_{(2)}H \cdots O_{(3)}$ or $O_{(2)}H \cdots O_{(3)}$ Conf. B	1.85	163.9
$O_{(2)}H \cdots O_{(3)}$ or $O_{(2)}H \cdots O_{(3)}$ Conf. A	1.79	164.5
$O_{(1)}H \cdots O_{(4)}$ or $O_{(4)}H \cdots O_{(1)}$	1.70	173.2
(*R,S*)-Atenolol		
$O_{(7)}$–H···$N_{(4)}$	2.49	119.7
$N_{(19)}$–H'···$O_{(18)}$	2.06	161.5
$N_{(19)}$–H''···$O_{(18)}$	2.04	173.9
Pindolol		
$N_{(4)}$–H···$O_{(9)}$	2.98	106
$N_{(4)}$–H···$O_{(7)}$ (intramolecular)	2.31	114
$N_{(15)}$–H···$O_{(7)}$	2.03	156
$O_{(7)}$–H···$N_{(4)}$	1.78	172
Betaxolol hydrochloride		
$N_{(4)}^{+}$–H_A···Cl^-	2.39	137.6
$N_{(4)}^{+}$–H_B···Cl^-	2.08	175.1
O···Cl^-	3.04	-

Table 3. Structural parameters (H···A distance and X-H···Y angle) characterizing the H-bonds existing in the crystalline structures of erythritol (Ceccarelli, et al. 1980), atenolol (Esteves de Castro, et al. 2007), pindolol (Chattopadhyay, et al. 1995) and betaxolol hydrochloride (Mairesse, et al. 1984).

Various attempts have been made to correlate the frequency shift of stretching vibration ($\Delta\nu$) with the enthalpy of the H-bond formation (Iogansen 1999, Stolov, et al. 1996). Although the accuracy of the values calculated from these empirical correlations may vary according to the H-bond under consideration, they give an approximate indication of the strength of these interactions in energetic terms. The values of ΔH presented in Table 4 were estimated by the Iogansen equation (Iogansen 1999).

Compound	ν /cm^{-1}	$\Delta\nu^a$ / cm^{-1}	ΔH / kJ mol^{-1}	Assignment
Erythritol	3234	418	27	ν [$O_{(2)}$H] or ν [$O_{(3)}$H] Conf. B
	3190	462	28	ν [$O_{(2)}$H] or ν [$O_{(3)}$H] Conf. A
	3124	528	30	ν [$O_{(1)}$H] or ν [$O_{(4)}$H]
(*R,S*)-Atenolol	3459	184	18	ν (OH)
	3345	204	19	ν_{ass} (NH_2)
	3292	74	11	ν [$N_{(4)}$H]
	3160	258	21	ν_s (NH_2)
Pindolol	3404	125	15	ν [$N_{(15)}$H]
	3309	45	9	ν [$N_{(4)}$H]
	3258	330	24	ν (OH)
Betaxolol hydrochloride	3263	110	14	ν (NH_A) ^{b}
	3126	499	29	ν (OH)
	2791	581	31	ν (NH_B) ^{b}

Table 4. Assignment of the bands found in the 4000-3000 cm^{-1} region of the infrared spectra of solid erythritol, atenolol, pindolol and betaxolol hydrochloride. [a] Values of the stretching vibration of the "free" XH group were obtained from the calculated spectra of the isolated molecules adopting the crystal conformation.[b] H_A and H_B refers to the hydrogen atoms of the NH_2^+ group.

6. Crystalline ability and hydrogen bonding

As it was previously described, the melt-cooling process can originate a crystalline or glassy state depending on the cooling rate. An example of this general behaviour is that of erythritol.

In spite of the structural similitude between the beta-adrenergic compounds, they exhibit a different behaviour in the melt-cooling process as far as crystallinity is concerned. Liquid atenolol becomes solid at a low supercooling and only a crystalline form is obtained, even for high cooling rates. One can say that liquid atenolol has a good crystal making ability. At the other extreme is betaxolol hydrochloride which vitrifies on cooling at scanning rates of tenths of a degree per minute, while the crystalline phase is only obtained upon heating the frozen solid. This is an example of a good glass forming liquid. In an intermediate position between crystal and glass former is pindolol. Upon cooling, at a temperature of ~143°C (26°C below the melting point), a phase separation of the viscous melt occurs, accompanied by a crystallization of one of the liquid phases; the other liquid phase crystallizes partially at ~75°C.

A crystal is an example of a supramolecule resulting from the self-recognition of the molecules present in the liquid (Redinha and Lopes Jesus 2011). Organic molecules such as those of the beta-adrenergic compounds have a long and flexible carbon chain (see Figure 1). In molecular terms, their packing occurs through intermolecular linking points. When these points are located at short distances the molecular fragments are constrained to occupy fixed positions and to adopt a certain conformation. The most important type of intermolecular interaction presented in organic crystals is hydrogen bonding. The gradual directional nature of this bond, its intermediate position between a covalent and a van der Waals interaction in respect to strength, and between covalent and ionic force as far as polarity is concerned, makes this interaction to play a key role in the molecular recognition process. The H-bond network pattern is fundamental in the type of solid formed.

In Table 3 are described the H-bond networks existing in the crystalline structures of the beta-adrenergics. The different distribution of H-bonds in atenolol and betaxolol hydrochloride clearly accounts for the difference observed on cooling from the melt. In the crystalline structure of atenolol each hydrogen atom of the acetamide group is linked to the $O_{(18)}$ atom of two neighbouring molecules. In addition, the $O_{(7)}H$ group is H-bonded to $N_{(4)}$. Weak interactions involving the phenyl group, $C_{(8)}$-H···π and π···π, were also proved to exist (Esteves de Castro, et al. 2007). That is, the only part of the atenolol molecule remaining free of relatively strong intermolecular forces is the terminal isopropyl group.

Regarding the betaxolol hydrochloride, the chloride ions establishes three charged-assisted H-bonds: two with the OH and NH^{+}_{2} groups of the same molecule and the other with the NH^{+}_{2} group of a neighbouring molecule (see Figure 13). The molecular fragment from $C_{(6)}$ to the cyclopropyl group is not constrained by any hydrogen bond and is free to move as in the liquid phase until it got trapped due to viscosity increasing. Thus, several conformations can be adopted by this moiety.

These two examples highlight the influence of the hydrogen bonding distribution on the tendency of the compound towards either crystal or glass formation.

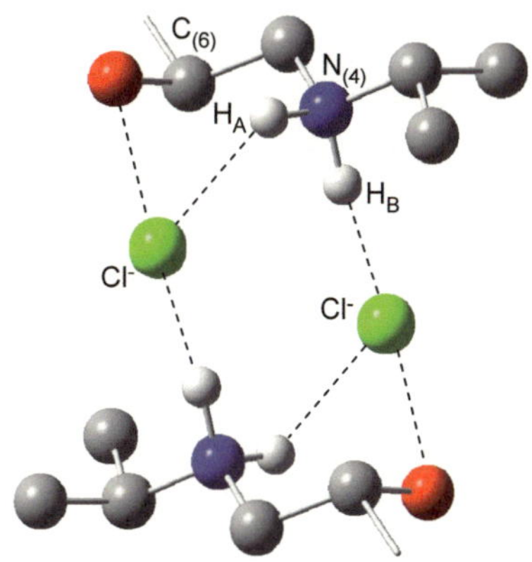

Fig. 13. Detail of the intermolecular hydrogen bonds between neighbour molecules in betaxolol hydrochloride. For better visualization some hydrogen atoms have been removed and only the molecular fragments involved in H-bonds are shown.

7. Partially disordered crystals

As it has been shown previously, the thermal methods provided evidence for the existence of more than one structure in the solids grown from melt (Table 2). Very likely, it will be rather difficult to obtain a perfect crystalline form in organic crystal of molecules of a certain size. Organic crystals may exhibit various types of imperfections in unknown amounts: vacancies in the crystal lattice, rotational or translational disorder and different orientations of the whole molecule or of some molecular fragments.

The crystalline structure of erythritol has been determined at 22.6K by neutron diffraction analysis (Ceccarelli, et al. 1980). This study reveals the existence of two conformations (A and B) (Jesus and Redinha 2009, Ceccarelli, et al. 1980), differing from each other in the positions of the two hydrogens connected to the middle oxygen atoms (see Figure 14). The occupancy percentages of the two conformations at 22.6 K are found to be 85% and 15%, respectively. Three types of hydrogen bonds have been identified: one of them is common to both conformations and connects the terminal OH groups, while the other two involve the middle OH groups of the conformations A or B, respectively (Ceccarelli, et al. 1980, Jesus and Redinha 2009). Since the existence of these two conformations does not affect the unit cell parameters, erythritol is an example of conformational isomorphism.

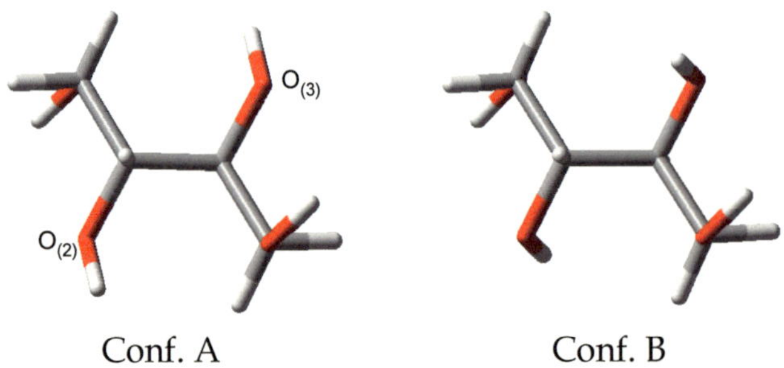

Fig. 14. Conformations of erythritol in the crystal lattice.

This situation is more complicated in the beta-adrenergic compounds. Let us consider the crystalline structures of (R,S)-atenolol and betaxolol hydrochloride. Crystals of (R,S)-atenolol of size suitable for X-ray diffraction were obtained by slow evaporation of the solvent from water-ethanol solutions. The compound crystallizes in the monoclinic crystal system (*C2/c* space group). Analysing the crystal geometry which is represented in Figure 15, three regions can be identified: the central part from $C_{(8)}$ to $C_{(16)}$ wherein the C and O atoms are in the plane defined by the benzene ring, the head consisting of the acetamide group with the C, N and O atoms in the same plane, making an angle of 81° with the central molecular backbone, and the tail extending from $C_{(6)}$ to the isopropyl group. The tail fragment exhibits two types of disorder: one is concerned with the position of the $C_{(5)}$ atom which can alternatively occupy positions 5a or 5b, Figure 15(A). The percentage occupancy of each site is 57 and 43%, respectively. Further thermal motion disorder arises from the flexibility of the tail fragment, as indicated by the ORTEP diagram given in Figure 15(B).
Betaxolol hydrochloride crystallizes in the triclinic system with P_1 symmetry (Mairesse, et al. 1984). As stated before the only detected H-bonds involve the chloride ion and the two functional groups of the molecular head. Besides these bonds, only loose van der Waals contacts occur between cyclopropyl groups. As can be seen from Figure 16(A) the structure presents disorder from $C_{(10)}$ to $C_{(19)}$. The cyclopropane group was left out of the structure refinement (Mairesse, et al. 1984). Basic betaxolol was also studied by X-ray diffraction (Canotilho, et al. 2008). This form crystallizes in the same crystal system and space group as the hydrochloride. It was found that the cyclopropylmetoxy group may acquire two alternative conformations (a and b with almost the same occupancy) resulting from the rotation around the $O_{(17)}$-$C_{(18)}$ bond as illustrated in Figure 16(B). The two cyclopropyl planes defined by each alternative conformation form an angle of 82°.

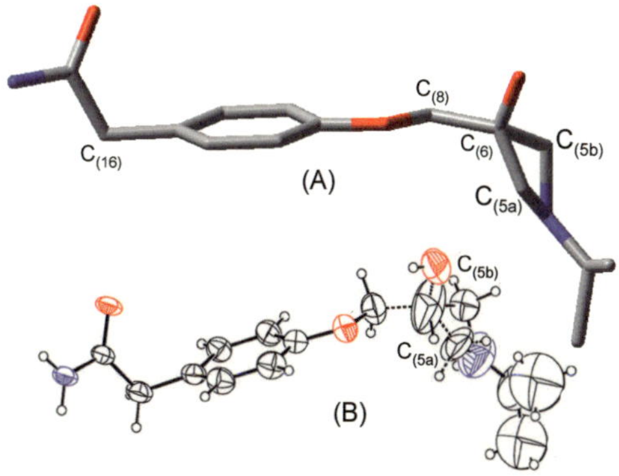

Fig. 15. (A) Geometry of (R,S)-atenolol in the crystalline structure showing the two alternative positions for the $C_{(5)}$ carbon atom; (B) ORTEP drawing (30% displacement ellipsoids) showing the thermal motion disorder exhibited by the tail fragment.

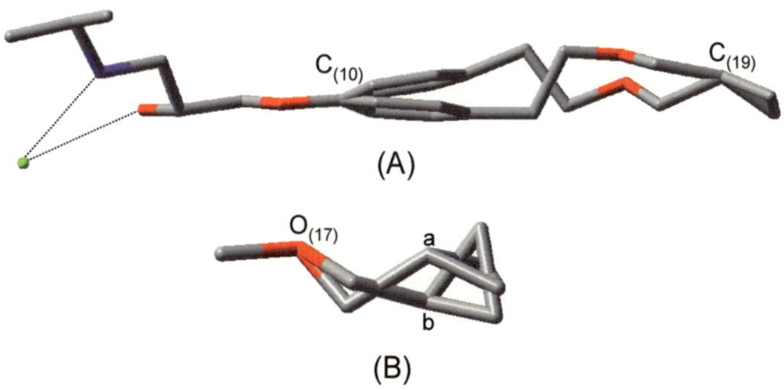

Fig. 16. (A) Geometry of betaxolol hydrochloride in the crystalline structure showing the disorder from C(10) to C(19) carbon atom; (B) two alternative conformations (a and b) adopted by the cyclopropylmetoxy group in the crystalline structure of basic betaxolol.

8. Conclusion

Grown from melt is a way to prepare solids with structures required for certain purposes. One of its greatest advantages relative to crystallization from solution is that no solvents are used. In this chapter the structure of the solid forms of compounds with pharmaceutical interest prepared in this way and the phase transitions occurring on cooling or heating were investigated. Compounds with different size and flexibility, as well as with different number and distribution of polar centres were selected for this study.

From the DSC results it was possible to conclude whether melt-cooling leads to crystalline or glassy states and to correlate the cooling rate with the crystallinity of the products obtained. In this respect it has been found that the three beta-adrenergic compounds exhibit quite distinct behaviours. To obtain crystalline compounds which produce liquids hard enough to be crystallized or just to improve the crystallinity of the solids obtained by cooling the melt, the frozen material has to be heated at a high temperature. The thermodynamic properties determined for phase transitions by DSC are clear "windows" to understand the crystalline structures. Statistical analysis of the fusion curves proved that the molecular crystals of the compounds under study are composed of a few well-defined (single-value melting point) structures.

Visual observation of the transformations under polarized light was found to be an excellent complement to the DSC technique. The mechanisms of phase transitions involving crystalline and glassy forms upon cooling or heating could be followed by PLTM. Moreover, the melt phase separation in pindolol that sometimes accompanies melt-cooling crystallization could be studied by this technique. Athermic transitions such as the solid-solid transformation occurring in erythritol were only detected by this technique.

Hydrogen bonding plays a fundamental role in molecular recognition and consequently it is a determinant interaction from the nucleation to the final crystal. The intermolecular H-bonds were studied in all compounds during the cooling and heating processes. These interactions were detected by the infrared spectra in the stretching vibration region. Difficulty of spectra interpretation caused by the strong overlap of the stretching bands concerning the polar groups involved in H-bonding was overcome by temperature lowering, deuteration and peak-fit analysis and a detailed table of the hydrogen bonds for all compounds was presented. Moreover, the combination of spectral and structural (X-ray or neutron diffraction) data allowed a full assignment of the experimental infrared bands.

With the data obtained by the thermal and spectroscopic methods it was possible to interpret the ability of the melts to originate crystals or glassy phases upon cooling.

A conclusion deserved to be emphasized is the proof of co-existence, as a rule, of more than one structure in the organic molecular crystals. To clarify the type of such structures, X-ray and neutron diffraction data taken from the literature or obtained by the authors were used. These data point out that those organic crystals are constituted by molecules differing themselves in the orientations of the groups (case of erythritol) or with parts of well-defined structure and others with different or disordered conformations (case of beta-adrenergics). It is understandable that these partially disordered crystals can originate DSC curves showing various close melting points. This conclusion should be borne in mind by those that are interested in the study of polymorphism because it is not be reliable to consider polymorphs based solely on the value of a certain property, for example fusion temperature. Indeed, the various structural forms evidenced by the DSC fusion curves that could be considered manifestation of polymorphism are in fact consequence of crystal disordering.

9. References

Aguiar, Armando J., John Krc, Arlyn W. Kinkel, and Joseph C. Samyn. "Effect of Polymorphism on the Absorption of Chloramphenicol from Chloramphenicol Palmitate." *Journal of Pharmaceutical Sciences* 56, no. 7 (1967): 847-53.

Bernstein, Joel. *Polymorphism in Molecular Crystals*. Oxford: Oxford University Press, 2002.

Breitenbach, Jörg. "Melt Extrusion: From Process to Drug Delivery Technology." *European Journal of Pharmaceutics and Biopharmaceutics* 54, no. 2 (2002): 107-17.

Brittain, H. G., ed. *Polymorphism in Pharmaceutical Solids*. 2 ed. Vol. 192, Drugs and the Pharmaceutical Sciences. New York: Informa Healthcare USA, Inc, 2009.

Burger, A., and R. Ramberger. "On the Polymorphism of Pharmaceuticals and Other Molecular Crystals. I." *Microchimica Acta* 72, no. 3 (1979): 259-71.

Canotilho, J., R. Castro, M. Rosado, S. Nunes, M. Cruz, and J. Redinha. "Thermal Analysis and Crystallization from Melts." *Journal of Thermal Analysis and Calorimetry* 100, no. 2 (2010): 423-29.

Canotilho, João, Ricardo A. E. Castro, Mário T. S. Rosado, M. Ramos Silva, A. Matos Beja, J. A. Paixão, and J. Simões Redinha. "The Structure of Betaxolol from Single Crystal X-Ray Diffraction and Natural Bond Orbital Analysis." *Journal of Molecular Structure* 891, no. 1-3 (2008): 437-42.

Ceccarelli, C., G. A. Jeffrey, and R. K. McMullan. "A Neutron-Diffraction Refinement of the Crystal-Structure of Erythritol at 22.6 K." *Acta Crystallographica Section B-Structural Science* 36, no. DEC (1980): 3079-83.

Chattopadhyay, Tapan K., Rex A. Palmer, and Daruka Mahadevan. "Molecular and Absolute Crystal Structure of Pindolol-1-(1h-Indol-4-Yloxy)-3-[(1-Methylethyl) Amino]-2-Propanol: A Specific Beta-Adrenoreceptor Antagonist with Partial Agonist Activity." *Journal of Chemical Crystallography* 25, no. 4 (1995): 195-99.

Chokshi, Rina, and Hossein Zia. "Hot-Melt Extrusion Technique: A Review." *Iranian Journal of Pharmaceutical Research* 3 (2004): 2-16.

Coulson, Christopher J. *Molecular Mechanisms of Drug Action*. 2 ed. London: Taylor & Francis, 1994.

Crowley, Michael M., Feng Zhang, Michael A. Repka, Sridhar Thumma, Sampada B. Upadhye, Sunil Kumar Battu, James W. McGinity, and Charles Martin. "Pharmaceutical Applications of Hot-Melt Extrusion: Part I." *Drug Development and Industrial Pharmacy* 33, no. 9 (2007): 909-26.

Dagnac, Thierry, Jean-Michel Guillot, and Pierre Le Cloirec. "Investigation of the Thermal Decomposition of Selected N,N-Dialkylamides at Low Temperature." *Journal of Analytical and Applied Pyrolysis* 42, no. 1 (1997): 53-71.

E.Donth. *Relaxation and Thermodynamics in Polymers: Glass Transition*. Berlin: VCH Pub, 1992.

Endo, Kosuke, Satoko Amikawa, Asami Matsumoto, Norio Sahashi, and Satomi Onoue. "Erythritol-Based Dry Powder of Glucagon for Pulmonary Administration." *International Journal of Pharmaceutics* 290, no. 1-2 (2005): 63-71.

Esteves de Castro, R. A., João Canotilho, Rui M. Barbosa, M. Ramos Silva, A. Matos Beja, J. A. Paixão, and J. Simões Redinha. "Conformational Isomorphism of Organic Crystals: Racemic and Homochiral Atenolol." *Crystal Growth & Design* 7, no. 3 (2007): 496-500.

Esteves de Castro, R. A., and J. S. Redinha. "Unpublished Work." 2002.

Gdanitz, R. "Ab Initio Prediction of Possible Molecular Crystal Structures." In *Theoretical Aspects and Computer Modeling of the Molecular Solid State*, edited by A. Gavevezzoti. Chichester: Wiley & Sons, 1997.

Gonnissen, Y., J. P. Remon, and C. Vervaet. "Development of Directly Compressible Powders Via Co-Spray Drying." *European Journal of Pharmaceutics and Biopharmaceutics* 67, no. 1 (2007): 220-26.

Hancock, Bruno C., and George Zografi. "Characteristics and Significance of the Amorphous State in Pharmaceutical Systems." *Journal of Pharmaceutical Sciences* 86, no. 1 (1997): 1-12.

Hilfiker, Rolf, ed. *Polymorphism in the Pharmaceutical Industry*. Weinheim: Willey-VCH, 2006.

Ioganseen, A. V. "Direct Proportionality of the Hydrogen Bonding Energy and the Intensification of the Stretching V(Xh) Vibration in Infrared Spectra." *Spectrochimica Acta, Part A: Molecular and Biomolecular Spectroscopy* 55, no. 7-8 (Jul 1999): 1585-612.

Jackson, K. A. "Theory of Crystal Growth." In *Treatise on Solid State Chemistry*, edited by N.B. Hannay. 50. New York: Plenum Press, 1975.

Jesus, A. J. L., M. T. S. Rosado, M. L. P. Leitão, and J. S. Redinha. "Molecular Structure of Butanediol Isomers in Gas and Liquid States: Combination of Dft Calculations and Infrared Spectroscopy Studies." *Journal of Physical Chemistry A* 107, no. 19 (May 15 2003): 3891-97.

Jesus, A. J. Lopes, and J. S. Redinha. "On the Structure of Erythritol and L-Threitol in the Solid State: An Infrared Spectroscopic Study." *Journal of Molecular Structure* 938, no. 1-3 (2009): 156-64.

Jesus, A.J.L., M.T.S. Rosado, I. Reva, R. Fausto, M.E. Eusébio, and J.S. Redinha. "Conformational Study of Monomeric 2,3-Butanediols by Matrix-Isolation Infrared Spectroscopy and Dft Calculations." *J. Phys. Chem. A* 110, no. 12 (2006): 4169-79.

Kelton, K . F . "Crystal Nucleation in Liquids and Glasses." In *Solid State Physics*, edited by H. Ehrenreich and D. Turnbull. 75–178. New York: Academic Press, 1991.

Kiselev, S. B. "Kinetic Boundary of Metastable States in Superheated and Stretched Liquids." *Physica A: Statistical Mechanics and its Applications* 269, no. 2-4 (1999): 252-68.

Klein, H., G. Schmitz, and D. Woermann. "Spinodal Decomposition in a Single-Component Fluid." *Physics Letters A* 136, no. 1-2 (1989): 73-76.

Kuhnert-Brandstatter, M. *Thermomicroscopy of Organic Compounds*. Amsterdam: Elsevier scientific publishing company, 1982.

Landau, L. D., and E.M. Lifshitz. *Statistical Physics*. Oxford: Pergamon Press, 1969.

Leităo, M., J. Canotilho, M. Cruz, J. Pereira, A. Sousa, and J. Redinha. "Study of Polymorphism from Dsc Melting Curves; Polymorphs of Terfenadine." *Journal of Thermal Analysis and Calorimetry* 68, no. 2 (2002): 397-412.

Lopes Jesus, A. J., Sandra C. C. Nunes, M. Ramos Silva, A. Matos Beja, and J. S. Redinha. "Erythritol: Crystal Growth from the Melt." *International Journal of Pharmaceutics* 388, no. 1-2 (129-35.

Mairesse, G., J. C. Boivin, D. J. Thomas, J. P. Bonte, D. Lesieur, and C. Lespagnol. "Structure Du Chlorhydrate De L'[Hydroxy-1-(R,S) Isopropylamino-2 Ethyl]-6 Dihydro-2,3 Benzoxazole-1,3 One-2, C12h16n2o3.Hcl." *Acta Crystallographica Section C* 40, no. 8 (1984): 1432-34.

Mollan, Matthew. "Historical Overview." In *Pharmaceutical Extrusion Technology (Drugs and the Pharmaceutical Sciences)*, edited by Isaac Ghebre-Sellassie and Charles Martin. New York: Marcel Dekker, Inc, 2003.

Munro, I. C., W. O. Bernt, J. F. Borzelleca, G. Flamm, B. S. Lynch, E. Kennepohl, E. A. Bär, and J. Modderman. "Erythritol: An Interpretive Summary of Biochemical, Metabolic, Toxicological and Clinical Data." *Food and Chemical Toxicology* 36, no. 12 (1998): 1139-74.

Myerson, Allan S. "Crystallization Basics." In *Molecular Modeling Applications in Crystallization*, edited by Allan S. Myerson. 55-105. New York: Cambridge University Press, 1999.

Nunes, S. C. C., M. E. Eusébio, M. L. P. Leitão, and J. S. Redinha. "Polymorphism of Pindolol, 1-(1h-Indol-4-Yloxyl)-3-Isopropylamino-Propan-2-Ol." *International Journal of Pharmaceutics* 285, no. 1-2 (2004): 13-21.

Ohmori, Shinji, Yasuo Ohno, Tadashi Makino, and Toshio Kashihara. "Characteristics of Erythritol and Formulation of a Novel Coating with Erythritol Termed Thin-Layer Sugarless Coating." *International Journal of Pharmaceutics* 278, no. 2 (2004): 447-57.

Redinha, J.S., and A.J. Lopes Jesus. "Molecular Recognition and Crystal Growth." In *Molecular Recognition: Biotechnology, Chemical Engineering and Materials Applications*, edited by Jason A. McEvoy. Chemical Engineering Methods and Technology: Novapublishers, 2011.

Repka, Michael A., Sunil Kumar Battu, Sampada B. Upadhye, Sridhar Thumma, Michael M. Crowley, Feng Zhang, Charles Martin, and James W. McGinity. "Pharmaceutical Applications of Hot-Melt Extrusion: Part Ii." *Drug Development and Industrial Pharmacy* 33, no. 10 (2007): 1043-57.

Repka, Michael A., Soumyajit Majumdar, Sunil Kumar Battu, Ramesh Srirangam, and Sampada B. Upadhye. "Applications of Hot-Melt Extrusion for Drug Delivery." *Expert Opinion on Drug Delivery* 5, no. 12 (2008): 1357-76.

Ropp, R.C. *Solid State Chemistry*. Amsterdan: Elsevier, 2003.

Sekiguchi, Keiji , and Noboru Obi. "Studies on Absorption of Eutectic Mixture. I. A Comparison of the Behavior of Eutectic Mixture of Sulfathiazole and That of Ordinary Sulfathiazole in Man." *Chemical & Pharmaceutical Bulletin* 9, no. 11 (1961): 866-72.

Silverman, Richard B. *The Organic Chemistry of Drug Design and Drug Action*. New York: Elsevier Academic Press, 2004.

Stenlake, J. B. *Foundations of Molecular Pharmacology*. Vol. 1, London: Athlone Press, 1979.

Stolov, Andrey A., Mikhail D. Borisover, and Boris N. Solomonov. "Hydrogen Bonding in Pure Base Media. Correlations between Calorimetric and Infrared Spectroscopic Data." *Journal of Physical Organic Chemistry* 9, no. 5 (1996): 241-51.

Threlfall, Terence L. "Analysis of Organic Polymorphs. A Review." *Analyst* 120, no. 10 (1995): 2435-60.

Turnbull, David, and Brain G. Bagley. "Transitions in Viscous Liquids and Glasses." In *Treatise on Solid State Chemistry*, edited by N. B. Hannay. 513-54. New York: Plenum Press, 1975.

Wiedemann, H. G., and G. Bayer. "Application of Simultaneous Thermomicroscopy/Dsc to the Study of Phase Diagrams." *Journal of Thermal Analysis and Calorimetry* 30, no. 6 (1985): 1273-81.

Wong, Joe, and C. Austen Angell. *Glass: Structure by Spectroscopy* New York Marcell Dekker, 1976.

Zhou, L. M., J. Sirithunyalug, E. Yonemochi, T. Oguchi, and K. Yamamoto. "Complex Formation between Erythritol and 4-Hexylresorcinol." *Journal of Thermal Analysis and Calorimetry* 59 (2000): 951-60.

11

Alternative Protein Crystallization Technique: Cross-Influence Procedure (CIP)

Ivana Nemčovičová[1,2] and Ivana Kutá Smatanová[3]
[1]*La Jolla Institute for Allergy and Immunology, La Jolla,*
[2]*Slovak Academy of Sciences, Institute of Chemistry, Bratislava,*
[3]*University of South Bohemia, IPB, Nové Hrady,*
[3]*Academy of Sciences of the Czech Republic, INSB, Nové Hrady,*
[1]*USA,*
[2]*Slovakia,*
[3]*Czech Republic*

1. Introduction

In general, the crystallization of proteins is a very complex process. Experiences of many scientists point out that majority of proteins is difficult to crystallize and even if a protein tends to crystallize relatively easily there are many parameters that must be taken into account. There are multiple reasons that point out the difficulty of protein crystal growth. Apparently, protein molecules are very complex (large, flexible molecules often composed of several subunits), relatively chemically and physically unstable (unfolding, hydration requirements, temperature sensitivity) and they have dynamic properties. If the solution changes, the molecule properties (e.g. conformation, charge and size) will change too. Furthermore, every macromolecule is unique in its physical and chemical properties since every amino acid sequence produces a unique three-dimensional structure having distinctive surface characteristics. Thus, conditions applied for one protein can only marginally apply to others (Giacovazzo, 2002; Lattman, 2008).

Therefore, finding of successful crystallization conditions for a particular protein remains a highly empirical process. During optimization a variable set of parameters is screened to determine appropriate conditions for nucleation and growth of single crystals suitable for X-ray diffraction analysis. In parallel to modern high-throughput approaches used in the protein crystallization, in recent years we performed basic research on physico-chemical properties and molecular interactions influencing crystal growth. Empirically, we have explored another tool useful for optimization strategy that was first described by Tomčová and Kutá Smatanová (2007). A new crystallization procedure modifying protein crystal morphology, internal packing and influencing crystal growth was tested particularly. For the first time the metal ion salts were added simultaneously to the protein drop and even to neighboring drops to allow a cross-influence effect of additives during crystallization experiment. The presence of metal ions significantly influences the crystal growth, as the modification of crystal morphology and internal packing were observed. This newly discovered cross-crystallization method (Tomčová & Kutá Smatanová, 2007; Tomčová et al., 2006) was called Cross-Influence Procedure (CIP).

This book chapter contains the brief introduction to protein crystallization that is given within the first paragraph. The second paragraph anticipates general principles of macromolecular crystallization by uniting the nucleation kinetics, crystal growth, and physical methods with the phase diagram. Advantages and disadvantages of each described crystallization technique are exclusively introduced. The main chapter describing the alternative crystallization techniques is taking advantage of discussing the importance of additives in protein crystallization, as well as, the novel approach to macromolecular crystallization and reporting the effects of CIP on crystallization of two different proteins. In addition, for the first time the detailed protocol of CIP is given within this chapter to help readers to perform their own cross-crystallization experiment by using selective additives. Until now there has been no monograph devoted exclusively to the role of cross-influence of additives and their general usage guideline in protein crystallization. The present text is an attempt to fill this gap. We have learned heavily on the work of numerous other authors during the writing and this text is primarily addressed both to specialists and to graduate students who are interested in looking for a mini-review of the modern alternative techniques used in macromolecular crystallization. Thus, this book chapter stands as a valuable guide to the alternative protein crystallization.

2. The basic principles of macromolecular crystallization

2.1 Introduction to crystal growth

Crystallographers try to grow the protein crystals by slow, controlled precipitation from aqueous solution under conditions, which do not evoke the protein denaturation. A number of substances cause proteins to precipitate. Ionic compounds, usually salts, precipitate proteins by a process called salting out (Weber, 1991; Stura 1991). Organic solvents also provoke precipitation, but they often interact with hydrophobic parts of proteins and thereby denature them. The water-soluble polymer polyethylene glycol (PEG) is widely used because it is a powerful precipitant and a weak denaturant (McRee, 1999).

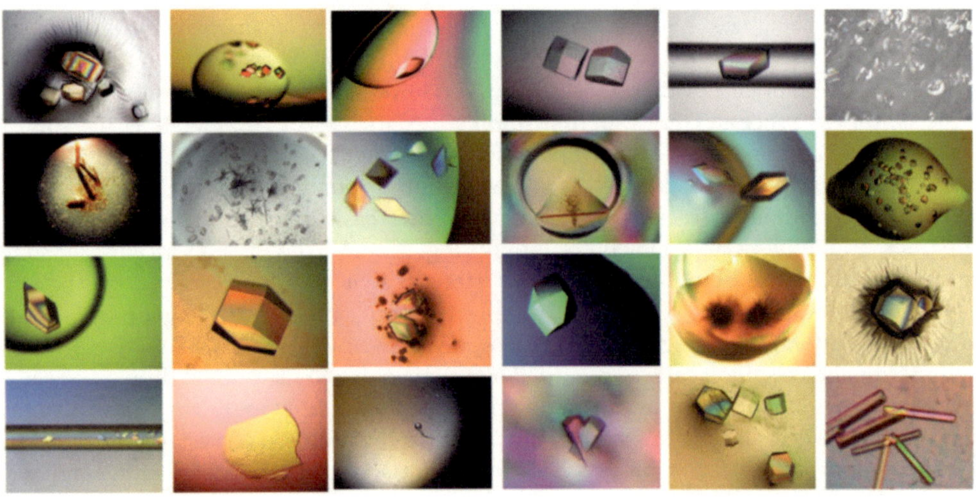

Fig. 1. Example of various protein crystals from the author's personal crystal gallery.

One simple way of causing slow precipitation is adding denaturant to an aqueous solution of protein until the denaturant concentration is slightly below than concentration of required precipitating the protein. In this case the water is allowed to evaporate slowly that gently raises the concentration of both protein and denaturant until precipitation occurs. Whether the protein forms crystals of specific shapes, a useless amorphous solid, depends on many properties of the solution, such as a protein concentration, temperature, pH, ionic strength, etc. Finding the exact conditions to produce good crystals of specific protein often requires many exhaustive trails, and in some cases, it is more art than science.
However, under certain circumstances, many molecular substances, including proteins, solidify to form crystals. In entering the crystalline state from solution, individual molecules of the substance adopt one or a few identical orientations. The resulting crystal is an orderly three-dimensional array of molecules, held together by non-covalent interactions or proteins in the crystal stick to each other primarily by hydrogen bonds through inverting water molecules (D. Voet & J. G. Voet, 1995). In terms of X-ray crystallography, the sample being examined is in the crystalline state, called crystal. A few macromolecular crystals are shown on Fig. 1.

2.2 The principle of crystallization in phase diagram

Crystallization of a protein is a multiparametric process in which the parameters are varied in the search for optimal crystallization conditions. The most common parameters are protein concentration, the nature and the concentration of the precipitant, pH, and temperature. Specific additives that affect the crystallization by manipulating of sample-sample and sample-solvent interactions to enhance or alter sample solubility can also be added in low concentration (see next paragraph).

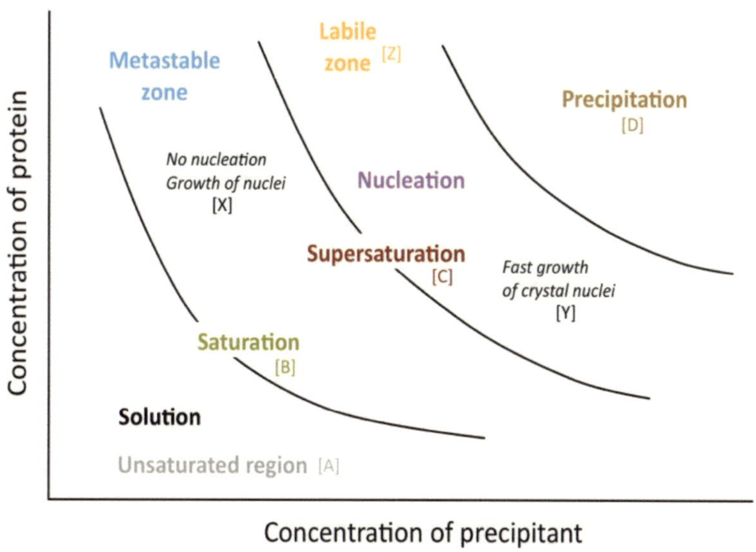

Fig. 2. A typical schematic solubility curve for a protein, as a function of the salt concentration or another parameter.

Two-dimensional solubility diagram (Fig. 2) is classically used to explain formation of crystal nuclei and their growth. The solubility curve (B) divides the concentration space into two areas - the undersaturated (A) and supersaturated zones (C). Each point on this curve corresponds to a concentration, at which the solution is in equilibrium with the precipitating agent. These correspond to the situation either at the end of the crystal growth process coming from a supersaturated solution or to a situation when crystal dissolution occurs in an undersaturated solution. In the area under the solubility curve, the solution is undersaturated (A) and the crystallization will never take place. Above the solubility curve lays the supersaturation zone (C); here, for a given concentration of precipitating agent, the protein concentration is higher than that at equilibrium. Depending on the kinetics to reach equilibrium and the level of supersaturation, this region may itself be subdivided into three zones (Table 1; Chirgadze, 1998; Stura, 1991; McPherson, 1990).

1.	The precipitation zone (D) is the zone, where the excess of protein molecules immediately separates from the solution to form amorphous aggregates.
2.	The nucleation zone (C) is the zone, where the excess of protein molecules aggregates in a crystalline form. Near the precipitation zone, crystallization may occur as a shower of microcrystals, which can be confused with a precipitate.
3.	A metastable zone (Z); a supersaturated solution may not nucleate for a long period, unless the solution is mechanically shocked or a seed crystal introduced. To grow well-ordered crystals of large size, the optimal conditions would have to begin with the formation of a preferably single nucleus in the nucleation zone just beyond the metastable zone. As the crystals grow, the solution would return to the metastable region and no more nuclei could occur. The remaining nuclei would grow, at a decreasing rate that would help to avoid defect formation, until equilibrium is reached.

Table 1. Schematic localization (see in Fig. 2) and a short description of the three zones of the phase diagram.

To obtain the best results, crystals should be grown at a lower level of supersaturation than is required for nuclei formation. To achieve crystal growth, supersaturation must be reduced to a lower level; maintaining a high supersaturation would result in the formation of many nuclei and therefore many small crystals (Y). The basic strategy is to bring the system into a state of limited degree of supersaturation by modifying the properties of the solvent through equilibration with a precipitating agent, addition of additives, or by altering some physical properties such as temperature, etc. The most important condition is that the crystals should grow slowly to reach a maximum degree of order in their structure. In practice this fundamental rule is not always obeyed. The supersaturation can be also maintained by introducing the suitable excipients to the crystallization system. However, the easiest and most common way to change the degree of supersaturation is by changing the temperature to lower degree and/or adding the polymers (as Glycerol; 10% should be good starting concentration) to slow down the crystallization process.

2.3 The nucleation kinetics

As was described in previous paragraph, crystallization takes place in supersaturated solutions, which are far from equilibrium. To attain equilibrium a solid state must be first created, a nucleus must be formed (see also next paragraph, "Formation of self-assembly"). The system proceeds to equilibrium through crystal growth. The formation of a nucleus is the rate-limiting step in the crystallization process, the kinetic determinant. It requires more energy than does any prior or subsequent step. If no adequate amount of energy ever becomes available, the system will remain in a metastable, nonequilibrium state of supersaturation (McPherson, 1990; Drenth, 2006). No nuclei will form, and no crystals will grow.

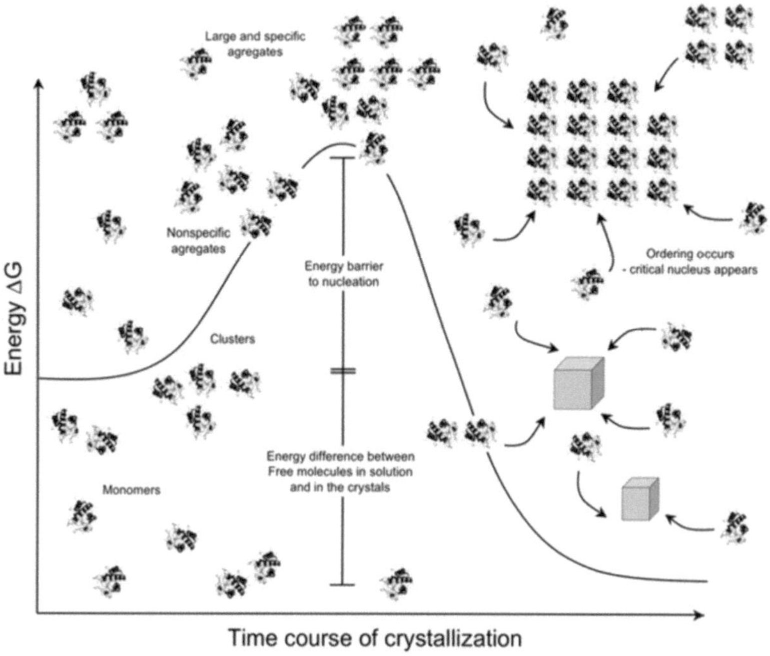

Fig. 3. The activation energy and nucleation barrier: Analogous to phase diagram for conventional chemical reactions, the formation of crystals from solution can be similarly represented. As shown here, molecules free in solution under conditions of supersaturation (left) are at a higher energy state with respect to those in the crystalline state (right). Figure originaly adapted from Drenth (2006) and modified by authors.

There are two approaches helping the system over the activation barrier: putting energy into system by some means, or effectively lowering the energy barrier (Fig. 3). Chemists do it by heating of their reactants and putting force energy into the system; biochemists do it by adding of biological catalysts (enzymes) to their reactants. Crystal growers utilize both approaches; they increase the supersaturation of their solution and later they seed the solution with preexisting crystals or with heterogeneous nuclei or introducing additive into crystallization system. The important point is that nucleation is difficult and improbable; growth of preexisting crystals is by contrast easier and more likely. Nucleation is characterized by a distinct series of events, and exactly these events do not need pertain to growth, which is the ordered addition of molecules to a growing lattice (McPherson & Weickmann, 1990; McPherson, 1991).

2.4 Crystal formation requires the self-assembling

Many self-assembly processes rely on the self-assembling nature of anorganic molecules, including complex species such as DNA and proteins; these methods are termed chemical or molecular self-assembly (Fink, 2005). Generally, the molecular self-assembly is the spontaneous organization of relatively rigid molecules into structurally, well defined aggregates, through weak, reversible interactions such as hydrogen bonds, ionic bonds and van der Waals bonds. The aggregated structure represents a minimum energy structure or equilibrium phase. Other simpler methods rely on geometric self-organization, in which hard spheres or hard rods will arrange themselves into two- and three-dimensional structures based on packing considerations. Self-assembly is scientifically interesting and technologically important for various reasons. Firstly the self-assembly is centrally important in life. The cell contains an astonishing range of complex structures such as lipid membranes, folded proteins, structured nucleic acids, protein aggregates, molecular machines, and many others that are formed by self-assembly. The second reason is that self-assembly provides routes to a range of materials with regular structures: molecular crystals, liquid crystals, and semicrystalline and phase-separated polymers are examples. In crystallization of proteins, the nuclei can be created by simple addition of the selective excipient that might form the self-assembly. The concepts of self-assembly historically have come from studying of molecular processes (Fink, 2005). A self-assembling system consists of a group of molecules or segments of a macromolecule that interact with one another (Fig. 4). These molecules or molecular segments may be the same or different. Their interaction leads from some less ordered state (a solution, disordered aggregate, or random coil) to a final high ordered state (a crystal or folded macromolecule). Self-assembly occurs when molecules interact with one another through a balance of attractive and repulsive interactions. These interactions are generally weak and non-covalent (van der Waals and Coulomb interactions, hydrophobic interactions, and hydrogen bonds) but relatively weak covalent bonds (coordination bonds) are recognized increasingly as appropriate for self-assembly. Complementarity in shapes among the self-assembling components is also crucial.

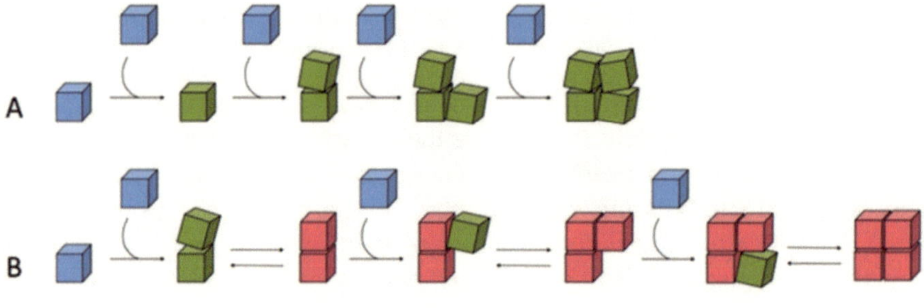

Fig. 4. Schematic illustration of the essential differences between irreversible aggregation (A) and ordered self-assembly (B). The aggregation occurs when there is a net attraction and an equilibrium separation between the components. The equilibrium separation normally represents a balance between attraction and repulsion. (A and B). (A) Components (shown in blue) that interact with one another irreversibly form disordered glasses (shown in green). (B) Components that can equilibrate, or adjust their positions once in contact, can form ordered crystals if the ordered form is the lowest-energy form (shown in red).

For self-assembly to generate ordered structures, the association must be either reversible or must allow the components to adjust their positions within an aggregate once it has formed (Fig. 4). The strength of the bonds between the components, therefore, must be comparable to the forces tending to disrupt them. For molecules, the forces are generated by thermal motion. Processes, in which collision between molecules leads to irreversible sticking, generate glasses, not crystals. The self-assembly of molecules normally is carried out in solution or at an interface to allow the required motion of the components. The interaction of the components with their environment can strongly influence the course of the process. The molecules must be also mobile for self-assembly to occur. In solution, thermal motion provides the major part of the motion required to bring the molecules into the contact. In nanoscale, mesoscopic, and macroscopic self-assembly systems, the components interact in ways that are analogous to those involving molecules. To design such systems, the first challenge often is assuring the mobility of the components, as they become larger than molecules (Fig. 4), Brownian motion rapidly becomes irrelevant, and gravity and friction become important (Fink, 2005).

2.5 Physical methods of protein crystallization

The most of the classical protein crystallization methods have been already well described in scientific literature. However, this paragraph is briefly summarizing the four of them, which are commonly employed to affect supersaturation in the crystallization of macromolecules: vapor diffusion, free interface diffusion, batch, and dialysis. Although each of these techniques achieves supersaturation of the particular macromolecule to be crystallized, the means by which supersaturation is achieved in each case varies to a great extent. A discussion on each of these techniques is presented below; moreover, we exclusively introduced their advantages and disadvantages to help readers choose the right technique.

2.5.1 Vapor diffusion

Vapor diffusion technique utilizes evaporation and diffusion of water between solutions of different concentration as a way to approach and achieve supersaturation of macromolecules. Typically, a solution containing a macromolecule is mixed in a ratio 1:1 with a solution containing the precipitant at the final concentration, which is to be achieved after vapor equilibration. The drop containing the 1:1 mixture of protein and precipitant (both of which have been diluted to 1/2 the original concentration by mixing with the other) is then suspended and sealed over the well solution, which contains the precipitant at the target concentration, as either a hanging or a sitting drop. Glass capillaries containing protein and precipitant at concentrations below that required for crystallization can also be vapor equilibrated against a well solution in a sealed test tube. DeMattei and coworkers have shown capillary based vapor equilibration to occur at rates up 102 times slower than drop based methods, resulting in improved crystals (DeMattei et al., 1992). The difference in precipitant concentration between the drop and the well solution is the driving force that causes water to evaporate from the drop until the concentration of the precipitant in the drop equals that of the well solution. Since the volume of the well solution is much larger than the volume of the drop (1-3 mL as compared to 1-20 μL) its dilution by the water vapor leaving the drop is negligible. Fowlis has demonstrated that the rate of vapor equilibration in normal gravity is dependent strictly on the rate of vapor diffusion of water in the space separating the drop and the well (Fowlis et al., 1988). Due to convection effects (caused by the increased concentration of the precipitant at the edge of the drop as water evaporates), the rate of diffusion of a water molecule in the suspended drop (in solution) is actually higher than that of the vaporized water molecule. In microgravity experiments, the rate of equilibration is based solely on the rate of diffusion of water in the drop until crystal nucleation increases the rate of equilibration (Provost & Robert, 1991). Many scientists report that the use of simple agarose gels can offset convection currents under normal gravity producing improved results with hanging drops (McPherson, 1999; Otalora et al., 2009; Van Driessche et al., 2008). Vapor diffusion is the optimal technique to use either when screening a large number of conditions (by varying the composition of each well solution) or when the lack of protein prevents the use of other methods. Furthermore, this method can be used to increase or decrease the concentration of protein in the equilibrated state relatively to its initial concentration. This is done, by varying the volume of protein mixed with the well solution when the drop is initially setup. Since the drop equilibrates so that the precipitant concentration matches that of the well solution, the final volume of the drop will always be equal to that of the initial amount of well solution mixed with the protein. One of the drawbacks to vapor equilibration is higher tendency to form smaller crystals than in other methods. This may be due to small drop volumes limiting the quantity of crystallizable solute present or creating a higher level of impurities as compared to other techniques, which utilize larger volumes. As crystals grow, the concentration of defective molecules increases relatively to perfect molecules (which are being selected for the crystal). When this factor is combined with the higher probability of impurities diffusing to the face of the crystal (due to the smaller volumes), the likelihood of inclusion of defects into the growing crystal increases (Giacovazzo et al., 2002). Thus, the production of X-ray quality crystals may be better suited to the use of batch, free interface diffusion, or dialysis techniques, which utilize larger solution volumes at equilibrium.

2.5.2 Free interface diffusion

Free interface diffusion is the method when layering of a low-density solution onto one of higher density, usually in the form of concentrated protein onto concentrated salt, can be used as the means of growing large crystals. Nucleation and crystal growth generally occur at the interface between the two layers (García-Ruiz et al., 2003; García-Ruiz & Morena, 1994), at which both the concentration of salt and the concentration of protein are at their highest values. The two solutions slowly intermix over time, and should be made up so that at equilibrium, (at which point in time both solutions are diluted to some fraction of their initial values), the concentration of the precipitant is still high enough to promote crystal growth. Since the solute to be crystallized must be concentrated, this method tends to consume fairly large amounts of protein.

2.5.3 Batch

In the batch method (Rayment, 2002; D'Arcy et al., 1996; Blow et al., 1994) concentrated protein is mixed with concentrated precipitant to produce a final concentration, which is supersaturated in terms of the solute macromolecule and therefore leads to crystallization. This can be done with up to milliliters amounts of solution and typically results in larger crystals due to the larger volumes of solute present and the lower chance of impurities diffusing to the face of the crystal. This technique is much expensive in terms of consumption of the soluble macromolecule, and thus should not generally be used to screen initial conditions for crystallization.

2.5.4 Dialysis

Dialysis utilizes diffusion and equilibration of small precipitant molecules through a semipermeable membrane as a means of slowly approaching the concentration, at which the macromolecule solute crystallizes. Initially, the solute is contained within the dialysis membrane that is than equilibrated against a precipitant solution. Equilibration against the precipitant in the surrounding solvent slowly achieves supersaturation for the solute within the dialysis membrane, eventually resulting in crystallization. Dialysis tubes can be used by itself, in the case of large amounts of protein being available, or tubes can be used to cover the opening of a dialysis button, allowing diffusion of the surrounding solvent in to the solute through the dialysis membrane. Dialysis buttons themselves come in a variety of sizes from 7-200 µL. The advantage of dialysis in comparison with other methods is in the simplicity with which the precipitating solution can be varied, simply by moving the entire dialysis button or sack from one condition to another. Protein can thus be continuously recycled until the correct conditions for crystallization are found (Carter et al., 1988). One drawback of this method is that it does not work at all with concentrated PEG solutions, as they tend to draw all the water out of the button or sack faster than PEG dialyzes across the membrane, thus resulting in precipitated protein.

3. Alternative crystallization techniques

3.1 Importance of additives in protein crystallization

Any foreign substance other than the crystallizing compound is considered as an additive. Thus, a solvent used for growth and any other compound deliberately added to the growth medium or inherently present in it, is an additive. Different terms, such as impurity, admixture, excipient, inhibitor and poison, are used in the literature for foreign substances other than the solvent used for obtaining supersaturated solutions (Stura et al., 1991; Sangwal, 1998; Cox & Weber, 1988). Irrespective of its concentration, a deliberately added impurity is called an additive, but by the term admixture we mean an impurity added in relatively large amounts (up to several percent). A surfactant may be any chemical compound active on the surface in changing its growth behavior. An impurity can accelerate or decelerate the growth process. The impurity that decelerates growth is called a poison or an inhibitor, while one that accelerates growth is named as a growth promoter (Sangwal, 1998). Foreign substances (called additives) present in the aqueous solution used for the crystallization of substances can be as diverse as simple ions of common bivalent metal salts, various proteinaceous compounds such as aspartic and glutamic acids, as well as natural compounds as protein cofactors or ligands. During the different phases of crystallization, the same additive can modify the crystallization behavior of both highly and sparingly soluble proteins. Thus, additives affect different processes involved during crystallization. Therefore, the understanding of interactions between additives and crystallizing phase is important.

3.1.1 Additive screening

In general, the additive screen commonly used in protein crystallography consists of a number of different small molecules that can affect the solubility and crystallizability of

biological macromolecules, including both soluble and membrane proteins. These small molecules can perturb and manipulate sample-sample and sample-solvent interactions, as well as perturb water structure, which can alter and improve both the solubility and crystallization of the sample. Additives can stabilize or create conformity by specific interaction with the macromolecules. There are numerous reports of the use of small salt additives to improve the quality and size of macromolecular crystals (Tomčová et al., 2006, 2007; Hartmut, 1991; Ducruix & Giege, 2000; Cudney, 1994; Sousa, 1995; Trakhanov & Quiocho, 1995). However, as the general protocol of additive usage has not been well defined yet, we are including this procedure in the following text.
The additive screen is usually designed to allow the rapid and convenient evaluation of number of unique additives and their ability to influence the crystallization of the sample. The screen is generally well compatible with plenty of popular crystallization reagents including in all of the commercial screens. The Hampton Research Additive Screens (Hampton Research, Aliso Viejo, CA, USA) can serve as a good example. Additives are commonly preformulated in deionized water and need to be equilibrated at room temperature until all the components are dissolved. This guide is describing the use of additive screen kit using the sitting drop vapor diffusion method with a 1 mL reservoir volume. Other methods such as hanging drop vapor diffusion crystallization or micro-batch may also be utilized as well as smaller reservoir and drop volumes. Two separate methods of setup are to be used for volatile (Table 3) and non-volatile additives (Table 2). The final concentration of the additive can be ranged from 5 to 20 mM. The other variants of additive composition in the drop and reservoir are listed in the paragraph of volatile buffers usage (see Fig. 8).

1.	Add 2 μL of sample into a sitting drop well.
2.	Add 1 μL (or 0.5 μL) of additive into the sample drop.
3.	Add 2 μL of the crystallization reagent (buffer, salt, etc) into the sample/additive drop.
4.	Add 1 mL of crystallization reagent into the reservoir.
5.	Seal the reservoir with tape or greased slides.

Table 2. Step-by-step setting of one drop-well and reservoir setup for non-volatile additives used in single crystallization experiment.

1.	Add 2 μL of sample into a sitting drop well.
2.	Add 2 μL of the crystallization reagent (buffer, salt, etc) into the sample drop.
3.	Add 900 μL of crystallization reagent and 100 μL of the volatile additive into the reservoir.
4.	Seal the reservoir with tape or grease and slides.

Table 3. Step-by-step setting of one drop-well and reservoir setup for volatile additives used in single crystallization experiment.

3.1.2 Natural additives – Cofactors and ligands

A cofactor is a non-protein chemical compound that is bound to a protein and it is required for its biological activity. These proteins are commonly enzymes, and cofactors can be considered as auxiliary molecules assisting in biochemical transformations. Cofactors are either organic or inorganic. They can also be classified depending on how tightly they bind to an enzyme (called also as coenzymes or prosthetic groups). For example, DNA and RNA binding proteins can be co-crystallized with oligonucleotides. Crystallization of certain macromolecule often requires presence of natural ligands in crystallizing solution. Some of these compounds can be added directly to the crystallization drop (co-crystallization), while other need to be co-incubated in advance to form a complex. Recent studies on ligand-protein crystallization show several advantages. The ligands, in general, stabilizing macromolecule in mother liquor and thus crystallization is less affected by protein aggregation (see also next paragraph). The stabilizing effect was also observed in crystallization of high affinity protein complexes (Brooijmans et al., 2002). However, this approach is somewhat specialized and clearly not universally applicable. The natural ligand may be unknown or not available in sufficient amounts. Crystallization with the natural ligand is not always useful in drug development, as the natural ligand occupies what is likely to be the most potent site for a drug.

3.1.3 Inhibitors and heavy-atoms ligands

It is the solvent content that makes the difference between a classical molecular crystal and a protein crystal: in the former, all the atoms can be described in terms of a regular lattice, whilst in the latter a crystalline array coexists with high portion of material in the liquid state (Brooijmans et al., 2002). The mother liquid, whose content can range approximately from 30 to 75 % (or more), has a strong influence on the behavior of this kind of crystal, making them very peculiar and creating some advantages along with some obvious disadvantages. The major disadvantage is that protein crystals are much less ordered than classical crystals, not only for the large amount of disordered material presents in the crystal itself, but also because surface groups of the macromolecule in contact with the solvent can show a great mobility. On the other hand, among the advantages is that the environment of the macromolecule in the crystal is not such different from that of the solution from which the crystal was obtained (the influence of the solvent on the conformation of the protein cannot be underestimated) and we can profit from the solvent in the preparation of heavy-atoms derivatives, as well as inhibitor soaking (Garman & Murray, 2003). Thus, another group of additives are considered inhibitors and heavy-atoms. A general guideline, how to co-crystallize or soak the existing crystal with the heavy atom, or an inhibitor, cannot exist for obvious reasons. However, there are some empirical rules, which should be followed (see The Hampton Research Heavy Atom Screens - Hampton Research, Aliso Viejo, CA, USA). The concentration may range from 0.05–50 mM, but should not be more than 100 mM, it depends on molar ratio between a molecule to be crystallized and the heavy atom or inhibitor to be introduced into the molecule. Therefore, a small stock solution of 10 mM is an appropriate working stock solution. Please note, the heavy atoms and inhibitors are only sparingly soluble in water. Addition of these reagents to macromolecular crystals (so-called soaking) can be accomphisehed by solubilizing the heavy atom or inhibitor in a carrier solvent that is less polar than water. Acetonitrile is a frequently used carrier solvent. Disolve these reagents in the carrier solvent and then pipet the solution with heavy atom or inhibitor into the crystal mother liquor so that the final carrier solvent concentration in the crystal mother liquor is 3 to 5% v/v. Important note here, the volatile reagents containing heavy atoms should be handeled in a fume hood. In some cases one can first derivate the protein with heavy atom then attempt crystallization. This procedure is less frequently used since the procedure may not produce derivated crystals, which are isomorphous with the native crystals because the positioning of the heavy atom or inhibitor molecules may disrupt intermolecular contact. Sometimes the presence of these additives with the native macromolecule in solution can change the solubility of the macromolecule which in turn can change the crystallization conditions. The native and derivated crystals now grow under different conditions and thus one must screen for new crystallization condition. Also, with the macromolecule free in solution it is possible that additional heavy atom sites are now introduced (which can complicate the phasing) since sites previously hidden by crystal contacts are now exposed. However, the method can be useful when one is trying to derivate the macromolecule with a heavy atom, ligand or inhibitor that is large enough to diffuse into and through the crystal's solvent channels.

3.1.4 Antibody and affibody fragments

While antibody-based therapeutics have become firmly established as frontline drugs, the use of antibodies as research tools in small molecule drug discovery is still in its infancy. This paragraph is focused on the use of antibody fragments as crystallization chaperones to aid the structural determination of otherwise uncrystallizable target proteins. Genetic engineering of proteins, specifically enhancing crystallization, is now common practice to troubleshoot proteins that crystallize with difficulty. For example, engineering an increase in the hydrophilic surface area of the molecule through the generation of a fusion protein has been used to enhance crystallization of several integral membrane proteins. Easy-crystallizing T4 lysozyme has been utilized successfully as a fusion partner by replacing the cytoplasmic loop in the target protein. However, enzymatic or genetic modification carries with it risk that the conformation adopted by the recombinant target may not be achievable with the native protein. Co-complexing a target protein with an auxiliary protein (antibody or affibody fragment), which acts as a chaperone, represents a particularly attractive option with wide applicability (Griffin & Lawson, 2011). Inclusion of the chaperone increases the probability of high-quality crystal formation by minimizing the target conformational heterogeneity through 'locking' or 'clamping' the target in a particular conformation (possibly previously unknown), masking inhibitory surfaces and extending facilitating surfaces. Formation of the crystal lattice based on contacts between antibody molecules is also very likely, in most circumstances, to be an advantage, so that sites of biological interest on target proteins are not occluded by crystal contacts. In addition, an antibody fragment with a previously characterized structure can facilitate molecular replacement phasing. While antibody fragments such as the Fab, Fv, single-chain Fv and single-domain camelid-derived VHH have proven ability to enhance solubility and stabilize target proteins, alternative scaffolds such as affibodies and the helical, but disulphide-free, designed ankyrin repeat proteins with randomized surface residue positions (DARPins) have also shown promisingly. DARPins and antibody fragments may be considered complementary in assisting the crystallography of proteins, as largely independent epitopes are defined by the two scaffolds.

Another promising approach, aimed at generalizing antibody mediated crystallization, is to engineer a tag binding epitope into a known loop region of a protein, thus facilitating the generic application of Fab fragments from well-known, and easily sourced, antibodies. If this approach prove to be successful with a wide range of targets, it will eliminate the need to generate or select new antibody-binding sites for each new protein, and thus has considerable attraction.

3.2 Influence of additives promoting crystal formation

3.2.1 Addition and effect of salts in protein crystallization

A protein's multiple acid-base groups make its solubility properties during crystallization dependent on the concentration and type of dissolved salts, the polarity of the solvent, the pH, and the temperature. Different proteins vary greatly in their solubility under given set of conditions. Certain proteins precipitate from solution under conditions, in which others remain quite soluble. This effect is routinely used as the basis for protein crystallization and greatly affected by addition of various excipients (additives described above).

Thus, the solubility of a protein in aqueous solution is a sensitive function of the concentration of dissolved salts. The salt concentration is usually expressed in terms of the ionic strength, which is defined by molar concentration and ionic charge. The use of this parameter to account for the effects of ionic charges results from theoretical consideration of ionic solution. However, protein solubility at a given ionic strength varies with the types of ions in solution. The order of effectiveness of these various ions on influencing protein solubility is quite similar for different proteins and it is apparently mainly due to the ions size and hydration. Thus, the solubility of a protein at low ionic strength generally increases with the salt concentration. The explanation of this salting in phenomenon is that as salt concentration of protein solution increases, the additional counter-ions more effectively shield the protein molecules multiple ionic charges and thereby increase the protein solubility. At high ionic strengths, the solubility of proteins, as well as those of most other substances, decreases. This effect, known as salting out, is primarily a result of the competition between the added salt ions and the other dissolved solutes for molecules of solvation. At high salt concentration many of the added ions are solvated that the amount of bulk solvent available becomes insufficient to dissolve other solutes. In thermodynamic terms, the solvents activity is decreased. Hence, solute–solute interactions become stronger than solute-solvent interactions and the solute precipitates. This is the basis for the most commonly used protein crystallization procedures.

Ammonium sulfate is the most commonly used reagent for salting out proteins during the crystallization because its high solubility (3.9 M in water at 0 °C) permits the achievement of solutions with high ionic strengths (D. Voet & J. G. Voet, 1995). Certain ions, notably I^-, ClO_4^-, SCN^-, Li^+, Mg^{2+}, Ca^{2+}, and Ba^{2+}, increase the protein solubility rather than salting them out. These ions also tend to denaturate proteins. Conversely, ions that decrease the solubility of proteins stabilize their native structures so that proteins having been salted out are not denaturated.

3.2.2 Additives influencing the pH

Proteins generally bear numerous ionisable groups, which have a variety of pK's. At a pH characteristic for each protein, the positive charges on the molecule exactly balance molecule's negative charges. At this pH, the protein's isoelectric point, pI, the protein molecule does not carry the net charge and is therefore immobile in an electric field. Thus, the solubility behavior, which is shared by most proteins, is easily explained.

Physicochemical considerations suggest that the solubility properties of uncharged molecules during the crystallization are insensitive to salt concentration. Therefore, a protein at its isoelectric point should not be subjected to salting in (D. Voet & J. G. Voet, 1995). Oppositely, as the pH is varied from a protein's pI, that is, as the protein's net charge increases, it should be increasingly subjected to salting in because the electrostatic interactions between neighboring molecules that promote aggregation and precipitation likewise increase. Hence, in crystallization solution of moderate salt concentration, the solubility of a protein as a function of pH is expected to be at a minimum at the protein pI and to increase about this point with respect to pH. Proteins vary in their amino acid compositions and therefore in their pI's. The solubility of the protein may be changed by adding salt as additive or enhancing the concentration of acid or base (changing pH-value of the solution). Consequently, the pH of the solution may affect not only the growth and dissolution rate (Langer, 1985; Kuznetsov & Hodorowicz, 1987), but also the different physical properties of the saturated solution like osmotic pressure, the density, the surface tension and the metastable region. Only a limited number of studies deal with the effect of pH level on the crystallization kinetics. Langer studied the effect of pH levels on the growth rate of various salts. His results show a maximum crystal growth in the neutral solution, with lower crystal growth rates in both acidic and alkaline solutions. There are several more or less satisfactory explanations of the effect of pH on crystallization from general point of view. A plausible explanation says that the presence of free acids or bases modifies the nature and the concentration of ions in solution. Mohameed and Ulrich (1996) explain the effect of pH on crystal growth in terms of a structure in a solution, namely of a hydration of ions. Most cations and OH^- ions are hydrated, the largest hydration enthalpy has the H^+ ion so that its presence in solution has stronger tendency of interaction with water molecules than, for example, the K^+ ion so that a competition of ions to acquire water molecules takes place. The K^+ ions have smaller chances to be fully hydrated and therefore they tend to drift towards the crystal surface rather than to remain in the solution.

3.2.3 Effect of organic solvents in protein crystallization

Water-miscible organic solvents, such as acetone and ethanol, are generally good protein precipitants because their low dielectric constants, reduce the solvating power of their aqueous solutions for dissolved ions such as proteins. The decreasing of the dielectric constant by organic solvents also magnifies the differences in the salting out behavior of proteins thus this effect can be helpful in crystallization. Some water-miscible organic solvents, such as dimethyl sulfoxide (DMSO) or N,N-dimethylformamine (DMF), are rather good protein solvents, although they have relatively high dielectric constant (D. Voet & J. G. Voet, 1995).

3.2.4 Crystal-additive interactions

During the nucleation and growth of crystals, different types of interactions are involved between growth species (molecule or ions) and the growing surface. These interactions include van der Waals, ionic and hydrogen bonding. Van der Waals interactions are important during the growth of simple organics compounds such as *n*-alkanes and ionic interactions during the crystallization of simple ionic crystals, while hydrogen bonding is crucial for both organic and inorganic crystals. During the growth in the presence of additives, similar chemical interactions occur at the interface between the additive species and the solid phase. Recent results on the associations between protein molecules in crystal lattices, crystal–solution surface energy, elastic properties, strength, and spontaneous crystal cracking are reviewed and discussed in Chernov (2003). Unlike that of colloids, which are crystallized at increasing particle density because of mutual particle repulsion, protein crystallization is driven by specific chemical attraction between macromolecules that result in the onset of intermolecular macrobonds within the crystal (see self-assembly paragraph, Fig. 4). Each of these macrobonds includes at least several interatomic (ionic) contacts and forms a patch on the macromolecular surface. An area occupied by each patch on a molecule is defined as the surface area accessible to a spherical water molecule probe rolling over the surface atoms and overlapping with a similar area on a neighboring molecule (Richards, 1977). For the same contact, the areas on the two molecules may be different due to pockets in each of the contact areas. This roughness of surface determinates the strength of intermolecular binding (Matsuura & Chernov, 2003). The total area from all contact patches on a macromolecular surface usually varies widely. However, the patches do not occupy specific areas on macromolecular surfaces that are capable only of making contacts. It follows that different pH, temperature, and solution composition, resulting in different polymorphs, activate different mutually compatible areas on molecular surfaces to produce contacts, thus enabling different packing and crystal structures.

From the formal, purely geometrical standpoint, a crystal lattice may be built out of any species of arbitrary shape. Choice between the endless varieties of possibilities is made by the strongest attractions between the patches mutually compatible under various conditions. We may speculate that the larger the difference between various possible contact systems, the easier to crystallize a perfect crystal. However, during crystal dissolution or growth affected by an additive, detachment or attachment of a molecule occurs in solution in order to create initial nuclei. Therefore, part of the energy spent for virtual separations returns from hydration (Chernov, 2003; see self-assembly paragraph).

3.3 Cross-influence procedure (CIP)

In parallel to modern high-throughput approaches, basic research on physico-chemical properties of proteins and their molecular interactions has increasingly gained on importance again in recent years. Physico-chemical properties of the crystallization of biological macromolecules are of particular interest for an efficient way to get high-quality crystals. To improve the success of any crystallization attempts significantly as well as to find new methods of predicting it we have explored another tool useful for optimization strategy that was firstly described in Tomčová and Kutá Smatanová (2006). A crystallization procedure was tested to modify crystal morphology, internal packing and also to influence a crystal growth. For the first time the metal ion salts were added simultaneously to the protein drop and even to neighboring drops to allow a cross-influence effect during crystallization experiment. This new crystallization procedure was verified and the effects of selected additives on crystallization of two different proteins; one well-known "model" protein thaumatin and one crystallographically unexplored di-heme cytochrome c4 from an anaerobic organism (Tomčová et al., 2006) using this procedure were tested. Thaumatin has been chosen for our study because its crystals are often obtained in many different crystallization conditions as well-shaped tetragonal bipyramids, thus it makes sense to study any changes in their internal packing and morphology by cross-crystallization method (Tomčová & Kutá Smatanová, 2007). The final results are summarized and briefly discussed in followed paragraph.

3.3.1 The principles of cross-crystallization method

The CIP is a procedure applied to standard vapor diffusion sitting and/or hanging drop method. This procedure is based on using a set of additives that influence the quality of crystal growth. In principle, the inclusion of other droplets (containing metallic compounds) against the same reservoir slightly changes the vapor pressure of water over the neighboring drop including protein. Products of reaction of metallic salts and precipitant in acidic buffer are often unstable, light sensitive complex compounds (Greenwood & Earnshaw, 2002; Hennig & Theopold, 1951; Nemčovič et al., 2008) that decompose into volatiles, which influence pH slightly. Whereas, used salts of metals (Ba, Co and Cd) have different chemical reactivity, volatile is released in sequence and thus a quality of crystal is improved.

As was describe previously (Tomčová et al., 2006), the Emerald BioStructures CombiClover Crystallization Plate (EBS plate, Emerald BioStructures, Bainbridge Island, WA, USA) with one central reservoir connected to four satellite drop-chambers (A, B, C, D) via dedicated vapor diffusion channels, was used in this procedure (Fig. 5). Each of drop-chambers A, B, C, D was filled with 0.5 µl of different additive (in this case; 5 mM chloride salts of copper, cadmium, cobalt and barium) and equal volume of the precipitating agent. A protein was added only into the drop-chamber B containing 5 mM cupric chloride. Additives and reservoir solution, and not protein, were placed to the three remaining drop-chambers to promote crystallization in the fourth drop-chamber. The range of protein concentration 5-20 mg/ml, a reservoir solution volume of 1.0 ml and drops consisting of 1 µl protein solution plus 1 µl of reservoir solution were used in each crystallization trial (Fig.5; Fig.7).

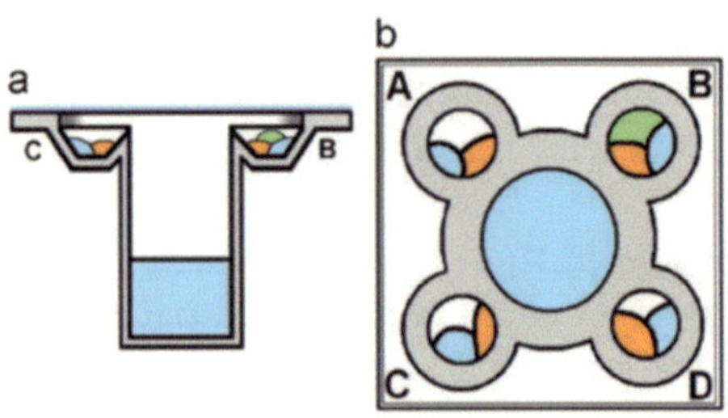

Fig. 5. Schematic side and top view of Emerald BioStructures Combi Clover Crystallization Plate (EBS plate) for sitting drop experiments. Blue color presents reservoir solution, red areas indicate each additives and green color represents protein-containing solution.

3.3.2 Hanging drop variant of CIP

The Hampton Research Linbro plate (Hampton Research, Aliso Viejo, CA, USA) with one central reservoir covered with 15 mm square cover slide was used as a hanging drop alternative in this method. All hanging drops with the same volumes and contents as sitting drops were placed on a siliconized cover slide (Fig. 6). This variant of cross-crystallization shows several disadvantages for protein crystal growth. Compared to the sitting drop method the drops are closer to each other. During crystallization mostly clustered twinned crystals growing on inside perimeter of a drop were observed (see next paragraph). Another variant of hanging drop application consists of the cover slip with drops above the EBS plate reservoir, which allows a hanging drop crystallization experiment to be conducted simultaneously with four sitting-drop experiments.

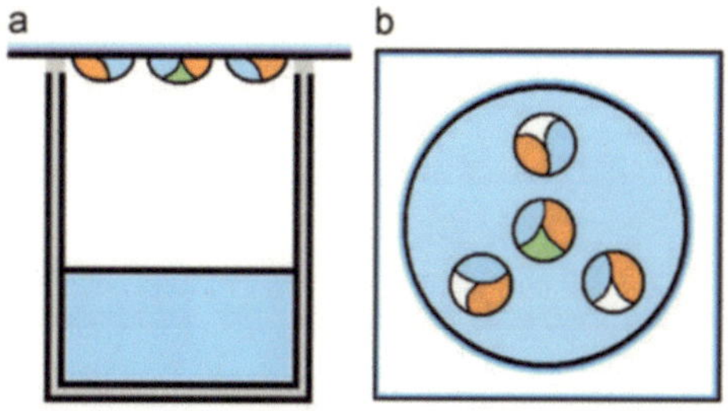

Fig. 6. Schematic side and top view of Hampton Research Linbro plate for hanging drop crystallization experiments. Blue color presents reservoir solution, red areas indicate each additives and green color represents protein-containing solution.

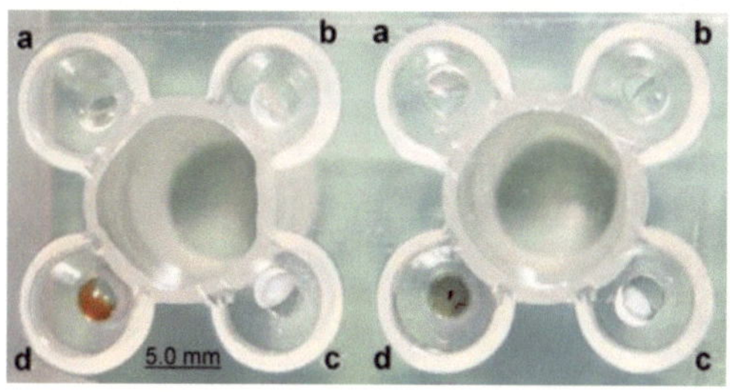

Fig. 7. Original picture of the EBS crystallization plate (with cytochrome crystals - in right d) used for performing cross-crystallization experiment.

3.3.3 The observed effect of CIP: Cytochrome crystallization

Initial crystallization conditions yielding crystals of Cyt c4 in not good quality (Table 4. A-E) were further optimized to improve diffraction quality and especially stability of the crystals (Table 4. F-J). Standard vapor diffusion method with combinations of additives was applied. Screening of pH values (from 6 to 5), NaCl concentration (from 0.1 to 0 M) and $(NH_4)_2SO_4$ concentration (from 1.7 M to 3.2 M) was used for optimization. In this optimization step, using of CIP improved quality of crystals by addition of additives (Table 4., Fig. 7). Deep red well-shaped cytochrome crystals grew within 3–4 days at 20 °C in the presence of 5 mM cupric chloride and ammonium sulphate in citric acid buffer at pH 5. Those crystals were not reproducible unless the other metal salts ($CdCl_2$, $BaCl_2$, $CoCl_2$) were present in the remaining drop chambers. Only metal salts and reservoir solution, and not protein, were required in the three remaining wells to promote crystallization in the fourth well. The CIP effect has been tested several times. The cytochrome crystals grew only in hexagonal prism form in all these cases. The same outer shape of crystals was observed even when a cytochrome was cross-crystallized by hanging drop (Table 4). Results described in Table 4 show that sodium chloride in combination with ammonium sulphate produce quasi crystals only and these are not good enough for diffraction experiments (drops A-C); presence of any of three used metal salts ($CdCl_2$, $CoCl_2$ and even $BaCl_2$) in protein solution makes heavy protein precipitation (drop D). Comparing drops E and F, it is clear that CIP causes a change in crystal formation what allow us to see the effect of CIP. Drops F-J show diffractable crystals in each successive optimization step.

NO CRYSTALS

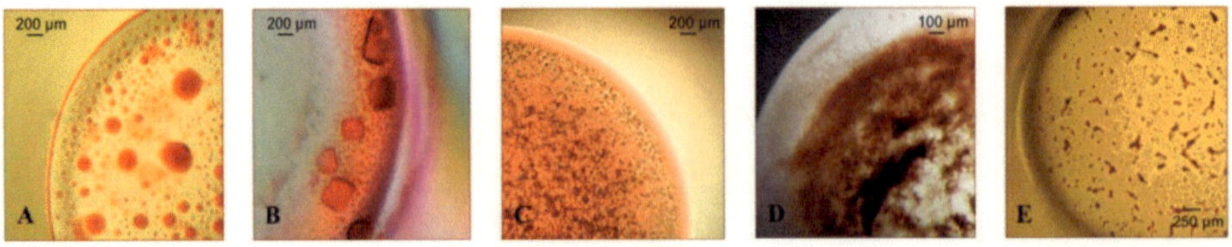

A	B	C	D	E
A prot:ppt = 2:8 2 mg/ml 1.0 M ammonium sulfate 0.1 M citric acid 6.0 0.2 M sodium chloride *phase separation 3d*	**B** prot:ppt = 2:8 2 mg/ml 1.7 M ammonium sulfate 0.1 M citric acid 6.0 0.2 M sodium chloride *quasi crystals 3-5d*	**C** prot:ppt = 2:8 2 mg/ml 2.5 M ammonium sulfate 0.1 M citric acid 6.0 50 mM sodium chloride *micro-crystals 2-3d*	**D** prot:ppt:add = 1:1:1 4 mg/ml 3.0 M ammonium sulfate 0.1 M citric acid 5.0 3.3 mM cadmium or cobalt chloride *heavy precipitation 0d*	**E** prot:ppt:add = 1:1:1 5 mg/ml 3.0 M ammonium sulfate 0.1 M citric acid 5.0 3.3 mM cupric chloride *phase separation 3d*

CRYSTALS

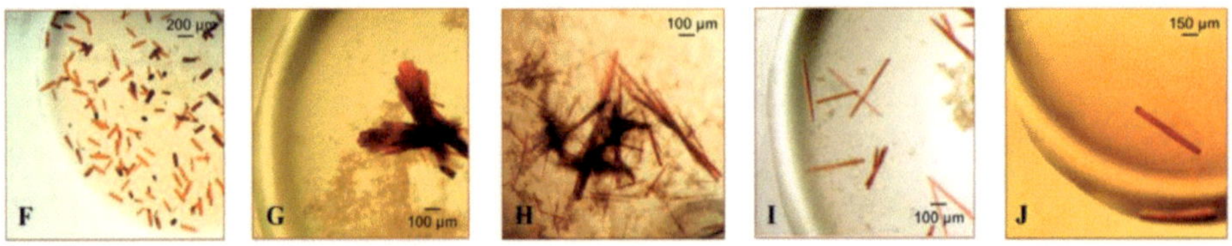

F	G-I	J
F prot:ppt:add = 1:1:1 5 mg/ml 3.0 M ammonium sulfate 0.1 M citric acid 5.0 3.3 mM cupric chloride + **cross influence** of $CdCl_2$, $CoCl_2$, $BaCl_2$ *small crystals 1-2d WD*	**G-I** prot:ppt:add = 2:½:½ 7 mg/ml - 10 mg/ml 3.2 M ammonium sulfate 0.1 M citric acid 5.0 1.6 mM cupric chloride + **cross influence** of $CdCl_2$, $CoCl_2$, $BaCl_2$ *rod clusters/ needles/ single crystal* *1-2d WD*	**J** prot:ppt:add = 1:½:½ 7.5 mg/ml 3.2 M ammonium sulfate 0.1 M citric acid 5.0 5 mM cupric chloride + **cross influence** of $CdCl_2$, $CoCl_2$, $BaCl_2$ *single crystal 3-4d HD*

d – days; HD - high diffraction ; WD - weak diffraction; ID - inferior diffraction

Table 4. Overview of cytochrome crystallization experiments: A, C, D - standard crystallization in hanging drop; B, E - standard crystallization in sitting drop; F, J - cross-crystallization in sitting drop; G-I - cross-crystallization in hanging drop.

3.3.4 The observed effect of CIP: Thaumatin crystallization

Previously thaumatin has been crystallized in four different forms: orthorhombic (1.75 Å), monoclinic (2.60 Å), tetragonal (1.75 Å) and hexagonal (1.60 Å) (McPherson, 1999; Lee et al., 1988; McPherson & Weickmann, 1990; Charron et al., 2004; Van der Wel et al., 1975). Orthorhombic and monoclinic crystal forms were obtained by using hanging drop vapor-diffusion method from polyethylene glycol as a precipitating agent (Lee et al., 1988; McPherson et al., 1990). The third, tetragonal crystal form was found in crystals grown by vapor diffusion method in the presence of 1 M sodium potassium tartrate containing 0.1 M ADA (sodium-N-2-acetamido-iminodiacetic acid) at pH 6.5 (McPherson, 1999). These crystals were similar to one grown from ammonium sulphate solution (Van der Wel et al., 1975). Thaumatin has also been crystallized in hexagonal crystal form from a tartrate and glycerol containing solution after shifting temperature from 293 K to 277 K (Charron et al., 2004; Ng et al., 1997). All these crystals grew as tetragonal bipyramids with dimensions commonly exceeding 0.5-1.0 mm (see Table 5, Fig. A1-2). In our experiments, thaumatin was crystallized using the standard sitting drop method from the polyethylene glycol (PEG) as a precipitating agent (Tomčová et al., 2007). Well-constructed tetragonal bipyramids were obtained from described crystallization conditions.

Thaumatin				
	Sitting drop experiments		Hanging drop experiments	
[no Cu, no CIP]	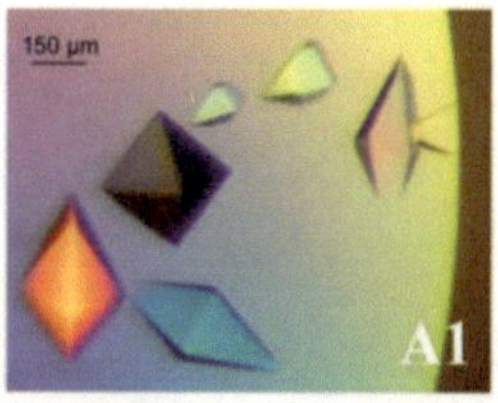	**A1** 35% PEG 3350 15% PEG 6K 0.1M TRIS 6.5 *single crystals 1-2d* *HD*	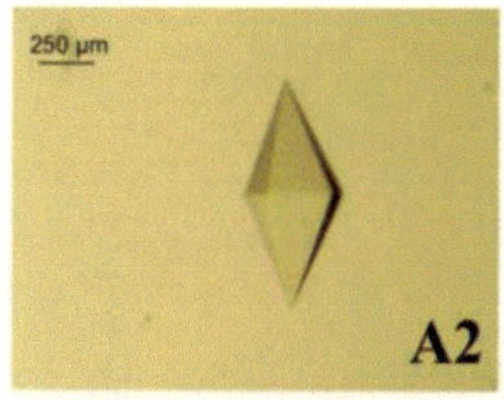	**A2** 35% PEG 3350 15% PEG 6K 0.1M TRIS 6.5 *single crystals 1-2d* *HD*
[+ Cu, no CIP]	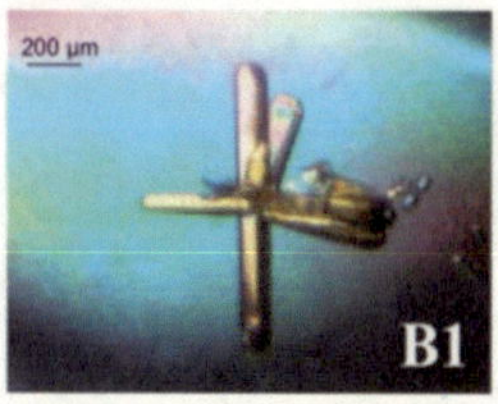	**B1** 35% PEG 3350 15% PEG 6K 0.1M TRIS 6.5 5mM cupric chloride *rod clusters 2d* *ID*	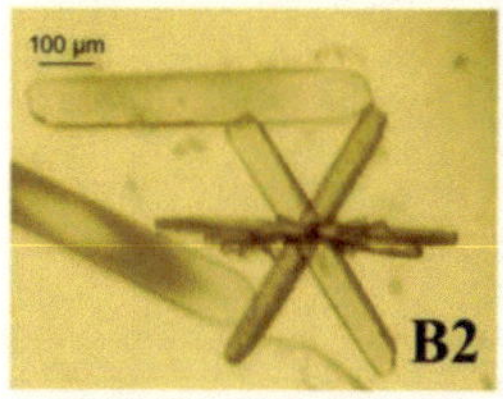	**B2** 35% PEG 3350 15% PEG 6K 0.1M TRIS 6.5 5mM cupric chloride *rod clusters 2d* *ID*
[no Cu, + CIP]		**C1** 35% PEG 3350 15% PEG 6K 0.1M TRIS 6.5 + **cross influence** of $CdCl_2$, $CoCl_2$, $BaCl_2$ *clusters 2d* *WD*	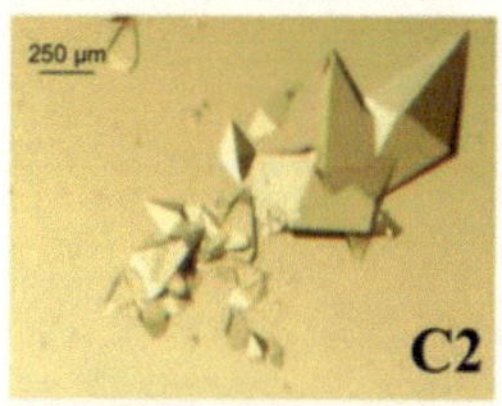	**C2** 35% PEG 3350 15% PEG 6K 0.1M TRIS 6.5 + **cross influence** of $CdCl_2$, $CoCl_2$, $BaCl_2$ *clusters 2d* *WD*
[+ Cu, + CIP]	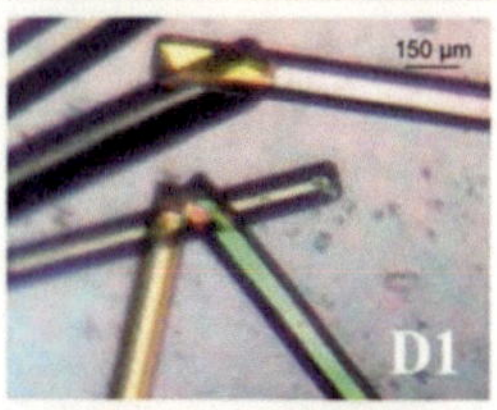	**D1** 35% PEG 3350 15% PEG 6K 0.1M TRIS 6.5 5mM cupric chloride + **cross influence** of $BaCl_2$, $CdCl_2$, $CoCl_2$ *single crystals 3d HD*	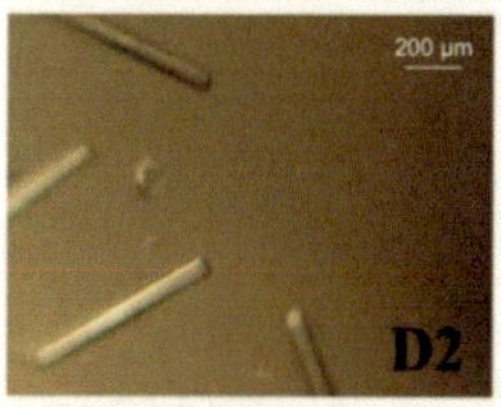	**D2** 35% PEG 3350 15% PEG 6K 0.1M TRIS 6.5 5mM cupric chloride + **cross influence** of $BaCl_2$, $CdCl_2$, $CoCl_2$ *single crystals 3d HD*
[special cases]		**E1** 35% PEG 3350 15% PEG 6K 0.1M TRIS 6.5 + **cross influence** of $BaCl_2$+$CuCl_2$, $CdCl_2$, $CoCl_2$ - *clusters 2d* *WD*		**E2** 35% PEG 3350 15% PEG 6K 0.1M TRIS 6.5 + **cross influence** of $BaCl_2$+$CuCl_2$, $CdCl_2$, $CoCl_2$ - *clusters 2d* *WD*
	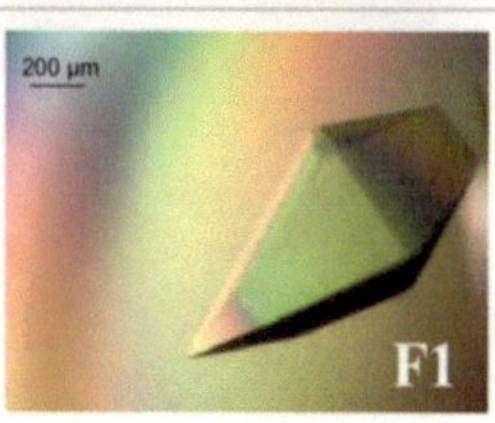	**F1** 35% PEG 3350 15% PEG 6K 0.1M TRIS 6.5 + **cross influence** of $CuCl_2$ *single crystal 1d* *HD*	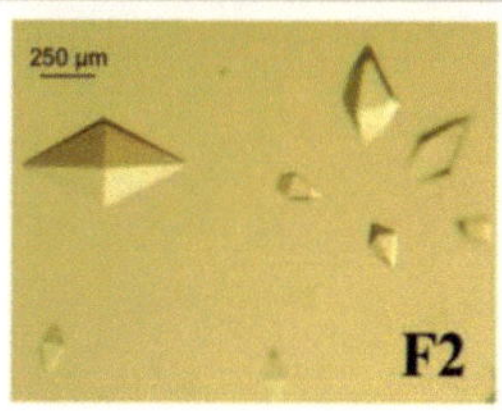	**F2** 35% PEG 3350 15% PEG 6K 0.1M TRIS 6.5 + **cross influence** of $CuCl_2$ *single crystal 1d* *HD*

d – days; HD - high diffraction ; WD - weak diffraction; ID - inferior diffraction

Table 5. Overview of thaumatin crystallization experiments: A1-B1 – standard crystallization in sitting drop; A2-B2 – standard crystallization in hanging drop; C1-F1 – cross-crystallization in sitting drop; C2-F2 – cross-crystallization in hanging drop.

Protein	Cytochrome		Thaumatin	
Crystallization method	Standard crystallization	Cross-crystallization	Standard crystallization	Cross-crystallization
Crystallization conditions in protein solution	3.2 M ammonium sulfate 0.1 M citric acid pH 5.0	3.2 M ammonium sulfate 0.1 M citric acid pH 5.0 5 mM cupric chloride	30-40% PEG 3350 15% PEG 6K 0.1 M TRIS pH 6.5	30-40% PEG 3350 15% PEG 6K 0.1 M TRIS pH 6.5 5 mM cupric chloride
Crystal outer shape	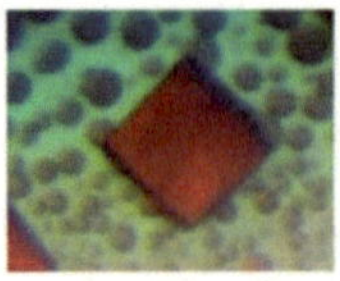 quasi crystals – plates 	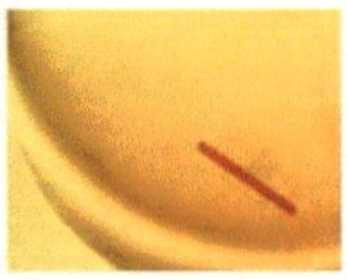 hexagonal prisms 	 tetragonal bipyramids 	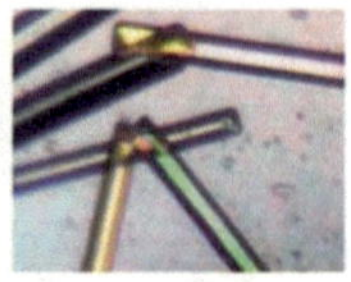 hexagonal prisms 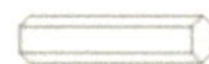
Crystal size (approx.)	300×300×5 μm	700×50×10 μm	550×300×300 μm	1000×100×100 μm
Crystal system		tetragonal	orthorhombic	tetragonal
Space group		$P4_12_12$	$P2_12_12_1$	$P4_12_12$
Unit cell parameters		a = b = 75.29 Å c = 37.12 Å	a = 44.3 Å b = 63.7 Å c = 72.7 Å	a = b = 58.6 Å c = 151.8 Å
Maximum resolution		**1.72 Å**	**1.51 Å**	**1.70 Å**
Resolution range low		15.227 Å	20.122 Å	28.020 Å
Number of unique reflections	no diffraction	11770	19968	41510
Completeness overall		99.69 %	97.50 %	98.10 %
Solvent content		26 %	40 %	56 %
Mosaicity		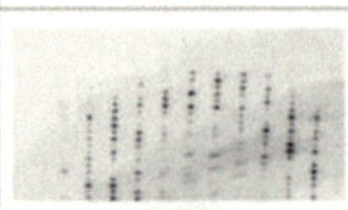 **0.58**	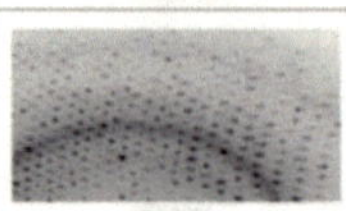 **0.89**	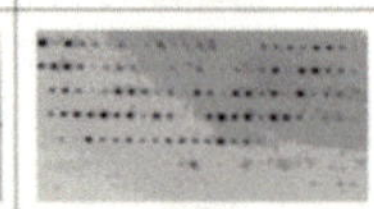 **0.44**
R_{merge} (%)		8.0	16.5	17.5
$I/\sigma(I)$		13.82	14.62	13.3

$R_{merge} = \sum|I_i - \langle I\rangle|/\sum\langle I\rangle$, where I_i is an individual intensity measurement and $\langle I\rangle$ is the average intensity for this reflection with summation over all data.

Table 6. Summary of crystallization and crystallographic statistics for both forms of cytochrome (plates and hexagonal prisms) and thaumatin (tetragonal bipyramids and hexagonal prisms). Crystal morphology and internal packing influenced by CIP are presented on the top of the table and X-ray diffraction statistics is listed below.

The effect of metal salt ions on cross-crystallization was tested. The most dramatic change in thaumatin crystal morphology and internal packing was observed when thaumatin was crystallized in hexagonal prisms shape (see Table 5, Fig. D1-2 and Table 6). From comparison of drops A1-2 with drops F1-2 and drops C1-2 with drops E1-2 illustrated in Table 5, it is clear there is no difference in crystal morphology, thus the cross-influence of single Cu^{2+} is minimal or probably was not evincible. This distinction in CIP composition allows us to conclude that several different metallic compounds (for example Co, Cd, Ba) added into satellite drops are necessary for effective CIP. The inclusion of CIP in drops D1-2 against drops B1-2 seems to reduce propensity of crystals to grow in rods and improve the diffraction quality while the inclusion of CIP realized simultaneously with addition of Cu^{2+} to protein (drops A1-2 against drops D1-2) shows more significant effect on crystal growth and morphology. In this case, cupric chloride caused the greatest change in crystal outer shape as it is present in Table 5 (drops A1-2 and drops D1-2).

3.3.5 Volatile buffers as crystallization inducers

Recently, the volatile buffers were found to be used in cross-crystallization experiments to induce crystallization of entire macromolecule (Tomčová et al., 2007). They are powerful in changing the pH and vapor pressure over the crystallization drop.

Reservoir volume	5.2 M Volatile buffer	Final drop concentration of volatile buffer	Drop pH
1000 µL	20 µL	0.1 M	3
500 µL	10 µL	0.1 M	3
100 µL	2 µL	0.1 M	3
75 µL	1.5 µL	0.1 M	3
50 µL	1 µL	0.1 M	3

Table 7. Using 5.2 M Acetic acid, the approximate final drop concentration will be 0.1 M Acetic acid. The pH of 0.1 M Acetic acid is approximately 3 but the actual final drop pH after addition of Acetic acid will depend upon the sample buffer and crystallization reagents in the drop.

Reservoir volume	5.2 M Volatile buffer	Final drop concentration of volatile buffer	Drop pH
1000 µL	20 µL	0.1 M	9
500 µL	10 µL	0.1 M	9
100 µL	2 µL	0.1 M	9
75 µL	1.5 µL	0.1 M	9
50 µL	1 µL	0.1 M	9

Table 8. Using 5.2 M Ammonium hydroxide, the approximate final drop concentration will be 0.1 M Ammonium hydroxide. The pH of 0.1 M Ammonium hydroxide is approximately 9 but the actual final drop pH after addition of Ammonium hydroxide will depend upon the sample buffer and crystallization reagents in the drop.

The volatile buffers, when added only to the reagent reservoir of a vapor diffusion experiment, can alter the pH of the crystallization drop by vapor diffusion of the volatile acid or base component from reservoir into the drop. This may be particularly useful when the sample is known to have pH dependent solubility and may be used to induce crystallization. For example, Acetic acid can be added to the reservoir to lower the pH of the drop. On the other hand, Ammonium hydroxide component can be added to the reservoir to raise the pH of the drop. Obviously, the final pH, the actual final volatile buffer concentration in the drop, rate and overall time of equilibration, will vary with drop and reservoir volume, geometry and temperature (Mikol, 1989; McPherson, 1990). The following table (Table 7-8) offers a general guideline when using a volatile buffer manipulates drop pH to induce crystallization (see also Hampton Research Volatile Buffers Usage; Hampton Research, Aliso Viejo, CA, USA). The volatile buffer may be added at the time of initial drop/reservoir set up (see Fig. 8). In this method, the initial drop pH will be that of the sample and crystallization reagent but it changes over time as the volatile buffer vapor diffuses from the reservoir to the drop. Alternatively, as a salvage method, to induce crystallization or improve the crystal, the volatile buffer can be added after the drop has fully equilibrated with the reagent reservoir.

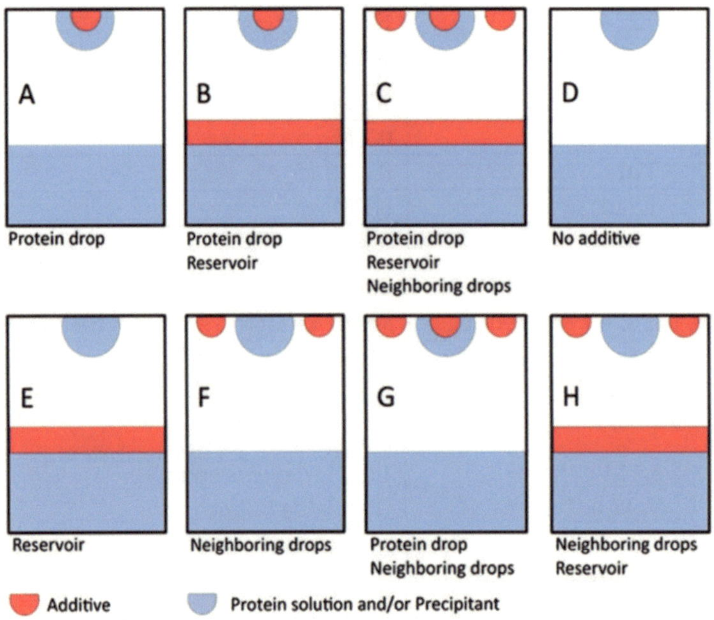

Fig. 8. The schematic view of the volatile buffer composition added at the same time to the hanging drop and reservoir. Each of the A-H crystallization experiment consists of reservoir solution and/or protein solution (shown in blue) and the placement of additive (shown in red). Additive can be volatile compound and can be placed in neighboring drops (C, F, G, H) or added to the reservoir (B, C, E, H).

4. Conclusion: New optimization tools in protein crystallization

In this chapter, we described cross-crystallization alias cross-influence procedure (CIP) in details and summarized several factors about CIP and selective additives that facilitate protein crystallization; through promotion of intermolecular contacts and gradually changing pH by the reaction of precipitating solution with divalent metallic compounds, stabilization of the protein with salts or changing the aggregation state with precipitating agents. In fact, any addition of a new substance into a crystallization mixture resulting in crystallization is usually classified as a new crystallization procedure and handled as a hot tip. From previous studies it was found that cupric ions in phosphate buffers have a tendency to produce heavy precipitate and even salt crystals (Lee et al., 1988; Jancarik & Kim, 1991; Jancarik et al., 2004).

Another example of an additive effect, which is explained on a molecular basis, is a formation of intermolecular contacts by intercalated divalent transition metal cations. Cadmium (in sulphate solutions) was long known as an inducing agent in crystallization of horse spleen ferritin (Trakhanov & Quiocho, 1995) and has been rediscovered as a useful agent to promote crystallization and increase diffraction quality in several cases. However, the specific morphology of thaumatin and cytochrome crystals may depend on factors such as a source of material used during crystal growth, chemicals presented in a crystallizing buffer in the mother liquor or on the mother liquor itself. Products of reaction MCl_2 (where M is metal) and ammonium sulphate with citric acid are unstable, light sensitive complex compounds like $C_6H_8O_7.nM.nNH_3$ (Greenwood & Ernshaw, 2002; Hennig & Theopold, 1951) decomposing into volatile ammonium that slightly influence pH. As Ba, Co and Cd have different reactivity, ammonium is released in sequence and thus quality of crystals is improved (Table 6). In a single crystal form the angles between the faces are constant (Drenth & Haas, 1992), but this is not true if the crystals belong to the different crystal forms such as tetragonal bipyramids and hexagonal prisms in the case of thaumatin. Their appearance depends on the use of metal salt cations (for example cupric chloride) with different chemical reactivity and partially on the buffer and the precipitating agent. We assume that added metal cations influence a character of evaporation in a protein drop and gradually change the pH even if they are only in the proximity of this protein drop. The influence of Cu^{2+} ions on cytochrome crystal growth appears to be specific, because no other successful combination of ion salts with cytochrome was found among these four salts singly or in pairs. A similar effect was observed even in thaumatin crystallization when conditions with cupric chloride produced thaumatin crystals with a different morphology. The combination of particular salts promoting crystallization by CIP can be reproduced with other metallic compounds as well, or even other volumes of the same drop in the remaining drop's chambers. Due to this fact, CIP can be appropriately used in any crystallization step to find or/and improve crystallization conditions. The presence of copper ions significantly influences the crystal growth, as the modification of crystal morphology and internal packing were observed. On the basis of our results and analyses of copper influence and CIP effect, we propose that the copper addition realized simultaneously with CIP provides a useful technique to modify crystal morphology and improve diffraction quality in protein crystallization and serves as a powerfull crystallization technique. In addition, for the first time the detailed

protocol of CIP and general guideline of additive usage was given within this chapter to help readers to perform their own cross-crystallization experiment. Thus, this book chapter stands as a valuable guide to modern alternative protein crystallization.
Perhaps we have stimulated your interest in crystallography itself, and have made you wonder if you might jump in and crystallize and determine the structure of that interesting protein you are studying. We are happy that we can encourage you by reiterating that crystallography, though still one of structural biology's more challenging callings, is faster and easier than ever before. Screening for crystal growth conditions does not require expensive equipment or chemicals and these chapter is giving you new ideas how to crystallize your protein of interest with no additional cost.

5. Acknowledgment

This research was supported by the Ministry of Education of the Czech Republic (LC06010, ME09016, COST LD11011, CZ.1.05/2.1.00/01.0024 and MSM6007665808) and by the Academy of Sciences of the Czech Republic (AV0Z60870520). We thank the EMBL for access to the X13 beamline at the DORIS storage ring of DESY in Hamburg. I. N. acknowledges the continuous support of the Slovak Academy of Sciences (Institute of Chemistry, Bratislava, Slovakia) and the La Jolla Institute for Allergy and Immunology (Department of Cellular Biology, La Jolla, California, USA).

6. References

Blow, D.M.; Chayen, N.E.; Lloyd, L.F.; & Saridakis, E. (1994). Control of nucleation of protein crystals. *Protein Science,* Vol. 3, Issue 10, pp. 1638-1643, DOI:10.1002/pro.5560031003

Brooijmans, N.; Sharp, K.A.; & Kuntz, I.D. (September 2002). Stability of Macromolecular Complexes. *Proteins: Structure, Function, and Genetics,* Vol. 48, Issue 4, pp. 645–653, DOI:10.1002/prot.10139

Carter, C.W.Jr.; Baldwin, E.T.; & Frick, L. (July 1988). Statistical design of experiments for protein crystal growth and the use of a precrystallization assay. *Journal of Crystal Growth,* Vol. 90, Issue 1-3, pp. 60-73, DOI:10.1016/0022-0248(88)90299-0

Charron, C.; Giege, R.; & Lorber, B. (January 2004). Structure of thaumatin in a hexagonal space group: comparison of packing contacts in four crystal lattices. *Acta Crystallographica Section D Biological Crystallography,* Vol. 60, Issue 1, pp. 83-89, DOI:10.1107/S0907444903022613

Chernov, A. (April 2003). Protein crystals and their growth. *Journal of Structural Biology,* Vol. 142, Issue 1, pp. 3-21, DOI:10.1016/S1047-8477(03)00034-0

Chirgadze, D. (1998). Available from http://www.xray.bioc.cam.ac.uk/xray_resources/ whitepapers/xtal-in-action/node1.html

Cox, M.J.; & Weber, P.C. (1988). An investigation of protein crystallization parameters using successive automated grid searches (SAGS). *Journal of Crystal Growth,* Vol. 90, Issue 1-3, pp. 318-324, DOI:10.1016/0022-0248(88)90327-2

Cudney, R.; Patel, S.; Weisgraber, K.; Newhouse, Y.; & McPherson, A. (July 1994). Screening and optimization strategies for macromolecular crystal growth. *Acta Crystallographica Section D Biological Crystallography,* Vol. 50, Issue 4, pp. 414-423, DOI:10.1107/S0907444994002660

D'Arcy, A.; Elmore, C.; Stihle, M.; & Johnston, J.E. (1996). A novel approach to crystallising proteins under oil. *Journal of Crystal Growth,* Vol. 168, issue 1-4, pp. 175-180, DOI:10.1016/0022-0248(96)00351-X

DeMattei, R.C.; Feigelson, R.S.; & Weber, P.C. (1992). Factors affecting the morphology of isocitrate lyase crystals. *Journal of Crystal Growth,* Vol. 122, Issue 1-4, pp. 152-160, DOI:10.1016/0022-0248(92)90238-E

Drenth, J. & Haas, C. (1992). Protein crystals and their stability. *Journal of Crystal Growth,* Vol. 122, Issue 1-4, pp. 107-109, DOI: 10.1016/0022-0248(92)90233-9

Drenth, J. (November 2006). *Principles of Protein X-Ray Crystallography.* (3rd Ed.), Springer Science and Business Media, ISBN 9780387337463, England

Ducruix, A.; & Giege, R. (January 2000). *Crystallization of Nucleic Acids and Proteins: A Practical Approach,* (2nd Ed.), Oxford University Press, ISBN 978-0199636785, USA

Fink, J.M. (2005). *Diploma Thesis: Self-organizing nanostructures,* Department of Applied Physics, University of Vienna, Austria

Fowlis, W.W.; DeLucas, L.J.; Twigg, P.J.; Howard, S.B.; Meehan, E.J.Jr.; & Baird, J.K. (1988). Experimental and theoretical analysis of the rate of solvent equilibration in the hanging drop method of protein crystal growth. *Journal of Crystal Growth,* Vol. 90, Issue 1-3, pp. 117-129, DOI:10.1016/0022-0248(88)90306-5

García-Ruiz, J.M; & Morena, A. (Jul 1994). Investigations on protein crystal growth by the gel acupuncture method. *Acta Crystallographica Section D Biological Crystallography,* Vol. 50, Part 4, pp. 484-90, DOI:10.1107/S0907444993014350

García-Ruiz, J.M. (2003) Counterdiffusion methods for macromolecular crystallization. *Methods in Enzymology,* Vol. 368, pp. 130-154, DOI:10.1016/S0076-6879(03) 68008-0

Garman, E.; & Murray, J.W. (November 2003). Heavy-atom derivatization, *Acta Crystallographica Section D Biological Crystallography,* Vol. 59, Issue 11, pp. 1903-1913, DOI:10.1107/S0907444903012794

Giacovazzo, C.; Monsaco, H.L.; Artioli, G.; Viterbo, D.; Ferraris, G.; Gilli, G.; Zanotti, G.; & Catti, M. (2002). *Fundamentals of Crystallography,* (2nd Ed.), Oxford University Press, ISBN 978-0198509585, Oxford, UK

Greenwood, N.N.; & Earnshaw, A. (2002). *Chemistry of the Elements.* (2nd Ed.), Pergamon Press, ISBN 9780080220567, The University of Michigan, USA

Griffin, L.; & Lawson, A. (September 2011). Antibody fragments as tool in crystallography. *Clinical and Experimental Immunology,* Vol. 165, Issue 3, pp. 285-291, DOI:10.1111/j.1365-2249.2011.04427.x

Hartmut, Michel (1991). *Crystallization of membrane proteins,* CRC Press, ISBN 978-0849348167, The University of Michigan, USA

Hennig, W.; & Theopold, W. (1951). Complex compounds of citric acid and calcium, *Zeitschrift fur Kinderheilkunde,* Vol. 69, Issue 1, pp. 55-61

Jancarik, J.; & Kim, S.-H. (1991). Sparse matrix sampling: a screening method for crystallization of proteins. *Journal of Applied Crystallography,* Vol. 24, pp. 409-411, DOI:10.1107/S0021889891004430

Jancarik, J.; Pufan, R.; Hong, C.; Kim S.H.; & Kim, R. (September 2004). Optimum solubility (OS) screening: an efficient method to optimize buffer conditions for homogeneity and crystallization of proteins. *Acta Crystallographica Section D Biological Crystallography,* Vol. 60, Issue 9, pp. 1670-1673, DOI:10.1107/S0907444904010972

Kuznetsov, D.A.; & Hodorowicz, S. (1978). On the effect of pH of solution on the kinetics of crystallization. *Kristall und Technik,* Vol. 13, Issue 12, pp. 1413-1416, DOI:10.1002/crat.19780131206

Langer, H. (1985). *Zum Stofftransport beim Kristallwachstum aus Lösungen,* Ph.D. Thesis, RWTH Aachen, Aachen 1985

Lattman, E.; Loll P.J.; & Loll P. (2008). *Protein crystallography: a concise guide,* Johns Hopkins University Press, ISBN 978-0801888069, Baltimore, USA

Lee, J.H.; Weickmann, J.L.; Koduri, R.K.; Ghoshdastidar, K.; Saito, K.; Blair, L.C.; Date, T.; Lai, J.S.; Hollenberg, S.M.; & Kendall, R.L. (July 1988). Expression of synthetic thaumatin genes in yeast. *Biochemistry,* Vol. 27, Issue 14, pp. 5101-5107, DOI: 10.1021/bi00414a023

Matsuura, Y.; & Chernov, A.A. (August 2003). Morphology and the strength of intermolecular contacts in protein crystals. *Acta Crystallographica Section D Biological Crystallography,* Vol. 59, Issue 8, pp. 1347-1356, DOI:10.1107/S0907444903011107

McPherson, A. (1990). Current approaches to macromolecular crystallization. *European Journal of Biochemistry,* Vol. 189, Issue 1, pp. 1-23, DOI:10.1111/j.1432-1033.1990.tb15454.x

McPherson, A. (1991). A brief history of protein crystal growth. *Journal of Crystal Growth,* Vol. 110, issue 1-2, pp. 1-10, DOI:10.1016/0022-0248(91)90859-4

McPherson, A. (August 1999). *Crystallization of Biological Macromolecules,* Cold Spring Harbor Laboratory Press, ISBN 978-0879696177, New York, USA

McPherson, A.; & Weickmann, J. (April 1990). X-ray analysis of new crystal forms of the sweet protein thaumatin. *Journal of Biomolecular Structure and Dynamics,* Vol. 7, Issue 5, pp. 1053-1060

McRee, D.E. (1999). *Practical Protein Crystallography,* (Hardcover), Academic Press Title, Elsevier, ISBN 978-0-12-486052-0

Mikol, V.; Rodeau, J.-L.; & Giegé, R. (1989). Changes of pH during biomacromolecule crystallization by vapor diffusion using ammonium sulfate as the precipitant. *Journal of Applied Crystallography,* Vol. 22, pp. 155-161, DOI:10.1107/S0021889888013433

Mohameed, H.A.; & Ulrich, J. (1996). Influence of the pH-value on the growth and dissolution rate of potassium chloride. *Crystal Research and Technology,* Vol. 31, pp. 27-31

Nemčovič, M.; Jakubíková, L.; Víden, I., & Farkaš, V. (2008). Induction of conidiation by endogenous volatile compounds in Trichoderma spp. *FEMS Microbiology Letters,* Vol. 284, Issue 2, pp. 231-236, DOI:10.1111/j.1574-6968.2008.01202.x

Ng, J.D.; Lorber, B.; Giege, R.; Koszelak, S.; Day, J.; Greenwood, A.; & McPherson, A. (November 1997). Comparative analysis of thaumatin crystals grown on earth and in microgravity. *Acta Crystallographica Section D Biological Crystallography*, Vol. 53, Issue 6, pp. 724-733, DOI:10.1107/S090744499700694X

Otalora, F.; Gavira, J.A.; Ng, J.D.; & García-Ruiz, J.M. (November 2009). Counterdiffusion methods applied to protein crystallization. *Progress in Biophysics and Molecular Biology*, Vol. 101, Issue 1-2, pp. 26-37, DOI:10.1016/j.pbiomolbio.2009.12.004

Provost, K. & Robert, M-C. (1991). Application of gel growth to hanging drop technique. *Journal of Crystal Growth*, Vol. 110, pp. 258-264, DOI:10.1016/0022-0248(91) 90894-B

Rayment, I. (February 2002). Small-scale batch crystallization of proteins revisited: an underutilized way to grow large protein crystals. *Structure*, Vol. 10, Issue 2, pp. 147-51, DOI:10.1016/S0969-2126(02)00711-6

Richards, F.M. (June 1977). Areas, volumes, packing and protein structure. *Annual Reviews of Biophysics and Bioengeneering*, Vol. 6, pp. 151-176, DOI: 10.1146/annurev.bb.06.060177.001055

Sangwal, K. (1998). Growth kinetics and surface morphology of crystals grown from solutions: Recent observations and their interpretations. *Progress in crystal growth and characterization of materials*, Vol. 36, Issue 3, Pages 163-248, DOI:10.1016/S0960-8974(98)00009-6

Sousa, R. (May 1995). Use of glycerol, polyols and other protein structure stabilizing agents in protein crystallization. *Acta Crystallographica Section D Biological Crystallography*, Vol. 51, Issue 3, pp. 271-277, DOI:10.1107/S0907444994014009

Stura, E.A.; Nemerow, G.R.; & Wilson, I.A. (1991). Strategies in protein crystallization. *Journal of Crystal Growth*, Vol. 110, pp. 1-12

Tomčová, I. & Kutá Smatanová, I. (August 2007). Copper co-crystallization and divalent metal salts cross-influence effect: A new optimization tool improving crystal morphology and diffraction quality. *Journal of Crystal Growth*, Vol. 306, Issue 2, pp. 383-389, DOI:10.1016/j.jcrysgro.2007.05.054

Tomčová, I., Branca, R.M.M., Bodó, G., Bagyinka, Cs.; & Kutá Smatanová, I. (August 2006). Cross-crystallization method used for the crystallization and preliminary diffraction analysis of a novel di-haem cytochrome c4. *Acta Crystallographica Section F*, Vol. 62, Part 8, pp. 820-824, DOI:10.1107/S1744309106027710

Trakhanov, S.; & Quiocho, F.A. (September 1995). Influence of divalent cations on protein crystallization. *Protein Science*, Vol. 4, Issue 9, pp. 1914-1919, DOI: 10.1002/pro.5560040925

Van der Wel, H.; Van Soest, T.; & Royers, E.C. (1975). Crystallization and crystal data of thaumatin I, a sweet-tasting protein from Thaumatococcus daniellii benth. *FEBS Letters*, Vol. 56, Issue 2, pp. 316-317

Van Driessche, A.E.S.; Otalora, F.; Gavira, J.A.; & Sazaki, G. (2008). Is agarose an impurity or an impurity filter? *In situ* observation of the joint gel/impurity effect on protein crystal growth kinetics. *Crystal Growth and Design*,Vol. 8, Issue 10, pp. 3623–3629, DOI:10.1021/cg800157t

Voet, D. & Voet, J.G. (1995). *Biochemistry,* (Hardcover), John Wiley & Sons, ISBN 0-471-58651-X, New York, USA

Weber, P.C. (1991). Physical principles of protein crystallization. *Advances in Protein Chemistry,* Vol. 41, pp. 1-36, 1991, DOI:10.1016/S0065-3233(08)60196-5

Part 5

Molecular Crystals

12

Optical Properties of Molecular Crystals: The Effect of Molecular Packing and Polymorphism

Silvia Tavazzi[1], Leonardo Silvestri[2] and Peter Spearman[1]
[1]*University of Milano Bicocca, Materials Science Department,*
[2]*School of EE&T, University of New South Wales,*
[1]*Italy*
[2]*Australia*

1. Introduction

Modern technologies rely not only on traditional inorganic semiconductors, but also on organic materials. In particular, organic semiconductors based on conjugated oligomers and polymers have technological applications owing to their compatibility with low-temperature processing, the relatively simple thin-film device fabrication, and the tunability of their electronic properties based on the richness of synthetic organic chemistry. Here, we are interested in molecular materials in their crystalline form, which play a role in the fields of both theoretical and experimental research. We will focus the attention on their UV-visible optical properties.

Compared to amorphous films, crystalline materials exhibit well defined polarization of the electronic states and interesting effects related to the anisotropic propagation of light. First of all, in order to maximize coupling to the crystal electronic transition moment, the pumping photon beams must be precisely aligned (Fichou et al. 1997). Emission is then observed with defined directions of propagation and of polarization. Optical gain can also result in amplified spontaneous emission (ASE), which is characterized by a spectrally narrowed emission upon increasing the excitation fluence above a threshold. Gain-narrowing has been demonstrated in a number of molecular crystalline materials, such as oligothiophenes, oligo(p-phenylene), oligo(p-phenylene vinylene), trans-1,4-distyrylbenzene, cyano derivatives, fluorene/phenylene co-oligomers, hydroxy-substituted tetraphenyl-imidazole, and 1,1,4,4-tetraphenyl-1,3-butadiene (Fichou et al. 1997, 1999; Hibino et al. 2002; Horowitz et al. 1999; Ichikawa et al. 2003, 2005; Losio et al. 2007; Nagawa et al. 2002; Park et al. 2005; Polo et al. 2008; Tavazzi et al. 2006c, 2007, 2008, 2010b; Xie et al. 2007; Zhu et al. 2003). Relatively low thresholds have been recently reported for p-sexiphenyl samples. Self-waveguided emission has been observed in needle-shaped crystals of this molecular species grown on potassium chloride (001) substrates (Cordella et al. 2007; Quochi et al. 2005; Yanagi et al. 2001; Yanagi & Morikawa 1999) with threshold down to 0.5 $\mu J/cm^2$. Random lasing from isolated nanofibres of various widths has also been reported and discussed by Quochi et al. (Quochi et al. 2005). For the same molecular species in the bulk crystalline form, ASE was reported (Losio et al. 2007) with different thresholds depending on the

pumping rates (ns: 0.885 mJ/cm^2, fs: 0.110 mJ/cm^2). The same authors also reported that exciton annihilation seems to be responsible for the increase of the ASE threshold in bulk crystals compared to nanofibers. The increase was attributed to differences in the exciton annihilation rate from the nanofibres to crystals, which, in turn, were attributed to the different molecular packing, namely to the different crystalline phase of bulk crystals with respect to nanofibres. Another example of emission gain narrowing has been reported to occur for single crystals of a thiophene/phenylene co-oligomer (Ichikawa et al. 2003; Nagawa et al. 2002). For these co-oligomers, thresholds as low as 0.027 mJ/cm^2 at room temperature have also been reported. Besides ASE and random lasing, some studies have also been reported on the coupling of excitons with photons in microcavity structures - which is a rapidly growing field of research, in particular as far as strong coupling phenomena are concerned, which may lead to novel laser devices such as the *polariton laser* (Klaers et al. 2010). Strong exciton-photon coupling has been demonstrated in microcavities of single crystal and thin film anthracene (Kena-Cohen et al. 2008; Kondo et al. 2009, 2008). More recently, Kéna-Cohen and Forrest observed polariton lasing at room temperature from a microcavity containing melt-grown anthracene single-crystal, with a threshold lower than the estimated threshold for conventional lasing in the same system (Kéna-Cohen et al. 2010). In order to better understand the origin of some gain-narrowed emission bands of molecular single crystals, the dependence of the excitation threshold on excitation lengths has been studied for poly(p-phenylene-vinylene) (Frolov et al. 1998) and for 5,5''-bis(4-biphenylyl)-2,2':5',2''-terthiophene (Hiramatsu et al. 2009; Matsuoka et al. 2010).
Here, we discuss three prototypical molecular crystals. They show remarkable changes in the energy, intensity, order, and polarization of the optical bands with respect to the single molecule. These species are: (i) 1,1,4,4-tetraphenyl-1,3-butadiene (TPB), a highly-efficient blue emitting material, (ii) dibenzo[d,d']thieno[3,2-b;4,5-b']dithiophene (DBTDT), a stable heteroacene reported in the literature for its good field effect mobility, (iii) N-pyrrole end-capped thiophene/phenyl co-oligomer (TPP), a molecule designed to explore the effect of an electron-rich aromatic ring, such as pyrrole, to end-cap oligothiophenes. Fig. 1 shows the structure of the three selected molecular species. TPB is taken into consideration to discuss possible remarkable differences between polymorphs of the same molecular species.

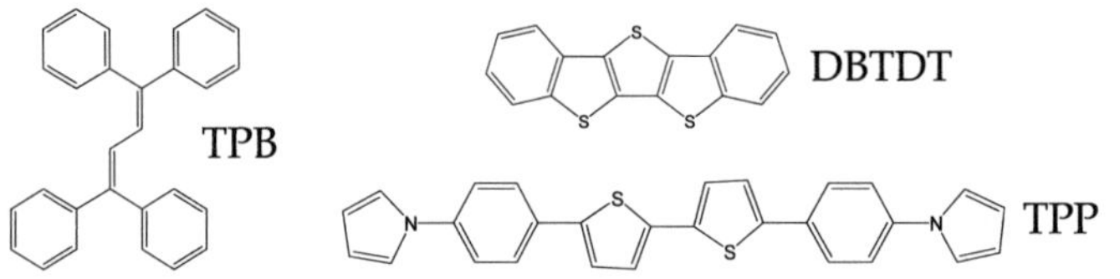

Fig. 1. Molecular structures of 1,1,4,4-tetraphenyl-1,3-butadiene (TPB), dibenzo[d,d']thieno[3,2-b;4,5-b']dithiophene (DBTDT), N-pyrrole end-capped thiophene/phenyl co-oligomer (TPP).

DBTDT is used as example to discuss the possible changes of the order of the excited states due to intermolecular interactions. Finally, TPP offers a prime example of the phenomenon of directional dispersion in which the molecular orientation within the crystal exposed face affects the absorption spectra. The chapter discusses some different techniques for the growth of molecular crystals (Section 2) and the microscopic and macroscopic theories of their optical properties, which is applied to the three selected species (Section 3). The general focus is on the UV-visible absorption properties, which are described and rationalised in the framework of the exciton theory of molecular materials. Some discussion on the emission properties is also given, but on a more qualitative level related to the molecular organization within the crystal. We emphasise that for a full description of the polarized emission spectra of molecular crystals, the excitonic bands should be studied as a function of momentum in the first Brillouin zone, while these details are here omitted. In contrast, for the interpretation of the absorption properties only the excitons at the centre of the Brillouin zone (k=0) need to be considered.

2. Growth of molecular crystals

2.1 Vapour-transport growth

The physical vapour growth of organic semiconductors is well described in the literature (Kloc et al. 1997; Kloc & Laudise 1998; Laudise et al. 1998; Siegrist et al. 1998). We have adopted a horizontal reactor arrangement, similar in design to that described by Laudise et al. (Laudise et al. 1998). The tube was placed in a three-zone furnace, which provided a suitable temperature gradient along the length of the tube under inert gas (nitrogen) at a flow rate of the order of a few tens of ml/min. The starting material was placed in a glass crucible at the position of the source temperature. For example, for the growth of TPB crystals the source was heated at 175°C in the first zone. The temperature descended to 135° at the end zone following a roughly linear temperature gradient. The nitrogen flux was 50 ml/min. Platelet crystals were obtained; few images are shown taken under the optical microscope (Fig. 2) and under the fluorescence microscope (Fig. 3).

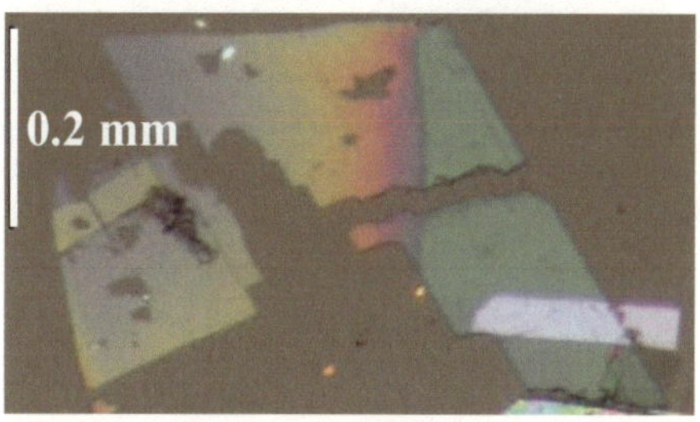

Fig. 2. Image under the optical microscope of α-TPB crystals grown by the vapour transport method. It was taken with polarizer and analyser parallel to each other and to a crystal principal axis. Coloured fringes are interference fringes in regions of different thickness.

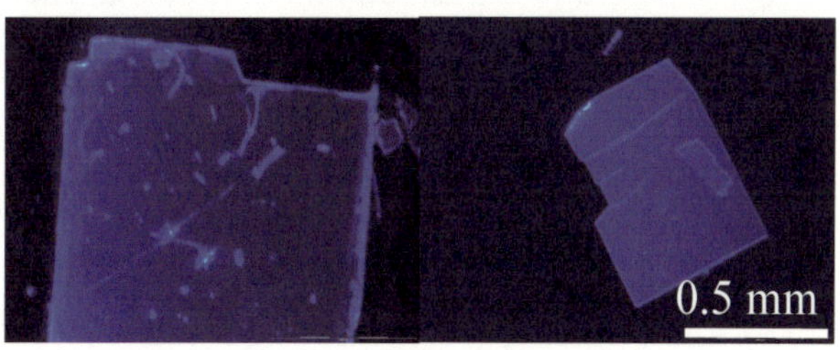

Fig. 3. Image under the fluorescence microscope of TPB crystals grown by the vapour-transport method.

2.2 Floating-drop growth

An alternative method for the growth of molecular crystals uses the technique of mixed solvents with different solubility in the extreme case where the molecule is completely insoluble in one of the solvents. The level of supersaturation in solution is typically lower than that of the vapour phase, so that the nucleation rate is reduced and the size of the crystal typically increases. To further reduce the nucleation, often induced by the vessel walls, the solution can be left on the surface of a denser and immiscible liquid. This method has been described elsewhere and called floating-drop technique (Adachi et al. 2003; Campione et al. 2005). For example, single crystals of DBTDT were obtained by such a floating-drop technique. Crystals grew as platelets with one accessible face for optical investigation. Images of a crystal are reported in Fig. 4 as taken under the optical microscope. The image on the left was taken with crossed polarizer and analyzer for a generic orientation of the sample. In this configuration, the emerging electric field is rotated with respect to the incident electric field and the angle of rotation depends on the wavelength of light and on the crystal thickness (birefringence). For this reason, the crystal is clearly distinguishable with respect to the dark background. The image on the right was taken with parallel polarizer and analyzer. The crystal appears slightly coloured (yellow due to the absorption of the complementary blue light).

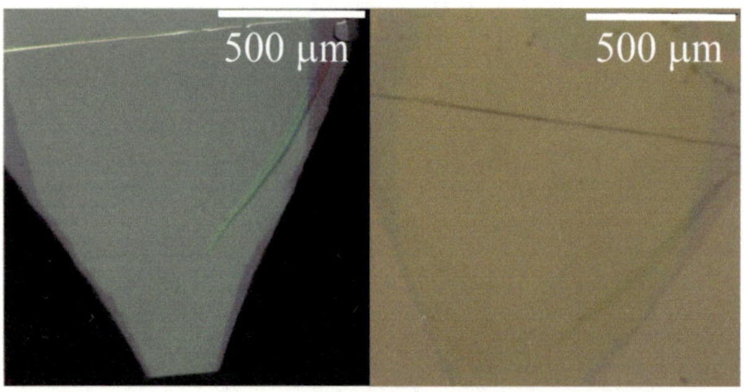

Fig. 4. Images of a DBTDT crystal grown by the floating-drop technique taken under the optical microscope with either crossed (left) or parallel (right) polarizer and analyzer.

2.3 Melt growth

When a molecular species possesses a well-defined melting point and remains stable at the corresponding temperature, crystals can be grown from the melted compound. This method is useful for incorporating crystalline materials into well-defined device geometries and several examples have been reported in the literature. For example, Hibino and co-workers reported the crystallization from the melt of anthracene and other oligomers (Hibino et al. 2002). They also observed ASE for many of them. The technique has also been adopted by Forrest's group in the fabrication of organic crystalline microcavities of anthracene grown by the melted compound (Kena-Cohen et al. 2008, Kena-Cohen et al. 2010). Another example has been reported (Liu & Bard 2000) on the growth and purification by zone-melting of organic single-crystals. Our group has similarly employed the melt-growth technique to grow crystals of the compounds TPB and TPP. Figure 5 displays an image taken under an optical microscope of a TPB crystal grown from the melted compound onto a quartz substrate. This substrate was initially functionalized with trichloro 3-bromopropylsilane followed by bromine substitution with dithiocarbamate. Large crystalline domains are observed. Other images of TPB grown from the melt between two quartz substrates are shown in Fig. 6. In this case, a fan-like distribution originating form a nucleation point is observed. The alternating ridge and valley bands are indicative of a rhythmic type of crystal growth as a result of the inability of the molten TPB to diffuse to the crystallization front. The image on the left (right) was taken with parallel (crossed) polarizer and analyzer.

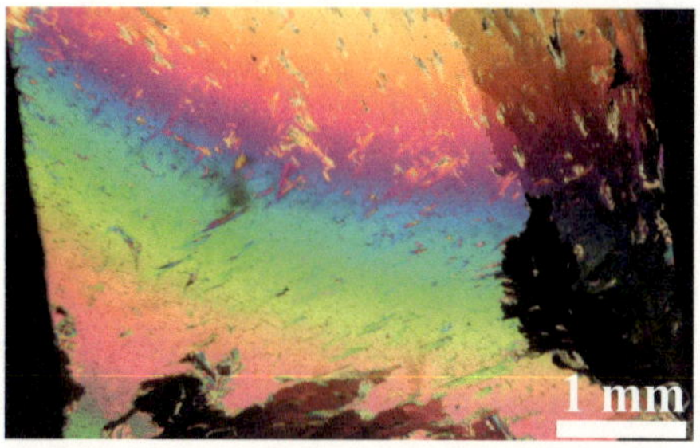

Fig. 5. Image taken under the optical microscope of TPB grown from the melted compound on a functionalized quartz substrate. The image was taken with crossed polarizer and analyzer for a generic orientation of the crystal, so that both interference and birefringence (but manly birefringence) contribute to give the observed fringes in regions of different thickness.

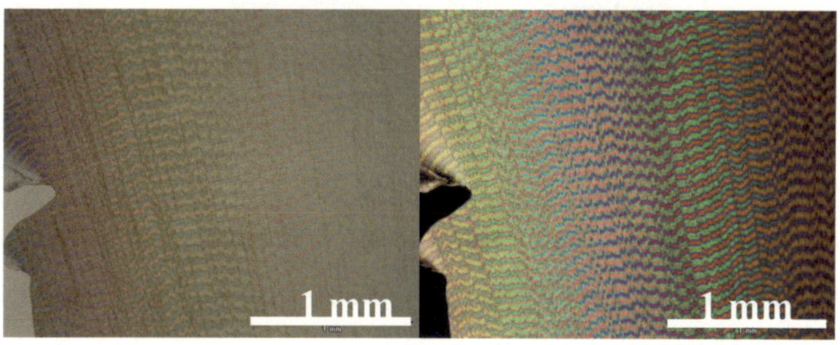

Fig. 6. Images under the optical microscope of a melt-grown TPB sample. They were taken with either parallel (left) or crossed (right) polarizer and analyzer for a generic orientation of the sample. Both interference and birefringence contribute to give the coloured fringes in regions of different thickness.

Another compound that is suitable for melt growth is TPP. Figure 7 shows the images taken under the fluorescence microscope at different temperatures during the growth of a sample of TPP from the melted compound. The starting polycrystalline powder was placed in the hot-stage apparatus and heated. The starting polycrystalline powder shows a disordered orange photoluminescence (inset of the first panel). The emission colour shifts to the green where the melting point (T=380°C) is approached (panel 1). Upon decreasing the temperature, crystallization takes place and abruptly the green regions clearly show orange borders and dark exposed surface (panels 2 and 3), which are even more evident when further decreasing the temperature (panel 4). We mention that a discrepancy between the nominal temperatures and the temperature of the sample is expected due to the presence of a quartz substrate between the material and the heating element.

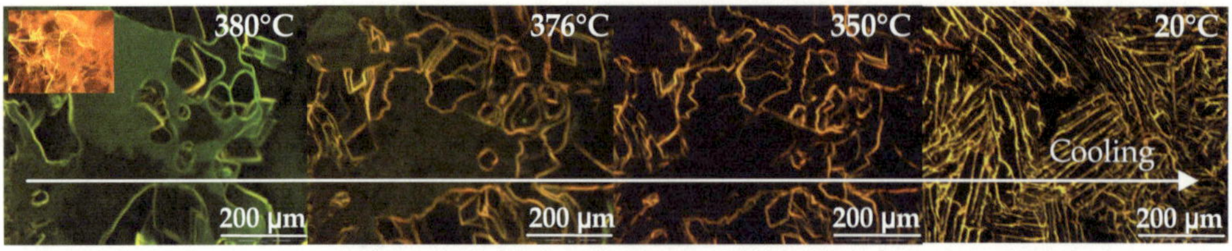

Fig. 7. Images under the fluorescence microscope of a sample of TPP grown from the melted material. Inset: starting polycrystalline powder.

Figure 8 shows a fluorescence microscope image of another melt-grown TPP crystal. The macroscopic order and the presence of cracks in this type of crystal are likely to be improved by adopting a slower and proper temperature profile during the growth. Nevertheless, these crystals exhibit self-waveguiding of the emitted light as a consequence of molecular packing and the melt technique is highly suited as a possible approach for device integration.

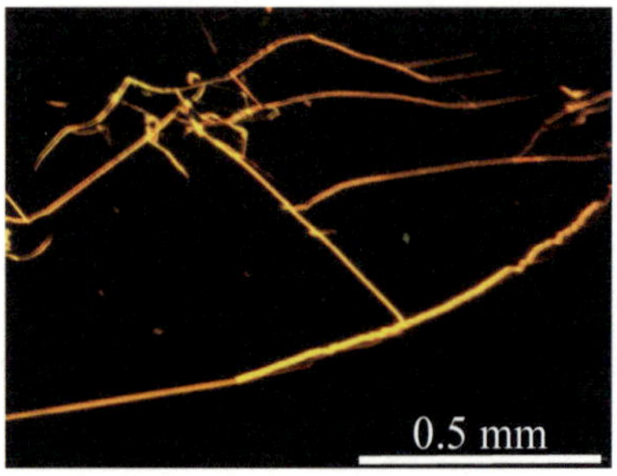

Fig. 8. Fluorescence microscope image of a crystal of TPP.

3. UV-visible optical properties of molecular crystals

3.1 Microscopic and macroscopic theory of the optical properties of molecular crystals

Organic molecular semiconductors are formed by molecules interacting via relatively weak van der Waals forces. Parallel molecular stacking, herringbone, and face-to-face arrangements are often found in the solid state. Absorption and emission spectra of molecular crystals are dominated by Frenkel excitons, which are very different from the Wannier-Mott excitons of conventional inorganic semiconductors. Molecular crystals are characterized by electronic energy bands separated by energy gaps. However, the band gap is strongly influenced by the molecular electronic transition between the highest occupied molecular orbital (HOMO) and the lowest unoccupied molecular orbital (LUMO) and it can be easily tuned to cover the whole UV-visible range by selecting the appropriate organic molecule. Molecular crystals are also characterized by narrow bands as a function of momentum **k** (typical bandwidths of tenths of an electronvolt), relatively large effective masses, and small dielectric constants. Due to these structural properties, they support Frenkel excitons, which consist of electron-hole pairs residing on the same molecule and have large binding energies (of the order of 1 eV). In addition, excitonic transitions in these materials typically show strong optical anisotropy, often associated with strong directional dispersion depending on the direction of the incoming light. By contrast, inorganic semiconductors, which are made of covalently bonded atoms, have relatively wide bands, small effective masses, and large dielectric constants. They therefore support Wannier-Mott excitons, characterized by a large electron-hole separation (much larger than the lattice constant) and by a small binding energy (of the order of 1 meV). Another striking difference between the two classes of semiconductors is that an optical excitation can easily create Frenkel excitons in molecular crystals, while it usually creates free carriers in inorganic semiconductors, Wannier-Mott exciton absorption lines appearing only at low temperatures just below the band edge.

In the following, we discuss the role that molecular packing plays in governing the UV-visible optical properties of molecular materials. In the framework of the Frenkel-Davydov theory, molecules in a crystal are treated as two level systems where the excited state corresponds to the presence of a Frenkel exciton. When interactions between different molecules are taken into account, optical properties are determined by the eigenstates of the whole crystal and they strongly depend on molecular packing. Since intermolecular interactions are much smaller than the molecular transition energy, coupling between crystal states with different numbers of excitons can be neglected. This is called the Heitler-London approximation. In the materials we study, molecular excitations occur in the UV-visible range of the spectrum and optical properties are mainly determined by crystal states with a single exciton. Within the above approximations we will now consider a molecular crystal made of N cells, each containing σ molecules. Crystal excitonic states are conveniently expressed in terms of delocalized states with a given wave vector **k**, defined as

$$\left|\mathbf{k},\alpha\right\rangle = N^{-1/2}\sum_{\mathbf{n}}\exp(i\mathbf{kn})\left|\mathbf{n},\alpha\right\rangle,$$

where

$$\left|\mathbf{n},\alpha\right\rangle$$

indicates a state in which the electronic excitation resides on molecule α in cell **n** while all other molecules in the crystal are in their ground state. The σ excitonic bands at each **k** can then be found by diagonalizing the crystal Hamiltonian

$$H = \hbar\omega_0 + D + \sum_{\mathbf{k},\alpha,\beta}\tilde{J}_{\alpha\beta}(\mathbf{k})\left|\mathbf{k},\alpha\right\rangle\left\langle\mathbf{k},\beta\right| \tag{1}$$

where $\hbar\omega_0$ is the molecular excitation energy, D is the gas-to-crystal shift and the resonance-interaction matrix is defined as

$$\tilde{J}_{\alpha\beta}(\mathbf{k}) = \sum_{\mathbf{n}}{}'\exp(i\mathbf{kn})J_{\alpha\beta}(\mathbf{n}) \tag{2}$$

$J_{\alpha\beta}(\mathbf{n})$ being the interaction energy between a molecule of type α in cell **0**=(0, 0, 0) and a molecule of type β in cell **n**. The primed summation indicates that self-interaction must be excluded, i.e. if α=β then **n**≠(0, 0, 0). Molecular interactions inside the crystal are often approximated to be (screened) interactions between point-dipoles located at the centre of mass of each molecule and corresponding to the dipole moment of the molecular transition. Such an approximation fails for molecules close to each other compared to their dimensions and it can be improved by using the transition charge distribution method, which consists in approximating the delocalized molecular transition dipole with a distribution of point charges located at atomic positions (Alessandrini et al. 2011, Markovitsi et al. 1995, Scholz et al. 2000, Vragovic & Scholz 2003). An even more accurate estimate of the molecular interactions can be obtained by quantum mechanical calculations involving several molecules, but such a refinement would be relevant only for nearest-neighbour or next-nearest-neighbour molecules and, in our test cases, would only slightly correct the absorption spectra. Particular care must be taken in computing the matrix elements $\tilde{J}_{\alpha\beta}(\mathbf{k})$, which, in the case of an infinite three-dimensional crystal, are non-analytic functions of **k** as **k**→0. If molecules are approximated by point-dipoles it is possible to evaluate the infinite sum (2) using Ewald's method (Philpott & Lee 1973), which allows to isolate the non analytic term corresponding to long-range molecular interactions. It is therefore possible to calculate Coulomb or mechanical exciton states by including or neglecting long-range interactions. Ewald's method can also be applied in conjunction with the transition charge distribution method, because even in that case interactions between distant molecules are well approximated by interactions between point-dipoles. Once excitonic bands have been computed, it is possible to compute absorption spectra through the macroscopic dielectric tensor ε_{ij}. Since the wave vector of the incident light is usually much smaller than the reciprocal lattice cell size, absorption is due to excitonic states with $k \approx 0$, which also correspond to poles in the macroscopic dielectric tensor. The latter can be explicitly computed as a function of the incident light energy, $\hbar\omega$, as

$$\varepsilon_{ij}(\hbar\omega) = \varepsilon_{\infty}\delta_{ij} + \frac{2}{\varepsilon_0 V}\sum_n \frac{dn,i\ dn,j\ En}{E_n^2 - (\hbar\omega)^2 - i\gamma\hbar\omega} \tag{3}$$

where ε_{∞} is the high frequency dielectric constant, V is the unit cell volume, γ is a damping factor which determines the width of the absorption line shape, E_n is the energy of the nth mechanical exciton state at $k=0$ and $d_{n,i}$ is the ith component of its dipole moment. We point out that since mechanical excitons enter equation (3), the limit k→0 is well defined and does not depend on the direction of **k**. Having the dielectric tensor, absorption and reflection spectra can be computed for a slab geometry by means of a transfer matrix method (Schubert 1996). In the case of more than one molecule per cell ($\sigma>1$) the Davydov splitting between excitonic bands at k=0 can be computed for each possible orientation of the crystal by simulating the absorption spectra of a very thin film. The above procedure splits the problem of computing absorption into two parts. First we compute the microscopic mechanical exciton states and then we consider the propagation of light in an anisotropic medium, including long-range interactions through the macroscopic dielectric tensor. This theoretical approach takes full account of all the crystal symmetries and the numerical calculations provide a fairly accurate description of the dispersion of the purely excitonic bands, allowing a quantitative comparison with experiments (Raimondo et al. 2006; Silvestri et al. 2009; Spearman et al. 2005; Tavazzi et al. 2006a, 2006b).

It is important to recognize that the properties of excitons in molecular crystals are strongly influenced by their coupling to intramolecular phonons, which give rise to the vibronic progressions typically observed in optical spectra. This is because in organic semiconductors electronic excitation is accompanied by significant nuclear rearrangements, which do not occur in inorganic semiconductors instead. The conventional treatment of exciton-phonon coupling identifies different regimes by comparing the nuclear relaxation energy to a measure of the intermolecular (excitonic) interactions, usually taken to be the free-exciton bandwidth or, in the case of more than one molecule per unit cell, the free-exciton Davydov splitting W. For a single harmonic intramolecular vibrational mode of energy $\hbar\omega_v$, the nuclear relaxation energy is given by $\lambda^2\hbar\omega_v$, where λ^2 is the Huang–Rhys factor. Strong, intermediate and weak excitonic coupling corresponds to the three situations: $\lambda^2\hbar\omega_v << W$, $\lambda^2\hbar\omega_v \approx W$ and $\lambda^2\hbar\omega_v >> W$, respectively.

In the weak excitonic coupling regime, a good approximation is obtained by replacing free excitons with vibronic excitons (or vibrons), in which the electronic excitation and $\tilde{\mu}$ vibrational quanta reside on the same molecule. Such an approximation ignores higher particle states in which the deformation of the lattice surrounding the excitation allows for ground-state vibrations to coexist on neighbouring molecules (Philpott 1971, Spano 2003). In this regime, also called vibronic coupling regime, crystal eigenstates can be expressed in terms of delocalized vibronic states

$$|\mathbf{k},\alpha,\tilde{\mu}\rangle = N^{-1/2}\sum_{\mathbf{n}}\exp(i\mathbf{kn})|\mathbf{n},\alpha,\tilde{\mu}\rangle,$$

where the additional quantum number $\tilde{\mu}$ indicates the number of vibrational quanta residing on molecule α in cell **n.** The corresponding vibrational Hamiltonian is

$$H_{vib} = \hbar\omega_0 + D + \sum_{\mathbf{k},\alpha,\tilde{\mu}} \tilde{\mu}\hbar\omega_v|\mathbf{k},\alpha,\tilde{\mu}\rangle\langle\mathbf{k},\alpha,\tilde{\mu}| + \sum_{\mathbf{k},\alpha,\beta,\tilde{\mu},\tilde{\nu}} S_{\tilde{\mu}0}S_{\tilde{\nu}0}\tilde{J}_{\alpha\beta}(\mathbf{k})|\mathbf{k},\alpha,\tilde{\mu}\rangle\langle\mathbf{k},\beta,\tilde{\nu}| \quad (4)$$

where

$$S_{\tilde{\mu}0}^2 = \exp\left(-\lambda^2\right)\lambda^{2\tilde{\mu}} / \tilde{\mu}!$$

are the Franck-Condon overlap factors. As it can be seen, in this coupling regime, the oscillator strength is redistributed among all the vibronic excitons with k=0 and each excitonic absorption line is replaced by a progression of vibronic replicas.

As the intermolecular coupling increases, the spectral centroid of each vibronic progression tends to shift and the Davydov splitting between different progressions increases. In the extreme case of strong excitonic coupling regime, all the oscillator strength of each progression is again concentrated in a single peak, corresponding to the free exciton state. There is essentially no nuclear relaxation subsequent to the electronic excitation because the excitation resonantly jumps to a neighbour before relaxation can occur. A more detailed description of the various exciton-phonon coupling regimes and of their spectral signatures in molecular aggregates can be found in the literature (Spano 2010).

3.2 TPB

TPB forms single crystals and can grow in at least three polymorphic forms (Baba et al. 2003; Girlando et al. 2010; Ino et al. 2000; Tavazzi et al. 2010b). The two most commonly obtained polymorphs have monoclinic structures known as α and β forms (Table 1). β-TPB is particularly interesting as it exhibits amplified spontaneous emission (ASE) from the widest crystal face (Tavazzi et al. 2010b). The essential crystallographic parameters of the two structures are reported in Table 1.

	α polymorph	β polymorph
Space group	$P2_1$	$P2_1/c$
a (Å)	6.259(2)	9.736(5)
b (Å)	22.164(4)	8.634(2)
c (Å)	7.362(3)	24.480(13)
α (deg)	90	90
β (deg)	96.349(4)	97.11(4)
γ (deg)	90	90
Z	2	4

Table 1. Crystal data for the α and β polymorphs of TPB (Girlando et al. 2010).

The electronic optical transitions of the isolated TPB molecule were calculated as reported elsewhere (Girlando et al. 2010; Tavazzi et al. 2010b). The lowest energy transition is found at about 3.6 eV for the conformations of both the α and the β polymorphs. In both cases, the lowest transition is by far the strongest one calculated in the interval 3-4.8 eV, and is strongly polarized along the longest inertial axis of the molecule (*L*), that is, along the butadiene skeleton, where the HOMO and LUMO are mostly localized. All the components of the transition dipole moment along the principal molecular axes *L*, *M*, *N* (where *M* and *N* are other inertial axes of the molecule, *N* being the shortest one) are reported in Table 2.

	α			β		
molecule	*L*	*M*	*N*	*L*	*M*	*N*
	-8.43	0.26	0.02	-8.22	0.73	0.12
crystals	*b*	*ac*		*b*	*ac*	
		a	*c**		*c*	*a**
	0	7.39	4.54	0	16.44	1.36
	8.19	0	0	0.50	0	0

Table 2. Components along the *L*, *M*, *N* molecular axes of the calculated dipole moments (Debyes) of the molecular electronic transition at lowest energy of α- and β- TPB and components along the unit-cell axes of the unit-cell excitonic transitions.

What is relevant is the relative orientation of the molecules within the crystal and especially with respect to the most developed crystal face to determine different UV-visible properties. The α polymorph has two molecules per unit cell, so that each molecular electronic transition gives rise to two excitonic bands. Both bands are optically allowed at **k**=0 (centre of the Brillouin zone). They are polarized in the *ac* plane and along the monoclinic *b* axis, respectively. For the molecular transition calculated at 3.6 eV, the projections of the unit-cell excitonic transition dipole moments at **k**=0 are reported in Table 2 with respect to the (*abc**) frame of reference (the most developed face of the crystals is typically along the *ac* plane). The two excitonic transitions have comparable intensities, and the direction of polarization of the *ac* component is predicted at 32° to the *a* axis. In β-TPB there are four molecules per unit cell, so that each molecular electronic transition gives rise to four excitonic bands. Only two of them are optically allowed, which are polarized in the *ac* plane and along the monoclinic *b* axis, respectively (Table 2). The most developed face of these crystals is typically along the *bc* plane. We notice that the *b* component of the β polymorph is negligible, so that the corresponding *ac* exciton takes almost the whole available oscillator strength. In particular, the latter is mostly polarized along *c*. Figure 9 summarises the resonant splitting of the electronic transitions from the single molecule to the crystalline α or β polymorphs. The different way that the excitonic oscillator strength is distributed in the two polymorphs arises from the relative orientation of the butadiene skeleton between individual molecules in the two unit cells. In the α phase the butadiene backbones are arranged in a sort of herringbone structure, while in the β-phase the butadiene skeletons are parallel to each other and parallel to the *c* axis. As a result, one exciton dominates the UV-visible absorption spectrum and has twice the intensity of the corresponding α-transitions.

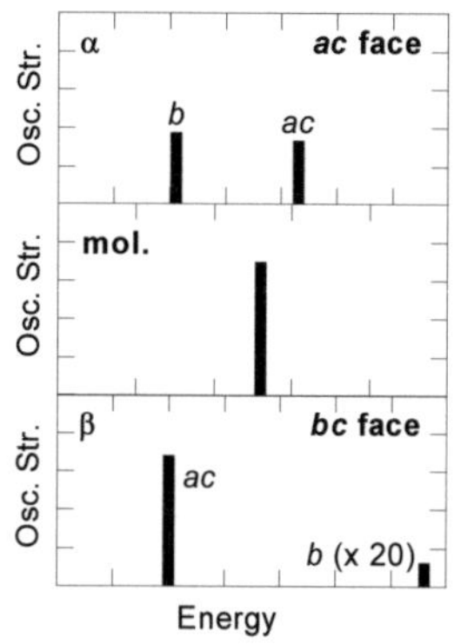

Fig. 9. Schematic picture of the modification of the first electronic transition of TPB from the isolated molecule (second panel) to the α and β crystals (first and third panels, where the labels indicate the polarization with respect to the unit-cell axes). The typical exposed face of the crystals of the two phases is also indicated.

The polarized absorption spectra of the exposed *ac* and *bc* crystal faces of α- and β-TPB, respectively, are reported in Fig. 10. The two continuous curves correspond to two orthogonal polarizations giving the maximum (//) and minimum (⊥) absorbance measured at normal incidence between 3 and 4 eV. The // spectrum saturates above about 3 eV. Despite the saturation, a band can be recognized. For the α polymorph, the maximum absorption at normal incidence is typically measured when the electric field of the incident light forms an angle of about 30° to one of the edges of the crystal. The comparison with the predicted value of 32° indicates that this edge corresponds to the *a* axis and the intense saturated band is attributed to the *ac* excitonic state of α-TPB. As expected, it becomes negligible in the ⊥ spectrum. For the β polymorph, the maximum absorption typically corresponds to one edge of the crystal. Indeed, the main peak can be attributed to the strong *ac* excitonic component, in particular to its *c* component (*bc* being the exposed face and the *c* axis lying along one edge of the crystal). At oblique incidence, another peak emerges centred at about 3.77 eV only for the α phase. It is attributed to the *b*-polarized excitonic component, which for this polymorph lies perpendicular to the exposed face. No emerging peaks were detected for the β phase, in agreement with the calculations which predict a negligible *b*-polarized component.

Up to now, we neglected possible effects due to vibronic coupling. The coupling of the electronic states with vibrational modes could induce shifts of the maxima of the optical bands resulting from a redistribution of the oscillator strength among the replicas. Fig. 11 shows the comparison between the simulated spectra (assuming weak excitonic coupling with one effective mode) for the polarization of the maximum absorption at normal incidence of the two polymorphs (i.e. at 32° to the *a* axis for the α polymorph and along the *c* axis for the β polymorph). The replicas at lowest energies are favoured in the β phase, indicative of J-type i interaction, whilst the *ac*-polarized spectra of the α polymorph shows a

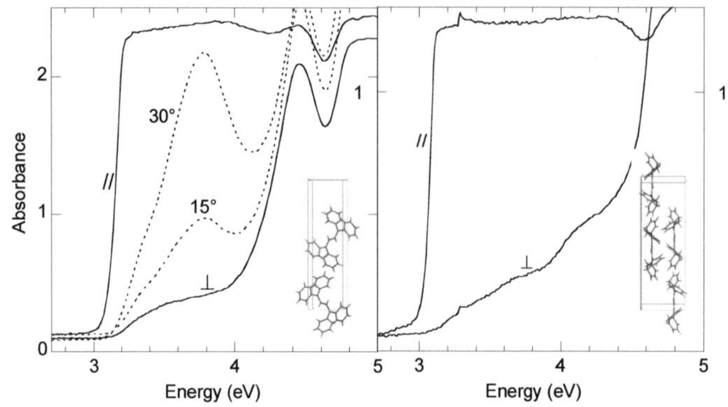

Fig. 10. Polarized absorbance spectra of monocrystals of α-TPB (left) and β-TPB monocrystals measured at normal incidence on the accessible face with two orthogonal polarizations corresponding to the minimum (⊥) and maximum (//) absorption between 3 and 4 eV and absorbance spectra (only for the α polymorph) taken at oblique incidence (angles of incidence 15° and 30°) with p-polarized light with plane of incidence defined by the direction of polarization of minimum absorption at normal incidence and the normal to the surface. Insets: sketch of the unit cell of the two polymorphs (left: α, right: β).

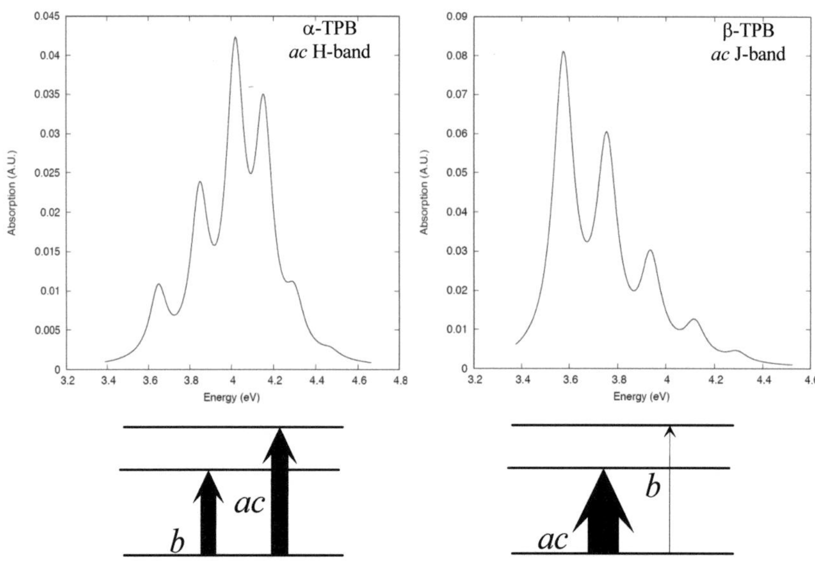

Fig. 11. Simulated absorbance spectra (vibronic approximation) corresponding to the polarization of the maximum absorption at normal incidence on the *ac* face of α-TPB (left) and on the *bc* face of β-TPB (right) (polarization direction at 30° to the *a* axis for the α polymorph and along the *c* axis for the β polymorph). The width of the arrows is proportional to the oscillator strength of the electronic transition.

shift of the maximum in absorption to higher energies due to the positive interaction energy for the upper transition, which is the one observed at normal incidence (H-type).

As far as the emission is concerned, the excitonic bands should be studied as a function of momentum **k** in the Brillouin zone. When the minimum of the excitonic band at the lowest energy is at k=0 or when the state at k=0 is thermally populated, the zero-phonon 0-0 emission transition is allowed and it is polarized as predicted by the excitonic model. For the α phase, the 0-0 emission transition is expected to be *b* polarized, in contrast to its vibronic replicas which are expected to be only partially polarized (Spano 2003, 2010). Since the crystals typically show the most developed face along the *ac* plane, the emitted light corresponding to the 0-0 transition is expected to be self-waveguided towards the borders, while at lower energy the emission from the replicas is expected both from the *ac* face and self-waveguided from the borders. For example, the image on the left of Fig. 3 was taken under the fluorescence microscope on a crystal with α crystal structure. In comparison, the 0-0 transition of the β polymorph is expected to be *ac* polarized, mainly along the *c* axis, which lies in the exposed *bc* face. Therefore, the 0-0 transition is expected to be detected when collecting the light from the face. The emission from the replicas is expected to be only partially polarized, but with much stronger intensity in the *c* polarization. Indeed, the molecular packing shown in the insets of Fig. 10 clearly indicates that in the β phase (and only in the β phase) the molecular transition moments (*L* polarized) are almost entirely parallel to the *c* axis (as also confirmed by the data in Table 2).

3.3 DBTDT

In order to circumvent the relatively low environmental stability of pentacene, heteroacenes, in which a carbocyclic ring or rings are replaced by a heteroaromatic group, have recently emerged (Anthony 2006, 2008; Bendikov et al. 2004; Gao et al. 2007; Gundlach et al. 1997; Jurchescu et al. 2004; Klauk et al. 2002; Li et al. 1998, Maliakal et al. 2004; Okamoto et al. 2005; Osuna et al. 2007; Sheraw et al. 2002; Wex et al. 2005; Xiao et al. 2005). DBTDT belongs to this family and here we use it to discuss an example of how the order of the electronic excited states changes due to intermolecular interactions. Okamoto and co-workers reported a UV-visible photophysical characterization and attribute the observed absorption bands of DBTDT to vibronic transitions of the single molecule (Okamoto et al. 2005). The electronic properties of the single DBTDT molecule and the structural properties in the solid state (Table 3) have been widely discussed also by Osuna et al., together with other thiophene- and selenophene-based heteroacenes studied by UV-visible-NIR spectroscopy (Osuna et al. 2007). The UV-visible absorption of a vacuum-deposited thin film of DBTDT has also been reported, which shows a characteristic red-shift of the lowest energy transition from solution to the solid state (Gao et al. 2007).

Space group	P*nma*
a (Å)	7.93
b (Å)	26.59
c (Å)	5.91
α (deg)	90
β (deg)	90
γ (deg)	90
Z	4

Table 3. Crystal data of DBTDT (Okamoto et al. 2005).

As discussed elsewhere (Alessandrini et al. 2011), the electronic optical transitions of the isolated molecule in vacuum and the corresponding directions of polarization were calculated using as input parameters the molecular geometries obtained from the crystallographic data. The components of the transition moments along the molecular axes (*L, M, N*, where *L* and *N* are the longest and shortest axes, respectively) are reported in Table 4 for the two lowest electronic transitions. The lowest energy band observed in solution (Okamoto et al. 2005; Osuna et al. 2007) is attributed to the transition calculated at lowest energy, which is also the strongest one. The calculated unit-cell dipole moments of the excitonic transitions at k=0 are also given in Table 4. Since the *L* molecular axes of all the four molecules in the unit cell are parallel to *b*, *L*-polarized molecular transitions give rise to a pure H aggregate and only one allowed *b*-polarized excitonic transition which is blue-shifted with respect to the molecular transition. By comparison, the molecular *M*-axes are organised in a herringbone fashion in the *ac* plane so that each *M*-polarized transition gives rise to two allowed excitonic transitions polarized along the *a* and *c* axes with significant projections along both of them.

molecule	*L*	*M*	*N*
	6.72	0	0
crystal	*a*	*b*	*c*
	0	13.44	0
molecule	*L*	*M*	*N*
	0	2.10	0
crystal	*a*	*b*	*c*
	1.93	0	3.73

Table 4. Components along the *L, M, N* molecular axes of the calculated dipole moments (Debyes) of the two molecular electronic transitions at lowest energy of DBTDT and components along the unit-cell axes of the corresponding excitonic transitions.

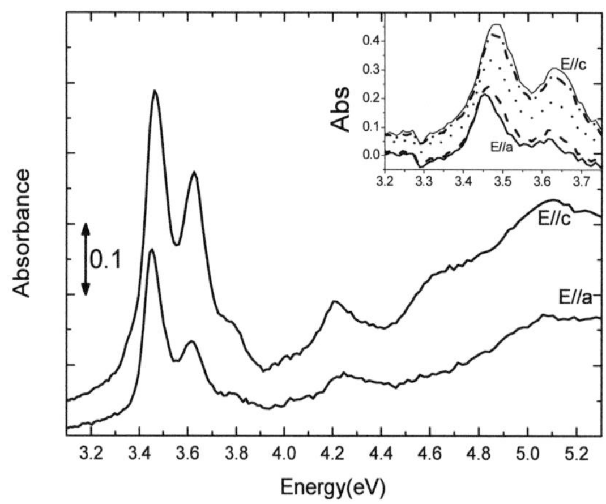

Fig. 12. Absorbance spectra of a DBTDT single crystal (50 nm thickness) taken at normal incidence with two orthogonal polarizations corresponding to the extremes high-energy shift (E//c) and low-energy shift (E//a) of the peaks between about 3.4 and 3.8 eV. Inset: spectra taken on the same crystal with intermediate polarizations.

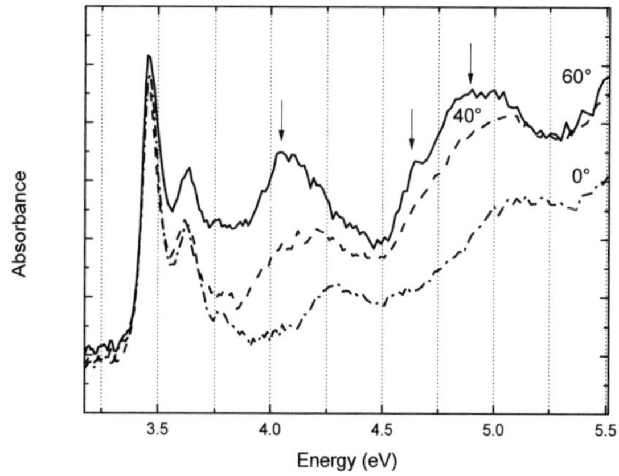

Fig. 13. Absorbance spectra of a DBTDT single crystal taken at different angles of incidence from 0° to 60° on the accessible face (*ac*) with *p* (*ab*) polarized light using the plane (*ab*) formed by the normal to the surface and the direction of polarization labelled as E//*a* in Fig. 12 as plane of incidence.

Polarized absorption spectra were taken at normal incidence to the most developed face of DBTDT crystals. Figure 12 shows the spectra of a monocrystal of about 50 nm in thickness. The reported spectra were measured with two orthogonal polarizations corresponding to the extremes of the low- and high-energy shift of the peaks between about 3.4 and 3.8 eV (the spectra taken with intermediate polarizations are shown in the inset in a restricted energy range). These two polarizations also correspond to the two directions of the electric field which give extinction of the transmitted light under crossed polarizer and analyzer. The spectra are dominated by two orthogonally-polarized vibronic series, with peaks at 3.45, 3.61 eV, and a shoulder at about 3.78 eV in one case, and at 3.46, 3.62 eV and a shoulder at about 3.78 eV in the other case. The stronger intensity is observed in the latter case. In Fig. 12, the two polarization directions are labelled E//a and E//c, since the comparison with the theoretical prediction (see below) allows deducing the corresponding polarization with respect to the axes of the crystal unit cell. Figure 13 shows the spectra taken at oblique incidence with p-polarized light (polarized in the plane of incidence) with plane of incidence formed by the normal to the surface and the direction of polarization labelled as E//a in Figure 12. Besides the replicas in the low-energy portion of the spectrum, at oblique incidence an emerging broad band is detected centred at about 4.07 eV, together with a further shoulder at about 4.6 eV and a structure at about 4.8 eV. The maximum intensity for these bands is detected for the highest angle of incidence; it decreases with decreasing angle of incidence, and eventually disappears at normal incidence.

The comparison between the experimental results and the theoretical predictions allows deducing that the accessible face, which is also the most developed one of the crystal, is the *ac* face. Indeed, two different excitonic transitions are detected in the normal incidence spectra, which are slightly shifted and orthogonally polarized. These bands are therefore attributed to the excitonic transitions (showing a vibronic progression) originating from the second *M*-polarized molecular electronic transition (calculated at 3.95 eV in this work and at 3.89 by Osuna et al. (Osuna et al. 2007). The labels E//*c* and E//*a* in Fig. 12 derive from this attribution. The Davydov splitting between the corresponding replicas of the *a* and *c* progressions is about 0.01 eV. Such a low value indicates that, for this electronic transition, exciton coupling is relatively weak ($\lambda^2\hbar\omega_v >> W$). At higher energy in Fig. 12, broad and weaker bands are detected at about 4.21 and 5.07 eV, and 4.21 and 5.08 eV, for the two polarizations. These bands are attributed to other excitonic transitions in the *ac* plane of *M* molecular origin. The emerging band at 4.07 eV in oblique incidence spectra is attributed to the *b*-polarized excitonic transition originating from the strongest transition of the single molecule at the lowest energy. This molecular transition has been calculated at 3.86 eV in this work and at 3.77 eV by Osuna et al. (Osuna et al. 2007) and has been observed at about 3.6 eV in the solution spectra (Osuna et al. 2007; Okamoto et al. 2005). The blue-shift observed in the crystal spectra is explained by the arrangement of the corresponding *L*-molecular transition moments forming a pure H aggregate. We underline that polarized spectroscopy on monocrystals shows that there is no correspondence between the replicas observed at lowest energy in the solid state (originating from the second molecular transition) and the strong electronic transition at lowest energy of the single molecule. The diagram in Fig. 14 schematically shows the modification of the first and second electronic states from the single molecule to the crystal. Numerical calculations based on the method described in section 3.1 predicted, in the case of the first molecular transition, a Davydov splitting of 0.44 eV between the dark exciton at lowest energy and the allowed *b*-polarized upper exciton while, for the second molecular transition, the Davydov splitting between the two allowed excitons was found to be less than 0.01 eV, thus confirming the weak excitonic coupling of the second *M*-polarized molecular electronic transition. This suggests that the lowest excitonic transition (dark) coming from the first molecular transition is expected to be at about the same energy as the excitonic transitions detected in the normal-incidence spectra, which stems from the second molecular *M* transition.

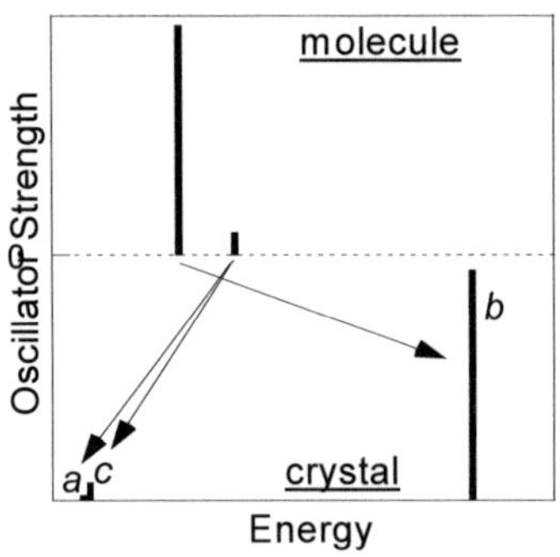

Fig. 14. Schematic picture of the modifications of the two lowest electronic transitions of DBTDT to the crystal excitons. The labels indicate the polarization of the excitonic states with respect to the unit-cell axes.

3.4 TPP

TPP is now taken into consideration to discuss another phenomenon, which may strongly influence the optical properties of molecular crystals, namely directional dispersion. Among thiophene/phenylene co-oligomers, this thiophene/phenyl co-oligomer end capped with pyrrole has been recently designed to explore the effect of an electron-rich aromatic ring, such as pyrrole, to end-cap oligothiophenes (Tavazzi et al. 2010a).

Space group	$P2_1$
a (Å)	5.77
b (Å)	7.43
c (Å)	24.72
α (deg)	90
β (deg)	91.76
γ (deg)	90
Z	2

Table 5. Crystal data for the α and β polymorphs of TPP (Tavazzi et al. 2010a).

The electronic optical transitions of the isolated molecule in vacuum were calculated (Tavazzi et al. 2010a) using as input parameters the molecular geometries obtained from the crystallographic data (Table 5). They were found at 2.95, 4.18, and 4.52 eV. In solution, the lowest absorption band was measured at about 3.1 eV (not shown here) and it is attributed to the lowest calculated transition. This transition is the strongest one with oscillator strength of 12.8 Debyes, which is quite high, compared to other conjugated molecules. For example, the lowest electronic transition of two of the most studied oligothiophenes, quaterthiophene and sexithiophene, have transition moments of about 10.6 and 10.4 Debyes, respectively (Petelenz & Andrzejak 2000; Silvestri et al. 2009; Spano et al. 2007). The components of the TPP transition at lowest energy with respect to the molecular inertial axes are listed in Table 6 and the results clearly indicate the strong polarization of the molecular transition at lowest energy along the long (*L*) molecular axis. In the crystal, each molecular electronic transition is predicted to give rise to two excitonic transitions, which are *b* and *ac* polarized. The components of the excitonic transitions were calculated and are provided in Table 6. There is a very weak *b* component and a strong *ac* component, mainly polarized along *c**, namely along the normal to the most developed face of the samples (the angle between the *ac* transition moment and the *c** axis is calculated to be about 7°). Indeed, the arrangement of the corresponding molecular transition moments in the unit cell gives rise to an almost pure H aggregate with almost parallel molecular transition moments.

Molecule	L	M	N
	-12.75	0.29	-0.10
crystal	*b*	*a*	*c**
	-0.18	0	0
	0	2.68	-17.84

Table 6. Components along the *L, M, N* molecular axes of the calculated dipole moment (in Debyes) of the molecular electronic transition at lowest energy of TPP and components along the unit-cell axes of the corresponding unit-cell excitonic transitions.

The measured absorption spectra taken in different configurations of a TPP single crystal are reported in Fig. 15. At normal incidence, the spectra correspond to two orthogonal polarizations giving the minimum and maximum absorption in the considered spectral range. These two polarizations also correspond to the two directions of the electric field which give extinction of the transmitted light under crossed polarizer and analyzer. The variable-angle measurements (at the angles of incidence indicated by the labels) were taken with *p* polarized light with plane of incidence formed by the normal to the surface and the direction of polarization of the maximum absorption measured at normal incidence. The spectra are dominated by a main broad band. The intensity and shape of this band strongly depends on the angle of incidence. The maximum intensity is detected for the highest positive angle of incidence (the centre of the band can be tentatively deduced to be 3.75 eV), it decreases with decreasing angle of incidence down to zero at about -15°, and finally it again emerges slightly shifted at higher energy (the maximum is found at about 3.85 eV). This indicates that the corresponding transition moment has a strong component along the normal to the surface and shows directional dispersion (Philpott 1969; Spearman et al. 2005; Tavazzi et al. 2006b; Weiser & Moller 2002). Considering that the band disappears at about -15°, the inclination of the transition moment to the normal to the surface can be approximately estimated from the experiment to be of the order of -10° (taking into consideration the effect of refraction at the air/crystal interface).

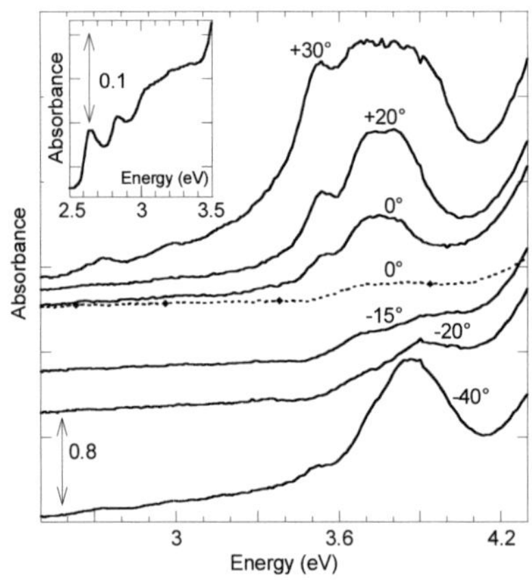

Fig. 15. Continuous lines: absorbance spectra of a TPP single crystal taken at different angles of incidence from +30° to -40° on the accessible face (*ab*) with *p*-polarized light and the plane formed by the normal to the surface and the direction of polarization of maximum absorption measured at normal incidence (*ac* plane) as plane of incidence. The portion of the spectrum between about 3.6 and 3.9 eV measured at +30° is affected by saturation. Dashed line with diamonds: normal-incidence spectrum with orthogonal (*b*) polarization). Inset: absorbance spectrum taken at normal incidence on a thicker crystal with *b* polarization.

The main band observed in Fig. 15 is attributed to the *ac* polarized exciton transition originating from the *L* molecular one, which is expected to be the strongest peak in the spectra. It reveals several structures (the lowest one at 2.73 eV, followed by shoulders at 2.97, 3.18 eV and by a peak centred at about 3.52 eV before the saturation region), which are easily detected for the highest angles of incidence and are attributed to vibronic replicas. The corresponding *b* excitonic component is not clearly observed in the spectra of Fig. 15, but it is recognizable in the spectrum (inset) taken in the same configuration on a thicker crystal with replicas of decreasing intensity at 2.65, 2.84, and 3.03 eV. On the basis of the calculated transition moments, the intensity ratio between the *ac* polarized and *b* polarized excitonic transitions is predicted to be of the order of 10^4. This ratio decreases for normal incidence spectra, where the intensity ratio between the *a*-component of the *ac* exciton and the *b* exciton is predicted to be about 200. Since the *ac* exciton transition dipole is nearly perpendicular to the *ab* plane, the predicted ratio is only a rough estimate: a change as small as 1° degree in the molecular dipole inclination can produce a reduction of a factor 10 in the estimated intensity ratio. In any case, these low values explain the difficulty to detect the *b* excitonic transition for the thin sample whose spectra are shown in Fig. 15. The results of the optical characterization also indicate that the accessible face is the *ab* face (and the spectra in Fig. 15 were taken with *ac* as plane of incidence). Moreover, the *b* axis is a material principal axis and *ac* is a principal plane, thus explaining the extinction of the transmitted light under crossed polarizer and analyzer (Spearman et al. 2005). The energy difference between the lowest *b*-polarized peak (the most intense replica of the progression) and the centre of the *ac* polarized band measured at oblique incidence for positive angles is about 1.2 eV. Therefore, TPP satisfies the conditions of the strong excitonic coupling $\lambda^2\hbar\omega_v << W$: the *ac* exciton is almost free, it possesses nearly all the available oscillator strength and it is considerably blue-shifted with respect to the molecular electronic transition. The *ac* polarized replicas in the spectral region between 2.7 and 3.6 eV are built on the lower *b* polarized excitonic band and obtain their oscillator strength through an intermolecular Herzberg–Teller coupling to the nearly free *ac* polarized exciton (Silvestri et al. 2009; Spano 2003, 2004; Spano et al. 2007; Tavazzi et al. 2006a). On the contrary, the *b* polarized exciton is very weakly allowed and it is more strongly coupled to phonons, showing a vibronic progression. On the basis of the calculations and the measurements in solution, other excitonic transitions are expected originating from the higher molecular electronic transitions (Tavazzi et al. 2010a). Indeed, in the *b*-polarized normal-incidence spectrum, a relatively weak structure is also detected between about 3.4 and 4.0 eV, showing poorly defined replicas. A similar band cannot be excluded in the *a*-polarized spectrum below the *ac* exciton of *L* origin.

Figure 16 summarises the Davydov splitting of the electronic optical transition at lowest energy from the molecule to the crystal. The structural and optical characterization indicates that the crystals show an almost H-type aggregation of the electronic transition moment at lowest energy with strong excitonic coupling, giving rise to a large splitting of about 1.2 eV between an intense and almost free excitonic state polarized along the normal to the crystal surface and a weak *b*-polarized component. This large splitting results from the combination

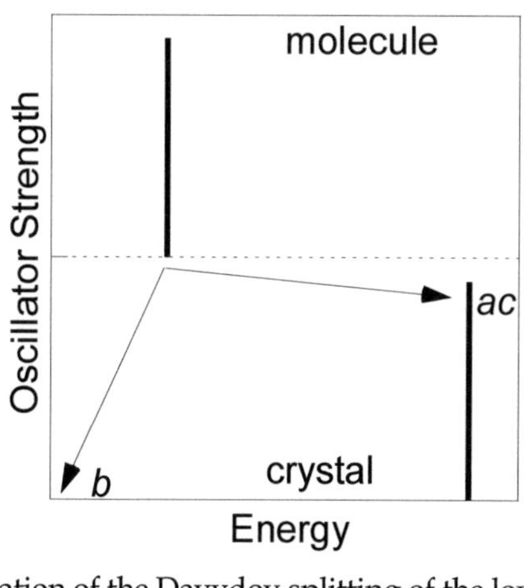

Fig. 16. Schematic representation of the Davydov splitting of the lowest electronic transition of TPP. The labels indicate the polarization of the excitonic states with respect to the unit-cell axes.

of the high oscillator strength and the steep inclination to the normal to the surface of the transition moment. Indeed, the splitting is demonstrated to be proportional to $(\hat{\mathbf{k}}_{inc}.\mathbf{d})^2$, where $\mathbf{k}_{inc}$ is the wave vector of the incoming light ($|\mathbf{k}_{inc}|$~0) and **d** is the excitonic dipole moment (Philpott 1969; Spearman et al. 2005; Tavazzi et al. 2006b; Weiser & Moller 2002). Thus, TPP is an example with strong directional dispersion (energy and shape of the absorption band are strongly dependent on the angle of incidence on the crystal face).

4. Conclusion

In this chapter, we have reported some relevant effects on the optical properties of molecular solids related to the packing of the molecules in the crystal lattice using the general theory of molecular excitons. Three prototypical examples have been taken into consideration: (i) 1,1,4,4-tetraphenyl-1,3-butadiene (TPB), (ii) dibenzo[d,d']thieno[3,2-b;4,5-b']dithiophene (DBTDT), (iii) N-pyrrole end-capped thiophene/phenyl co-oligomer (TPP).

First of all, we have described few methods, which are often used for the crystal growth. The physical vapour growth requires a crucible for the starting material, a tube under inert-gas flow, and a furnace. An alternative method is the floating-drop technique, which requires two liquids, the molecule of interest being completely insoluble in one of them. The size of the grown crystals is typically relatively large. The third method that we have described is the melt growth, which is applicable when the molecular species possesses a well-defined melting point and remains stable at the corresponding temperature. Future opto-electronics device structures may exploit the crystalline form of molecular materials, but limiting factors are the control of their growth and device integration. A promising technique to overcome this barrier is to grow crystals from the melted compound, which can be incorporated into well-defined device geometries. As an example, we have shown the evolution during the TPP melt growth of the colour of the emission observed under a fluorescence microscope. The emission is green and isotropic when the compound is melted, but, upon decreasing the temperature, crystallization takes place and abruptly the green regions clearly show orange borders and dark exposed surface (with self-waveguiding of the emitted light).

The discussion of the UV-visible optical properties of molecular crystals is based on the Frenkel-Davydov exciton theory. First of all, we have mentioned a few important differences between organic and inorganic semiconductors in the framework of solid-state physics. The main differences are the different interaction energies among the molecules/atoms, the different energy dispersions of the electronic bands in the Brillouin zone as a function of momentum (and thus the different effective masses of the charge carriers), the formation of different types of excitons (either Frenkel-Davydov or Wannier-Mott), the differences in the packing of the constituent molecules/atoms, with possible strong optical anisotropy in the case of molecular solids, the different interactions of the respective excitons with vibrations/phonons. The Frenkel-Davydov theory underlines these aspects. It has been described and used to deduce the dielectric tensor of the three mentioned materials.

As far as TPB is concerned, it has been taken into consideration to discuss possible remarkable differences between polymorphs of the same molecular species. Indeed, a common consequence of different packing of the same molecular entity is polymorphism, i.e. the occurrence of different crystalline phases. We have discussed that the different redistribution of the excitonic oscillator strength for the two polymorphs of TPB arises from the relative orientation of the butadiene skeleton between individual molecules in the two unit cells. In the α phase the butadiene backbones are arranged in a sort of herringbone structure, while in the β phase they are parallel to each other and almost parallel to one of the unit-cell axis. As a result, only one excitonic state dominates the UV-visible absorption spectrum, which has twice the intensity of the corresponding transitions in the α polymorph. These differences produce, in turn, differences in the exciton-vibration interaction. As discussed in the chapter, the coupling of the electronic states with vibrational modes can induce shifts of the maxima of the optical bands resulting from a redistribution of the oscillator strength among the replicas. For example, for TPB the replicas at lowest energies are favoured in the β phase (J-type interaction), while the normal-incidence spectra of the α polymorph shows a shift of the maximum in absorption to higher energies due to the positive interaction energy for the upper transition (H-type interaction).

DBTDT has been used as example to discuss the possible changes of the order of the excited states due to intermolecular interactions. The optical bands have been attributed to excitonic transitions showing a vibronic progression. However, there is no correspondence between the replicas observed at lowest energy in the solid state (originating from the second molecular transition) and the strong electronic transition at lowest energy of the molecule in solution (originating, as expected, from the first molecular transition). The remarkable blue-shift in the crystal of the first molecular transition is explained by the arrangement of the corresponding molecular transition moments forming a pure H aggregate.

Finally, TPP offers a prime example of the phenomenon of directional dispersion in which the molecular orientation within the crystal lattice affects the absorption spectra. This crystal shows an almost H-type aggregation of the electronic transition moment at lowest energy with strong excitonic coupling, giving rise to a large splitting at normal incidence of about 1.2 eV among the two allowed excitons. This large splitting results from the combination of the high oscillator strength of the molecular transition and the steep inclination to the normal to the surface of the transition moment since the splitting is proportional to the scalar product of the wave vector of the incoming light and the excitonic transition moment. Therefore, the energy and shape of the absorption band are strongly dependent on the angle of incidence on the exposed crystal face.

The chapter is mainly focused on the UV-visible absorption properties. A full description of the polarized emission spectra of molecular crystals requires the study of the excitonic bands in the whole first Brillouin zone (not only at the center of the Brillouin zone as for the interpretation of the absorption properties). However, some discussion on the emission properties has been given. For example, we have mentioned which are the expected directions and polarizations of either the zero-phonon emission transition or its replicas. We underline that future device structures may exploit the crystalline forms of molecular solids to provide highly directional emission useful in laser applications (vertical surface emitting geometries) or edge emission.

5. Acknowledgment

Fondazione Cariplo is acknowledged for financial support.

6. References

Adachi, H., Takano, K., Morikawa, M., Kanaya, S., Yoshimura, M., Mori, Y. & Sasaki, T. (2003) 'Application of a two-liquid system to sitting-drop vapour-diffusion protein crystallization', *Acta Crystallographica Section D-Biological Crystallography,* 59, 194-196.

Alessandrini, L., Braga, D., Jaafari, A., Miozzo, L., Mora, S., Silvestri, L., Tavazzi, S. & Yassar, A. (2011) 'Optical Properties of Dibenzo d,d ' thieno 3,2-b;4,5-b ' dithiophene Monocrystals: The Effect of Intermolecular Interactions', *Journal of Physical Chemistry A,* 115(3), 225-231.

Anthony, J. E. (2006) 'Functionalized Acenes and Heteroacenes for Organic Electronics', *Chemical Reviews,* 106(12), 5028-5048.

Anthony, J. E. (2008) 'The larger acenes: Versatile organic semiconductors', *Angewandte Chemie-International Edition,* 47(3), 452-483.

Baba, K., Kasai, H., Okada, S., Oikawa, H. & Nakanishi, H. (2003) 'Fabrication of organic nanocrystals using microwave irradiation and their optical properties', *Optical Materials,* 21(1-3), 591-594.

Bendikov, M., Wudl, F. & Perepichka, D. F. (2004) 'Tetrathiafulvalenes, oligoacenenes, and their buckminsterfullerene derivatives: The brick and mortar of organic electronics', *Chemical Reviews,* 104(11), 4891-4945.

Campione, M., Ruggerone, R., Tavazzi, S. & Moret, M. (2005) 'Growth and characterisation of centimetre-sized single crystals of molecular organic materials', *Journal of Materials Chemistry,* 15(25), 2437-2443.

Cordella, F., Quochi, F., Saba, M., Andreev, A., Sitter, H., Sariciftci, N. S., Mura, A. & Bongiovanni, G. (2007) 'Optical gain performance of epitaxially grown para-sexiphenyl films', *Advanced Materials,* 19(17), 2252.

Fichou, D., Delysse, S. & Nunzi, J. M. (1997) 'First evidence of stimulated emission from a monolithic organic single crystal: alpha-octithiophene', *Advanced Materials,* 9(15), 1178.

Fichou, D., Dumarcher, V. & Nunzi, J. M. (1999) 'One- and two-photon stimulated emission in oligothiophenes single crystals', *Optical Materials,* 12(2-3), 255-259.

Frolov, S. V., Vardeny, Z. V. & Yoshino, K. (1998) 'Cooperative and stimulated emission in poly(p-phenylene-vinylene) thin films and solutions', *Physical Review B,* 57(15), 9141-9147.

Gao, J., Li, R., Li, L., Meng, Q., Jiang, H., Li, H. & Hu, W. (2007) 'High-performance field-effect transistor based on dibenzo[d,d']thieno[3,2-b;4,5-b']dithiophene, an easily synthesized semiconductor with high ionization potential', *Adv. Mater. (Weinheim, Ger.),* 19, 3008-3011.

Girlando, A., Ianelli, S., Bilotti, I., Brillante, A., Della Valle, R. G., Venuti, E., Campione, M., Mora, S., Silvestri, L., Spearman, P. & Tavazzi, S. (2010) 'Spectroscopic and Structural Characterization of Two Polymorphs of 1,1,4,4-Tetraphenyl-1,3-butadiene', *Crystal Growth & Design,* 10(6), 2752-2758.

Gundlach, D. J., Lin, Y. Y., Jackson, T. N., Nelson, S. F. & Schlom, D. G. (1997) 'Pentacene organic thin-film transistors - molecular ordering and mobility', *IEEE Electron Device Lett.,* 18, 87-89.

Hibino, R., Nagawa, M., Hotta, S., Ichikawa, M., Koyama, T. & Taniguchi, Y. (2002) 'Emission gain-narrowing from melt-recrystallized organic semiconductor', *Advanced Materials*, 14(2), 119.

Hiramatsu, T., Matsuoka, N., Yanagi, H., Sasaki, F. & Hotta, S. (2009) 'Gain-narrowed emissions of thiophene/phenylene co-oligomer single crystals', *Physica Status Solidi C*, 6(1), 338.

Horowitz, G., Kouki, F., El Kassmi, A., Valat, P., Wintgens, V. & Garnier, F. (1999) 'Structure-dependent fluorescence in sexithiophene single crystals', *Advanced Materials*, 11(3), 234.

Ichikawa, M., Hibino, R., Inoue, M., Haritani, T., Hotta, S., Araki, K., Koyama, T. & Taniguchi, Y. (2005) 'Laser oscillation in monolithic molecular single crystals', *Advanced Materials*, 17(17), 2073.

Ichikawa, M., Hibino, R., Inoue, M., Haritani, T., Hotta, S., Koyama, T. & Taniguchi, Y. (2003) 'Improved crystal-growth and emission gain-narrowing of thiophene/phenylene co-oligomers', *Advanced Materials*, 15(3), 213.

Ino, I., Wu, L. P., Munakata, M., Kuroda-Sowa, T., Maekawa, M., Suenaga, Y. & Sakai, R. (2000) 'Bridged silver(I) complexes of the polycyclic aromatic compounds tetraphenylethylene and 1,1,4,4-tetraphenyl-1,3-butadiene', *Inorganic Chemistry*, 39(24), 5430-5436.

Jurchescu, O. D., Baas, J. & Palstra, T. T. M. (2004) 'Effect of impurities on the mobility of single crystal pentacene', *Applied Physics Letters*, 84(16), 3061-3063.

Kena-Cohen, S., Davanco, M. & Forrest, S. R. (2008) 'Strong exciton-photon coupling in an organic single crystal microcavity', *Physical Review Letters*, 101(11), 116401.

Kena-Cohen, S. & Forrest, S.R. (2010) 'Room-temperature polariton lasing in an organic single-crystal microcavity', *Nature Photonics*, 4, 371.

Klaers, J., Schmitt, J., Vewinger, F. & Weitz, M. (2010) 'Bose-Einstein condensation of photons in an optical microcavity', *Nature*, 468(7323), 545-548.

Klauk, H., Halik, M., Zschieschang, U., Schmid, G., Radlik, W. & Weber, W. (2002) 'High-mobility polymer gate dielectric pentacene thin film transistors', *Journal of Applied Physics*, 92(9), 5259-5263.

Kloc, C. & Laudise, R. A. (1998) 'Vapor pressures of organic semiconductors: α-hexathiophene and α-quaterthiophene', *J. Cryst. Growth*, 193, 563-571.

Kloc, C., Simpkins, P. G., Siegrist, T. & Laudise, R. A. (1997) 'Physical vapor growth of centimeter-sized crystals of α-hexathiophene', *J. Cryst. Growth*, 182, 416-427.

Kondo, H., Tongu, K., Yamamoto, Y., Yamamoto, S. & Kurisu, H. (2009) 'Cavity polariton dispersion of a single-crystalline anthracene film embedded in a microcavity', *Phys. Status Solidi C*, 6, 284-287.

Kondo, H., Yamamoto, Y., Takeda, A., Yamamoto, S. & Kurisu, H. (2008) 'Optical responses in single-crystalline organic microcavities', *J. Lumin.*, 128, 777-779.

Laudise, R. A., Kloc, C., Simpkins, P. G. & Siegrist, T. (1998) 'Physical vapor growth of organic semiconductors', *J. Cryst. Growth*, 187, 449-454.

Li, X. C., Sirringhaus, H., Garnier, F., Holmes, A. B., Moratti, S. C., Feeder, N., Clegg, W., Teat, S. J. & Friend, R. H. (1998) 'A highly pi-stacked organic semiconductor for thin film transistors based on fused thiophenes', *Journal of the American Chemical Society*, 120(9), 2206-2207.

Liu, C. Y. & Bard, A. J. (2000) 'In-situ regrowth and purification by zone melting of organic single-crystal thin films yielding significantly enhanced optoelectronic properties', *Chemistry of Materials,* 12(8), 2353-2362.

Losio, P. A., Hunziker, C. & Guenter, P. (2007) 'Amplified spontaneous emission in para-sexiphenyl bulk single crystals', *Applied Physics Letters,* 90(24), 241103.

Maliakal, A., Raghavachari, K., Katz, H., Chandross, E. & Siegrist, T. (2004) 'Photochemical stability of pentacene and a substituted pentacene in solution and in thin films', *Chemistry of Materials,* 16(24), 4980-4986.

Markovitsi, D., Germain, A., Millie, P., Lecuyer, P., Gallos, L., Argyrakis, P., Bengs, H. & Ringsdorf, H. (1995) 'Triphenylene Columnar Liquid Crystals: Excited States and Energy Transfer', *J. Phys. Chem.,* 99, 1005-17.

Matsuoka, N., Hiramatsu, T., Yanagi, H., Sasaki, F. & Hotta, S. (2010) 'Characterization of Gain-Narrowed Emission from Biphenyl-Capped Thiophene Single Crystals', *Japanese Journal of Applied Physics,* 49(1), 01AD05.

Meinardi, F., Borghesi, A., Cerminara, M., Sassella, A., Tavazzi, S., Tubino, R., Gurioli, M., Mura, A. & Bongiovanni, G. (2001) 'The origin of radiative emission of quaterthiophene ultra-thin films', *Synth. Met.,* 121, 1355-1356.

Nagawa, M., Hibino, R., Hotta, S., Yanagi, H., Ichikawa, M., Koyama, T. & Taniguchi, Y. (2002) 'Emission gain narrowing from single crystals of a thiophene/phenylene co-oligomer', *Applied Physics Letters,* 80(4), 544-546.

Okamoto, T., Kudoh, K., Wakamiya, A. & Yamaguchi, S. (2005) 'General synthesis of thiophene and selenophene-based heteroacenes', *Organic Letters,* 7(23), 5301-5304.

Osuna, R. M., Ponce Ortiz, R., Okamoto, T., Suzuki, Y., Yamaguchi, S., Hernandez, V. & Lopez Navarrete, J. T. (2007) 'Thiophene- and selenophene-based heteroacenes: Combined quantum chemical DFT and spectroscopic Raman and UV-vis-NIR study', *Journal of Physical Chemistry B,* 111(26), 7488-7496.

Park, S., Kwon, O. H., Kim, S., Choi, M. G., Cha, M., Park, S. Y. & Jang, D. J. (2005) 'Imidazole-based excited-state intramolecular proton-transfer materials: Synthesis and amplified spontaneous emission from a large single crystal', *Journal of the American Chemical Society,* 127(28), 10070-10074.

Petelenz, P. & Andrzejak, M. (2000) 'Vibronic interpretation of the low-energy absorption spectrum of the sexithiophene single crystal', *Journal of Chemical Physics,* 113(24), 11306-11314.

Philpott, M. R. (1969) 'Dipole Davydov splittings in crystalline anthracene, tetracene, naphthalene, and phenanthrene', *J. Chem. Phys.,* 50, 5117-28.

Philpott, M. R. (1971) 'Theory of coupling of electronic and vibrational excitations in molecular crystals and helical polymers', *Journal of Chemical Physics,* 55(5), 2039.

Philpott, M. R. & Lee, J. W. (1973) 'Some remarks on calculation and use of retarded and static dipole sums in molecular exciton theory', *Journal of Chemical Physics,* 58(2), 595-602.

Polo, M., Camposeo, A., Tavazzi, S., Raimondo, L., Spearman, P., Papagni, A., Cingolani, R. & Pisignano, D. (2008) 'Amplified spontaneous emission in quaterthiophene single crystals', *Applied Physics Letters,* 92(8), 083311.

Quochi, F., Cordella, F., Mura, A., Bongiovanni, G., Balzer, F. & Rubahn, H. G. (2005) 'One-dimensional random lasing in a single organic nanofiber', *Journal of Physical Chemistry B,* 109(46), 21690-21693.

Raimondo, L., Campione, M., Laicini, M., Moret, M., Sassella, A., Spearman, P. & Tavazzi, S. (2006) 'Absorbance spectra of polycrystalline samples and twinned crystals of oligothiophenes', *Appl. Surf. Sci.*, 253, 271-274.

Scholz, R., Kobitski, A. Y., Kampen, T. U., Schreiber, M., Zahn, D. R. T., Jungnickel, G., Elstner, M., Sternberg, M. & Frauenheim, T. (2000) 'Resonant Raman spectroscopy of 3,4,9,10-perylene-tetracarboxylic-dianhydride epitaxial films', *Physical Review B,* 61(20), 13659-13669.

Schubert, M. (1996) 'Polarization-dependent optical parameters of arbitrarily anisotropic homogeneous layered systems', *Phys. Rev. B: Condens. Matter,* 53, 4265-74.

Sheraw, C. D., Zhou, L., Huang, J. R., Gundlach, D. J., Jackson, T. N., Kane, M. G., Hill, I. G., Hammond, M. S., Campi, J., Greening, B. K., Francl, J. & West, J. (2002) 'Organic thin-film transistor-driven polymer-dispersed liquid crystal displays on flexible polymeric substrates', *Appl. Phys. Lett.*, 80, 1088-1090.

Siegrist, T., Kloc, C., Laudise, R. A., Katz, H. E. & Haddon, R. C. (1998) 'Crystal growth, structure, and electronic band structure of α-4T polymorphs', *Adv. Mater. (Weinheim, Ger.),* 10, 379-382.

Silvestri, L., Tavazzi, S., Spearman, P., Raimondo, L. & Spano, F. C. (2009) 'Exciton-phonon coupling in molecular crystals: Synergy between two intramolecular vibrational modes in quaterthiophene single crystals', *Journal of Chemical Physics,* 130(23), 234701.

Spano, F. C. (2003) 'The fundamental photophysics of conjugated oligomer herringbone aggregates', *J. Chem. Phys.,* 118, 981-994.

Spano, F. C. (2004) 'Temperature dependent exciton emission from herringbone aggregates of conjugated oligomers', *J. Chem. Phys.,* 120, 7643-7658.

Spano, F. C. (2010) 'The Spectral Signatures of Frenkel Polarons in H- and J-Aggregates', *Acc. Chem. Res.,* 43, 429-439.

Spano, F. C., Silvestri, L., Spearman, P., Raimondo, L. & Tavazzi, S. (2007) 'Reclassifying exciton-phonon coupling in molecular aggregates: Evidence of strong nonadiabatic coupling in oligothiophene crystals', *Journal of Chemical Physics,* 127(18), 184703.

Spearman, P., Borghesi, A., Campione, M., Laicini, M., Moret, M. & Tavazzi, S. (2005) 'Directional dispersion in absorbance spectra of oligothiophene crystals', *J. Chem. Phys.,* 122, 014706/1-014706/6.

Tavazzi, S., Borghesi, A., Gurioli, M., Meinardi, F., Riva, D., Sassella, A., Tubino, R. & Garnier, F. (2003) 'Absorption and emission properties of α,ω-dihexyl-quaterthiophene thin films grown by organic molecular beam deposition', *Synth. Met.,* 138, 55-58.

Tavazzi, S., Borghesi, A., Papagni, A., Spearman, P., Silvestri, L., Yassar, A., Camposeo, A., Polo, M. & Pisignano, D. (2007) 'Optical response and emission waveguiding in rubrene crystals', *Physical Review B,* 75(24), 245416.

Tavazzi, S., Campione, M., Laicini, M., Raimondo, L., Borghesi, A. & Spearman, P. (2006a) 'Measured Davydov splitting in oligothiophene crystals', *J. Chem. Phys.,* 124, 194710/1-194710/7.

Tavazzi, S., Laicini, M., Raimondo, L., Spearman, P., Borghesi, A., Papagni, A. & Trabattoni, S. (2006b) 'Evidence of polarized charge-transfer transitions by probing the weak dielectric tensor components of oligothiophene crystals', *Appl. Surf. Sci.,* 253, 296-299.

Tavazzi, S., Miozzo, L., Silvestri, L., Mora, S., Spearman, P., Moret, M., Rizzato, S., Braga, D., Diaw, A. K. D., Gningue-Sall, D., Aaron, J.-J. & Yassar, A. (2010a) 'Crystal Structure and Optical Properties of N-Pyrrole End-Capped Thiophene/Phenyl Co-Oligomer: Strong H-type Excitonic Coupling and Emission Self-Waveguiding', *Crystal Growth & Design,* 10(5), 2342-2349.

Tavazzi, S., Raimondo, L., Silvestri, L., Spearman, P., Camposeo, A., Polo, M. & Pisignano, D. (2008) 'Dielectric tensor of tetracene single crystals: The effect of anisotropy on polarized absorption and emission spectra', *Journal of Chemical Physics,* 128(15), 154709.

Tavazzi, S., Silvestri, L., Miozzo, L., Papagni, A., Spearman, P., Ianelli, S., Girlando, A., Camposeo, A., Polo, M. & Pisignano, D. (2010b) 'Polarized Absorption, Spontaneous and Stimulated Blue Light Emission of J-type Tetraphenylbutadiene Monocrystals', *Chemphyschem,* 11(2), 429-434.

Tavazzi, S., Spearman, P., Silvestri, L., Raimondo, L., Camposeo, A. & Pisignano, D. (2006c) 'Propagation properties and self-waveguided fluorescence emission in conjugated molecular solids', *Organic Electronics,* 7(6), 561-567.

Vragovic, I. & Scholz, R. (2003) 'Frenkel exciton model of optical absorption and photoluminescence in alpha-PTCDA', *Physical Review B,* 68(15).

Weiser, G. & Moller, S. (2002) 'Directional dispersion of the optical resonance of pi-pi* transitions of alpha-sexithiophene single crystals', *Physical Review B,* 64(5), 045203.

Wex, B., Kaafarani, B. R., Kirschbaum, K. & Neckers, D. C. (2005) 'Synthesis of the anti and syn Isomers of Thieno[f,f']bis[1]benzothiophene. Comparison of the Optical and Electrochemical Properties of the anti and syn Isomers', *J. Org. Chem.,* 70, 4502-4505.

Xiao, K., Liu, Y. Q., Qi, T., Zhang, W., Wang, F., Gao, J. H., Qiu, W. F., Ma, Y. Q., Cui, G. L., Chen, S. Y., Zhan, X. W., Yu, G., Qin, J. G., Hu, W. P. & Zhu, D. B. (2005) 'A highly pi-stacked organic semiconductor for field-effect transistors based on linearly condensed pentathienoacene', *Journal of the American Chemical Society,* 127(38), 13281-13286.

Xie, W., Li, Y., Li, F., Shen, F. & Ma, Y. (2007) 'Amplified spontaneous emission from cyano substituted oligo(p-phenylene vinylene) single crystal with very high photoluminescent efficiency', *Applied Physics Letters,* 90(14), 141110.

Yanagi, H. & Morikawa, T. (1999) 'Self-waveguided blue light emission in p-sexiphenyl crystals epitaxially grown by mask-shadowing vapor deposition', *Applied Physics Letters,* 75(2), 187-189.

Yanagi, H., Ohara, T. & Morikawa, T. (2001) 'Self-waveguided gain-narrowing of blue light emission from epitaxially oriented p-sexiphenyl crystals', *Advanced Materials,* 13(19), 1452.

Zhu, X. H., Gindre, D., Mercier, N., Frere, P. & Nunzi, J. M. (2003) 'Stimulated emission from a needle-like single crystal of an end-capped fluorene/phenylene co-oligomer', *Advanced Materials,* 15(11), 906.

13

Parity Violation in Unstirred Crystallization from Achiral Solutions

Marian Szurgot
Technical University of Łódź,
Center of Mathematics and Physics,
Poland

1. Introduction

Dissymetry of the atomic arrangement in molecules or crystals of chiral substances leads to the existence of left (L) and right-handed (D) forms being a mirror image of the other. The two structural forms of crystals, i.e. L and D enantiomers have mirror macro- and micromorphology and exhibit opposite sense of optical rotation. The formation and identification of enantiomers or racemates, and determination of enantiomeric purity is very important to chiral drugs manufacturing, since L and D enantiomers of a large number of drugs exhibit quite different therapeutic, toxic and adverse side effects. The different biological activity of chiral compounds is caused by the preferred chirality of chemistry of life and breakage of mirror symmetry in nature is a general phenomenon. Origin, mechanisms and conditions of formation of chiral asymmetry for various physical, chemical and biological objects and processes are extensively studied. The progress in understanding of the spontaneous breakage of chiral symmetry in crystallization processes of simple substances can help in better establishing the causes of chiral asymmetry in nature.

Single enantiomers or racemates can be formed during crystallization. Unequal amounts of L- and D-forms can be produced in mirror-symmetric chemical reactions through a phenomenon called spontaneous symmetry breaking. Breakage of chiral symmetry has been discovered by Kondepudi and co-workers in their studies of sodium chlorate crystallization from stirred solution (Kondepudi et al., 1990). The phenomenon has been investigated both experimentally and theoretically (McBride & Carter 1991; Kondepudi et al., 1995 a,b; Kondepudi et al., 1993; Kondepudi et al., 1995 a,1995 b; Kondepudi & Sabanayagam, 1994; Qian & Botsaris, 1998, 2004). At present, efforts are made to establish the experimental conditions under which the phenomenon occurs, to determine all important parameters controlling its kinetics and mechanisms, and to develop a reliable theory (Veintemillas-Verdaguer et al, 2007; Viedma 2004; Cartwright et al., 2004; Martin et al. , 1996; Metcalfe & Ottino, 1994; Bushe et al., 2000). Our results for sodium chlorate have shown that even in unstirred solutions there is a finite probability of creation of large enantiomeric excess (Szurgot & Szurgot, 1995a) and pure enantiomeric forms (Szurgot & Szurgot, 1995b), what means that chiral symmetry is broken in sodium chlorate unstirred crystallization.

The aim of the paper was to prove that parity violation really takes place in unstirred solutions, to establish the effect of first crystals and effect of growth conditions (supersaturation, temperature, and crystallizer size) on chiral nucleation of sodium bromate. Sodium bromate crystallizes in the cubic structure (space group $P2_13$, class 23). A large number of nucleation experiments have been carried out to establish various dependencies and to describe the statistics of chiral nucleation. Presented results describe chiral nucleation of sodium bromate but the same types of dependencies have been established also for other compounds.

2. Experimental

Sodium bromate solutions were prepared of 99.0% pure AR grade $NaBrO_3$ powder dissolved in warm distilled water using magnetic, teflon coated stirrer and filtered with a glass filter. Then they were transformed to several flat-bottomed crystallizers and left for crystals to grow. Large flat-bottomed crystallizers of 2 l capacity and of 23 cm in diameters and Petri flasks of 3.5, 6, 8, 9, 11, 13 and 14 cm in diameter and of volume of 20-100 ml were used. Simultaneously about 10-20 crystallization runs were conducted. Slow solvent evaporation at a constant temperature between -2 and 109 ^oC created the supersaturation necessary for nucleation and crystal growth. Crystals nucleated spontaneously at the air/solution interface and immediately on the flat bottom of crystallizers. First events of spontaneous nucleation or nucleation stimulated by L, D, and both L and D seeds occurred approximately 4-40 h after preparation of solution. Crystallizations were conducted at -2, 0, 1.5, 4, 6.5, 8, 10,14, 21,25, 31, 33, 40, 50, 56, 58, 60, 62, 64, 66, 68, 70, 82, 88, 90, 92, 94, 96, 98, 100, 103, 105, and 109 ^oC (boiling point). Temperature of nucleation and crystal growth varied about $\pm$ 0.2 ^OC. Crystals were identified and counted before and after completed crystallization, i.e. when the whole solvent evaporated and the whole material crystallized. The smallest of the crystals were about 100 μm and the largest about 2 cm in size. Nucleation and crystal growth were conducted in closed and in open crystallizers. In open crystallizers the whole surface of the solution evaporated freely, whereas in closed crystallizers evaporation was hindered since only the small slit was open to remove the molecules of the solvent. As a result supersaturation necessary for nucleation and crystal growth in open crystallizers was much higher than in closed crystallizers.

Chirality of the crystals was determined with a polariscope or polarizing microscope by optical activity. Laevorotatory (-)-$NaBrO_3$ crystals have the same absolute structural configuration as isomorphous sodium chlorate of the opposite sign of rotation, i.e. dextrorotatory (+)-$NaClO_3$ crystals (Vogl et al., 1995). In this paper (-)-$NaBrO_3$ enantiomers have been denoted as L forms, and (+)-$NaBrO_3$ enantiomers as D forms. Therefore sodium bromate D crystals rotated light clockwise, and L crystals counterclockwise. When white light was used to D crystals placed between polarizers the colour of the crystals changed from blue to orange when the analyser were rotated clockwise, and the L crystals from orange (or yellow) to blue.

As-grown surfaces of crystals were etched in formic acid for 1-10 s at room temperature. The crystals were then rinsed in acetone and dried between filter paper. As-grown and etched surfaces of crystals were examined under optical Zeiss microscopes in reflection and in transmission.

3. Results and discussion

3.1 Chiral nature of micromorphology of as-grown and etched surfaces

Sodium bromate crystals nucleated at the air/solution interface and immediately on the glass bottoms of the crystallizers and grew on {100} and {111} surfaces on the crystallizer bottoms. Macromorphology of high-temperature (growing above 85 °C) forms of sodium bromate was cubic (dominated by {100} faces) with small {111}, $\{1\bar{1}1\}$, {110} and {210} surfaces which occasionally appeared on the habit. Low temperature forms (growing below 40 °C), and medium temperature forms (growing between 40 and 85 °C) were tetrahedral with {111} and $\{1\bar{1}1\}$ as dominant surfaces and minor {100}, {110} and {210} surfaces. Figure 1 presents morphology of high-temperature L and D forms.

Fig. 1. L-$NaBrO_3$, and D-$NaBrO_3$ crystals seen under crossed polarizers. Note: L (levorotatory) crystal is nearly white, and D (dextrorotatory) crystal is nearly black. In white light left crystal is blue and right is orange-brown. Growth temperature 86 °C. Field of view: 2.9 cm x 2 cm.

As grown {100} surfaces of low-temperature $NaBrO_3$ crystals exhibits elliptical growth hillocks which are dislocation growth centres. The hillocks are present on both enantiomorphous forms and depending on the type of enantiomer their major axes are inclined about +13° or -13° to the <100> edge of the crystal (Fig. 2).

Etching of L and D crystals of $NaBrO_3$ in formic acid reveals well-defined dislocation pits. The morphology of etch pits formed on {100} surfaces of both enantiomorphous forms is illustrated in Figure 3. The shape and orientation of growth hillocks and etch pits on {100} surfaces of $NaBrO_3$ crystals are similar to hillocks and pits revealed on $NaClO_3$ crystals as they belong to the same symmetry class (Szurgot & Szurgot, 1995a, Szurgot, 1995). Chiral nature of etch pits and growth hillocks reflects chirality of the internal structure of crystals. Etch pits formed on L crystals are mirror images of the pits formed on D crystals. As a result, L and D enantiomers of sodium bromate can be distinguished by the shape of growth and dissolution centers. Calculations of surface entropy factor (α factor) for sodium bromate crystals have shown that in the whole temperature region of growth from aqueous solution, i.e. between 271 and 381 K surface entropy factor varies between 5.6 and 4.8 for low and high-temperature limits. This means that the presence of growth hillocks as dislocation growth centers is expected at low supersaturations. Values of α factor for sodium bromate are somewhat higher than those calculated for sodium chlorate (4.8-4.2) (Szurgot & Szurgot, 1994). This explains why etch pits and growth hillocks on sodium bromate crystals are more polygonal in comparison with the corresponding growth hillocks and etch pits formed by the same etchant on {100} surfaces of sodium chlorate.

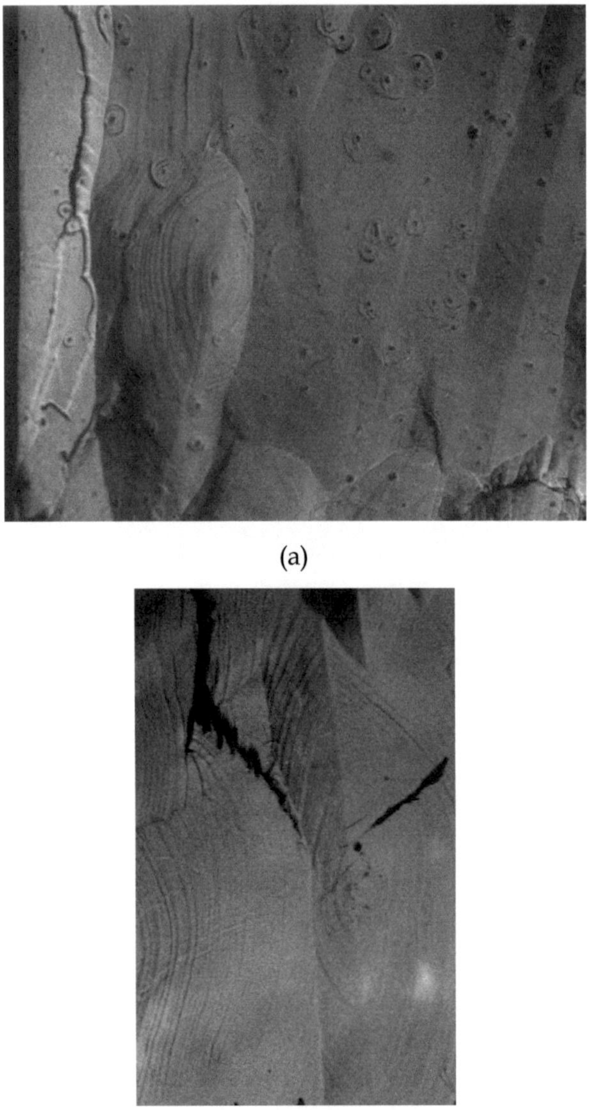

(a)

(b)

Fig. 2. Growth hillocks on (100) surfaces of dextrorotatory and laevorotatory crystals of sodium bromate. (a) (+)-$NaBrO_3$, and (b) (-)-$NaBrO_3$ crystal. Growth temperature 25 °C. Field of view: (a) 4 mm x 3.3 mm, (b) 1.8 mm x 2.8 mm.

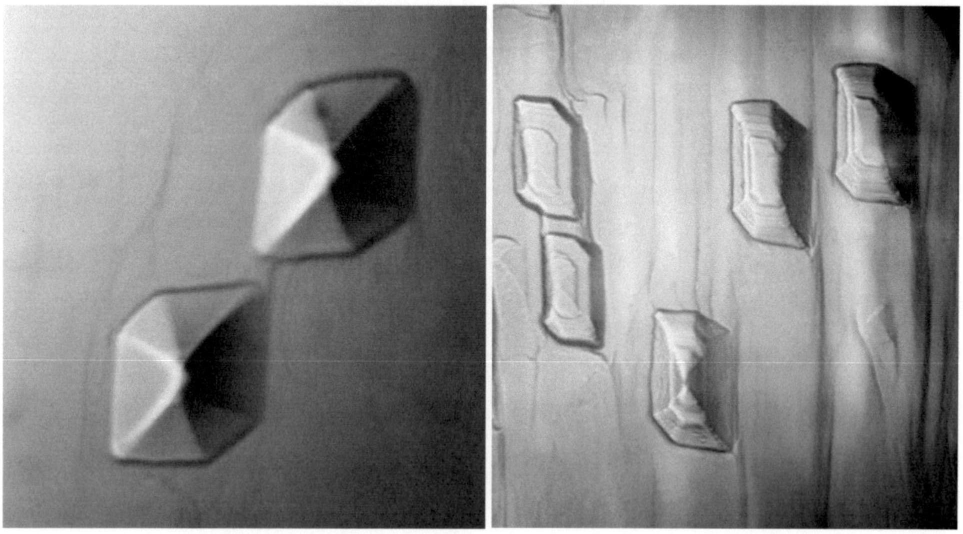

(a) (b)

Fig. 3. Dislocation etch pits on (100) surfaces of dextrorotatory and laevorotatory sodium bromate crystals.(a) (+)-$NaBrO_3$, and (b) (-)-$NaBrO_3$ crystal. Etched in formic acid: (a) 3s, (b) 5s. Field of view: (a) 0.2 mm x 0.2 mm, (b) 0.3 mm x 0.32 mm.

3.2 Convection effects on surface morphology and distribution of inclusions

Convection effects on crystal growth embrace rate and kinetics of growth and nucleation, macromorphology, micromorphology and formation of defects. In the case of buoyancy-driven convection due to Earth's gravity various defects such as inclusions, dislocations, and growth bands are preferably formed at top sectors of the crystals, and at upper parts of side faces (Szurgot & Sangwal, 1987; Wilcox, 1983). Lateral segregation of impurities is another effect caused by the convection flow which has been recently revealed by etch topography (Szurgot, 2010). Differences in supersaturations in various parts of crystals growing from solution, transient instabilities in convection flow, creation of starvation zones depleted in solute in the laminar boundary layer and in the closed wake at the rear, are responsible for the observed effects. Differences in supersaturations in various parts of growing crystals are responsible for different thickness of growth layers in various places of the same surfaces. Szurgot established that growth layers present on vertical, side faces of potassium bichromate crystals growing in aqueous solution are thicker in upper parts of crystals than layers situated in the lower parts of the same faces (Szurgot, 1991). Figure 4 illustrates the same effect on a side, vertical or slightly inclined with respect to the vertical (111) surface of sodium bromate crystal grown at low temperature. The preferred capture inclusions by the upper, top faces, and by upper parts of vertical faces of two sodium bromate crystals grown at high temperature region is shown in Figure 5. Both figures prove that buoyancy-driven convection participates in growth of sodium bromate crystals.

Fig. 4. Micromorphology of a vertical, side (111) surface. U, L-upper and lower parts of a crystal. Arrow indicates direction of gravitational field. Growth temperature: 25 °C. Field of view: 2.7 mm x 4 mm.

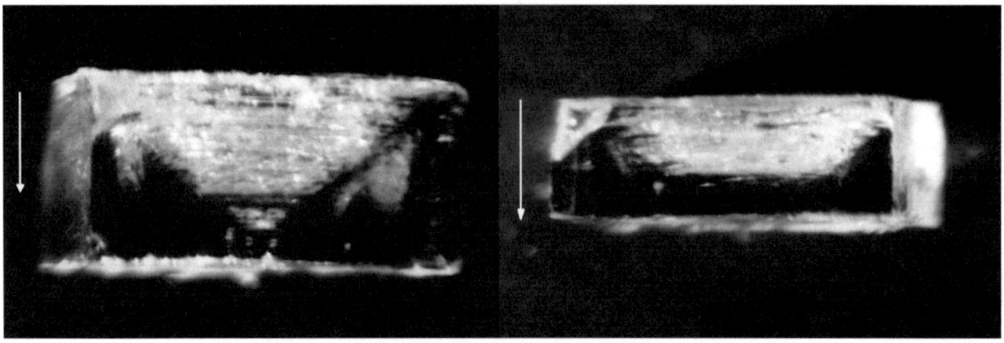

(a) (b)

Fig. 5. Distribution of inclusions in $NaBrO_3$ crystals grown by self-nucleation at the bottom of growth vessel from unstirred solution at 90 °C. The planes of contact with the bottom are (100). Arrows indicate direction of gravitational field. Note: inclusions are captured by the upper, horizontal, top (100) face, and by upper parts of vertical (010) faces. Front planes: (001). Field of view: (a), (b) 4 mm x 2.5 mm.

3.3 Statistics of nucleation of chiral forms. Unimodal, bimodal and trimodal distribution of enantiomeric excess

The enantiomeric excess ee is a dimensionless coefficient used to characterize the formation of enantiomers in crystallization. It expresses the relative excess of L-forms in the whole population of enantiomers. This order parameter can be determined by the formula

$$ee = (N_L - N_D)/(N_D + N_L), \quad (1)$$

where N_L and N_D are the numbers of L- and D-forms.
If ee = -1 then exclusively D-crystals, if ee = 1 then exclusively L-crystals are formed but if D- and L-crystals nucleate with the same probability then ee = 0. Since our identification of sodium bromate enantiomers has been performed by the direction of rotation of light N_L is the number of laevorotatory (-)- $NaBrO_3$ crystals, and N_D is the number of dextrorotatory (+)-$NaBrO_3$ crystals.
Statistics of distributions of enantiomeric excess ee has been the main goal of our studies. It is based on an analysis of large number of crystallization runs (over 4300 crystal growth runs). Figures 6-11 and Table 1 present experimental data on the occurrence and relative participation of enantiomorphous forms of sodium bromate spontaneously nucleated at various temperatures and at various supersaturations. It may be noted from the data that in unseeded, unstirred solutions ee assumes all possible values between -1 and 1, and depending on the growth conditions various unimodal, bimodal and trimodal distributions are created. This means that in the case of sodium bromate uni-, bi- and the trimodal distributions of enantiomers are formed in unstirred solutions rather than monomodal, as it was reported previously for sodium chlorate (Kondepudi et al., 1990; Kondepudi et al., 1995b; Martin et al., 1996). This proves that in the case of spontaneous nucleation of sodium bromate enantiomers chiral symmetry is broken.

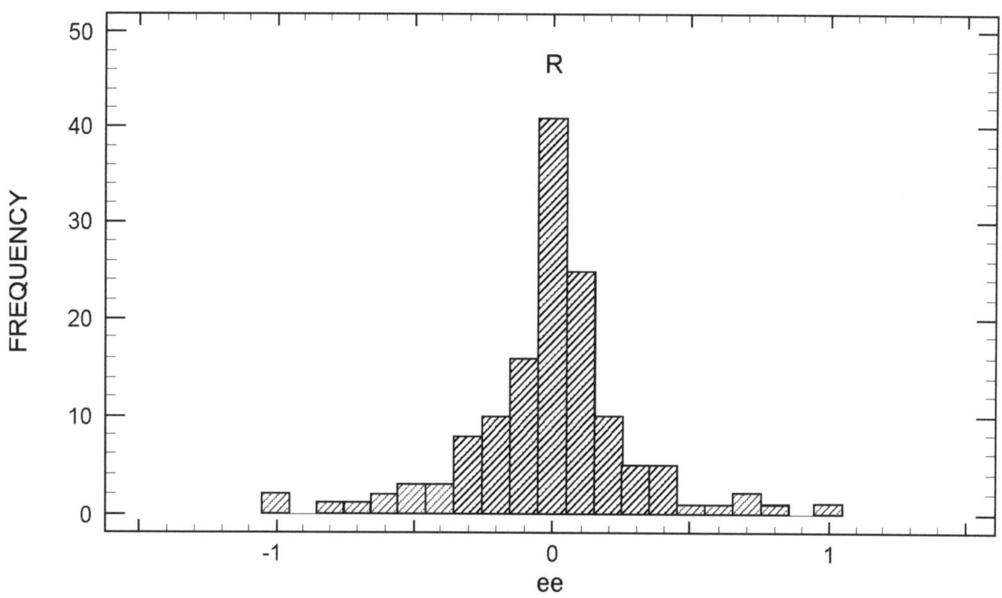

Fig. 6. Unimodal distribution of enantiomeric excess ee for $NaBrO_3$ crystals nucleated at 100 °C in unstirred solutions in open crystallizers diameter of 6.2 cm. Note: high probability of nucleation of racemates (50:50 mixture) (P = 0.30), and close to zero probability nucleation of pure L (PL = 0.007) and pure D forms (PD = 0.014). Ratio: PL/PD = 0.5, PR/(PL+PD) = 14.3. Number of growth runs: 138, high supersaturation.

Experiments of unstirred crystallization of sodium chlorate enantiomers performed by various researchers revealed that only unimodal ee distribution is obtained in which only one peak occurs, i.e. peak for ee = 0 which corresponds to the same probability of creation of

L and D-crystals (Kondepudi et al., 1990; Kondepudi et al., 1995b; Martin et al., 1996). Our data show that unimodal distributions of ee with peaks at ee = 0 (or at $-0.05<ee<0.05$) which are here assigned as R peaks (R since ee = 0 means racemate 50:50 mixture) are formed only in certain conditions of crystallization: in open, large crystallizers at the lowest, and at the highest temperatures and at high supersaturations. Figure 6 is an example of such an unimodal distribution of enantiomeric excess ee for $NaBrO_3$ crystals nucleated at 100 ^oC in unstirred solutions in open crystallizers diameter of 6.2 cm. It is seen that the probability PR of nucleation of racemate 50:50 mixture is high PR = 0.30, whereas probability of nucleation of pure L (PL, for ee = 1) and pure D (PD, for ee = -1) forms are close to zero (PL = 0.007, PD = 0.014). In this case supersaturation is high, ratio PL/PD = 0.5, and ratio PR/(PL+PD) = 14.3. Domination of formation of racemates over pure enantiomers is here evident.
Figure 7 reveals bimodal distribution of enantiomeric excess ee for $NaBrO_3$ crystals nucleated at 103 ^oC in unstirred solutions in small, closed crystallizers diameter of 3.5 and 6.2 cm. Peak R is absent, instead peaks D and L are present. Peak D presents the largest enantiomeric excess of D forms (ee between -1 and -0.95), and peak L shows the largest excess of L-forms (ee between 0.95 and 1). Probabilities of nucleation of pure L forms, and pure D forms are high (PL = 0.39, and PD = 0.35), and the probability of nucleation of racemates (50:50 mixture) is close to zero (PR = 0.02). In this case supersaturation is low, ratio PL/PD = 1.11, and (PL+PD)/PR = 37. Since in the class ee is between -0.95 and -1 about 90-95 % of sodium bromate crystals were pure D enantiomers (ee = -1), and in the class between 0.95 and 1 were pure L enantiomers (ee = 1) our PL and PD are assigned as the probabilities of nucleation of pure enantiomers. According to the author's knowledge it is the first evidence for a such bimodal ee distribution which is typical of stirred crystallization rather than unstirred crystallization. The distribution proves that in the case of spontaneous nucleation of sodium bromate enantiomers chiral symmetry is broken.

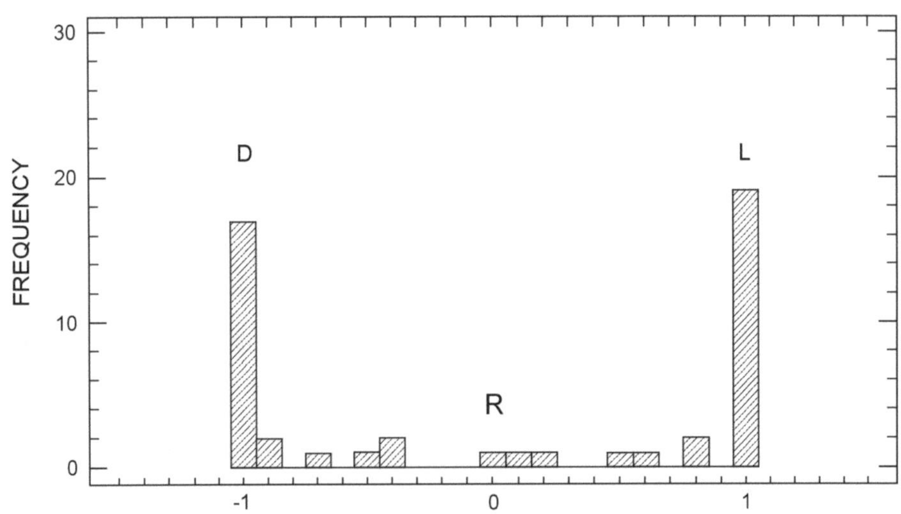

Fig. 7. Bimodal distribution of enantiomeric excess ee for $NaBrO_3$ crystals nucleated at 103 ^oC in unstirred solutions in small, closed crystallizers diameter of 3.5 and 6.2 cm. Note: high probability of nucleation of pure L (PL = 0.39) and pure D (PD = 0.35) forms, and close to zero probability nucleation of racemates (50:50 mixture) (PR = 0.02). Ratio: PL/PD = 1.11, (PL+PD)/PR = 37. Number of growth runs: 49, low supersaturation.

Figure 8 and Table 1 reveal another bimodal distribution of enantiomeric excess ee for $NaBrO_3$ crystals in which high peaks D and L are present and have practically the same height. In this case crystals nucleated at different temperatures in the range between 0 and 105 °C in unstirred solutions in small, closed crystallizers with a different size. Number of growth runs is here large: 727. It is seen from the figure that number of events for which racemates are formed (ee = 0) is so low that peak R may be neglected as a separate peak (PR = 0.02). Probabilities of nucleation of pure L forms, and pure D forms are high and comparable with those presented in Figure 7 (PL = 0.43, and PD = 0.41). Ratio PL/PD = 1.05, and (PL+PD)/PR = 42. These results show that bimodal distributions of enantiomeric excess ee are common, occur in the whole temperature range used for unstirred crystallization and parity violation in unstirred crystallization is a real phenomenon. The phenomenon occurs in closed, small crystallizers for which supersaturation is low due to low evaporation rate. Experiments at boiling point (109 °C) in closed crystallizers revealed bimodal distribution of enantiomers with high probability of nucleation of pure L and D forms: 0.41 and 0.43, respectively. In this case vigorous movement of the growth medium takes place due to boiling and such a distribution is expected since it corresponds to intensive stirring. We can notice that probabilities PL and PD are the same as those noted in unstirred solution (Figs. 8 and 7).

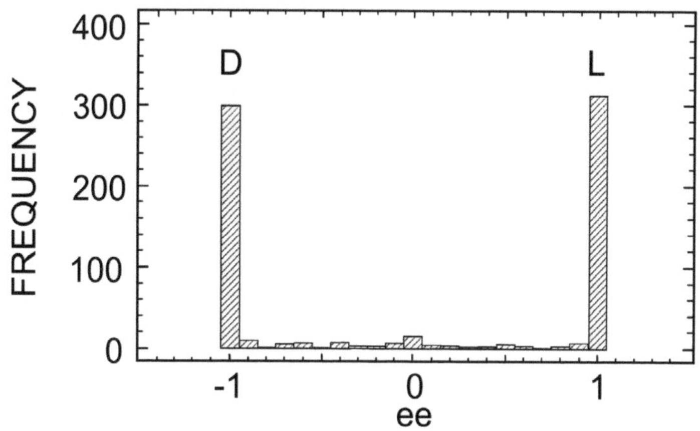

Fig. 8. Bimodal distribution of ee for crystals nucleated in closed, small crystallizers at low supersaturation at different temperatures. Probability of nucleation: PD = 0.41, PR = 0.02, PL = 0.43. Ratio: PL/PD = 1.05, (PL+PD)/PR = 42. Number of growth runs: 727.

Figure 9 reveals another type of bimodal distribution of ee. Here sodium bromate crystals nucleated spontaneously at 58 °C. The distribution presents results for 112 crystallization runs. It is seen from the figure that only R and L peaks are present in the distribution. Peak R is low and peak L is high. Peak D is absent. Probability of nucleation and probability ratios for this case are following: PD = 0.009, PR = 0.05, PL = 0.70, PL/PD = 7.8, (PL+PD)/PR = 14.2. Figures 7 and 9 show that distributions of ee and probabilities of nucleation are affected by temperature, and figures 6 and 7 reveal strong effect of evaporation rate, representing supersaturation.

Growth conditions	Total number of crystallization runs and probabilities PD, PR, PL of nucleation				Number of peaks and type of ee distribution
(° C)	Total	pure L form	Racemate	pure D forms	
Closed crystallizers (Tg = 0-105 °C)	727	PL = 0.43	PR = 0.02	PD = 0.41	2 (L,D) bimodal
Open crystallizers (Tg = 0-105 °C)	3312	PL = 0.16	PR = 0.16	PD = 0.05	3 (L, R, D) trimodal
Open crystallizers HT region (Tg = 98-105 °C)	423	PL = 0.002	PR = 0.16	PD = 0.01	1 (R) Unimodal

Table 1. Spontaneous nucleation of $NaBrO_3$ enantiomers in unstirred aqueous solutions under various conditions. (For L forms ee = 1 or 0.95 ≤ ee ≤ 1, for D forms ee = -1 or -1 ≤ ee ≤ -0.95, R – racemate 50:50 mixture (equal number of L- and D-crystals ee ≈ 0 i.e. -0.05<ee<0.05, HT-high- temperature growth, Tg-growth temperature).

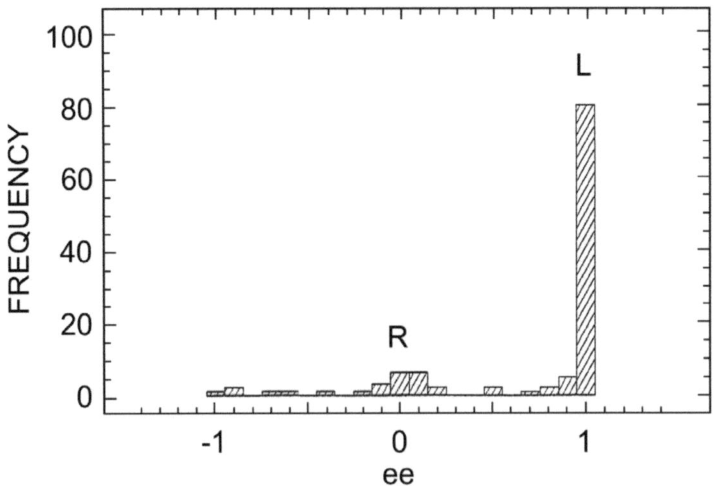

Fig. 9. Bimodal distribution of ee for crystals nucleated spontaneously at 58 °C. Probability of nucleation and probability ratios: PD = 0.009, PR = 0.05, PL = 0.70, PL/PD = 7.8, (PL+PD)/PR = 14.2. Number of growth runs: 112.

Trimodal distributions of enantiomeric excess ee are formed when sodium bromate crystals nucleated in closed and open crystallizers. Figure 10 shows an example of ee distribution when crystals nucleated spontaneously at constant temperature 105 °C in small, medium and large crystallizers. During nucleation and crystal growth in open crystallizers operating supersaturation was high, and in closed crystallizers low. The distribution presents results for 354 crystallization runs. It is seen from the figure that three peaks are present: peak D, peak R and peak L. The peaks are high and well defined. Probability of nucleation and probability ratios for this case are following: PD = 0.23, PR = 0.18, PL = 0.21, PL/PD = 0.91, (PL+PD)/PR = 2.4. This results deal with crystallization at high temperature, but trimodal distributions were obtained at any temperature. Trimodal distributions of ee are in fact combined bimodal distributions (peaks D and L), and unimodal distributions contributing to the central peak R.

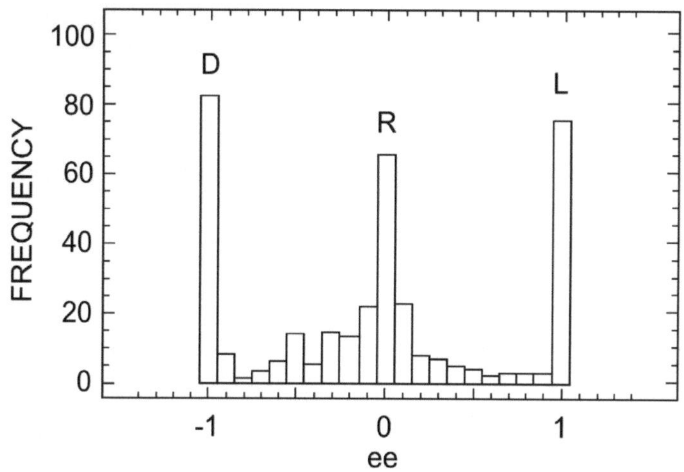

Fig. 10. Trimodal distribution of ee for crystals nucleated spontaneously in unstirred solution in open and closed crystallizers at 105 °C. Probability of nucleation: PD = 0.23, PR = 0.18, PL = 0.21. Ratio: PL/PD = 0.91, (PL+PD)/PR = 2.4. Number of growth runs: 354, low and high supersaturations.

Figure 11 shows trimodal ee distribution when crystals nucleated spontaneously at constant, but various temperatures (between -2 and 105 °C), in small, medium and large, open and closed crystallizers. The distribution includes a large number of nucleation runs (2201 runs). Operation supersaturation was low, medium and high. The figure reveals that, as previously, three peaks are present in the distribution, but peak D is smaller than peak R, and smaller than peak L, which is the highest of the three. Probability of nucleation and probability ratios for these results are following: PD = 0.06, PR = 0.12, PL = 0.23, PL/PD = 3.8, (PL+PD)/PR = 2.4. The results show that crystallization at low, medium and high temperatures and at various supersaturations leads inevitably to the trimodal ee distributions.

Bimodal and trimodal ee distributions with peaks D and L prove that for certain growth conditions the probability of the creation of exclusively L- and D- crystals is high in unstirred crystallization, and probability of creation of racemates is low. This means that it is not true that only unimodal distributions of enantiomeric excess with exclusively peaks R are created in unstirred crystallization. Chiral symmetry is broken in spontaneous nucleation of sodium bromate from unstirred solutions.

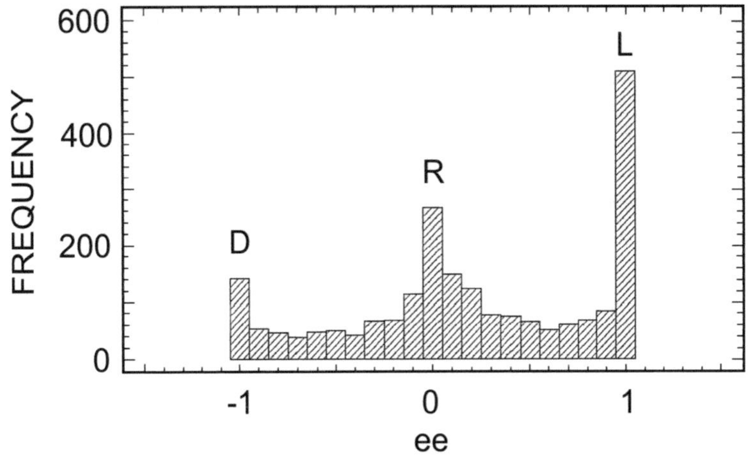

Fig. 11. Trimodal distribution of ee for crystals nucleated spontaneously in unstirred solution. Distribution includes a large number of nucleation runs conducted at various temperatures, various supersaturations with various crystallizer sizes. Probability of nucleation: PD = 0.06, PR = 0.12, PL = 0.23. Ratio: PL/PD = 3.8, (PL+PD)/PR = 2.4. Number of growth runs: 2201.

Comparison of probabilities for both trimodal distributions presented in Figures 10 and 11 shows that PR and PL are practically the same, but PD decreased. This means that L enantiomers of sodium bromate are preferentially formed with respect to D enantiomers. Figures 7 and 8 reveal that asymmetry between D and L enantiomers is not created during nucleation in closed crystallizers since ratio PL/PD for nucleation in closed crystallizers varies between 1.05 and 1.11, and both values of PL/PD are close to one, whereas PL/PD resulting from figure 11 is about four.

In strirred crystallization, in forced convection regime the bimodal distribution of enantiomeric excess is preferentially obtained since the high efficiency of cloning of the single parent crystal by the collision breeding e.g. by breaking off the whiskers from the crystal surface of the parent crystal. In unstirred crystallization when the mechanisms connected with free convection operate as it was suggested by Szurgot and Szurgot (Szurgot & Szurgot, 1995b), and supported by the Cartwright and coworkers (Cartwright et al., 2004) the efficiency of cloning of the first crystal is expected to be much lower due to slow velocity of movement of the solution. Much time is needed prior to the next crystal will nucleates in unstirred solution. When small, closed crystallizers are used in unstirred crystallization then for a small volume of the solution and at low supersaturation it is possible that nearly all secondary crystals will acquire the handedness of the first crystals (Figs 7, 8). In this case, in a single growth run in a small crystallizer either L pure or D pure enantiomers will be created and bimodal distribution will result for a large number of experiments.

In unstirred crystallization it is highly probable that two or more initial crystals will control the secondary nucleation, rather than the single parent crystal. In the case of two initial crystals which may be either LL, DD or LD pairs the trimodal distribution will result (Figs 10, 11). In large crystallizers due to the huge volume of the mother solution, and at additionally high supersaturation, which are reachable especially in open crystallizers, three, or larger number of initial crystals is formed and the racemate peak (peak R at ee = 0) will play an important role. Unimodal distribution is thus preferentially formed (Fig 6), or multimodal distribution with the dominating racemate peak.

Figure 12 presents dependence of ratio PL/PD on temperature for sodium bromate crystals nucleated in open and closed crystallizers. Diamonds present experimental points for nucleation in open crystallizers, and squares nucleation in closed crystallizers. It is seen from the figure that in the case of nucleation in closed crystallizers PL/PD is independent of temperature in the whole temperature range and is nearly constant. Values of PL/PD are spread in the narrow range between 0.95 and 1.1 which means that they are very close to one. Since PL ≈ PD, nucleation in closed crystallizers leads to symmetric distributions of enantiomeric excess. In the case of nucleation in open crystallizers at higher supersaturation PL/PD values are spread in the wide range between 0 and 20. On an average PL/PD increases with the increasing temperature what can be approximated by the linear function PL/PD = 0.035 t + 1.454, where t is temperature in Celsius degrees. At most temperatures PL > PD, at selected temperatures PL = PD, and at certain temperatures PL/PD = 0 what means that PL = 0, and L forms are not created. Values of PL/PD are scattered between two curves, lower limit is the straight line PL/PD = 0, the upper limit is curve UL. It is seen that spreading PL/PD values is relatively narrow at low temperatures and wide at high temperatures.

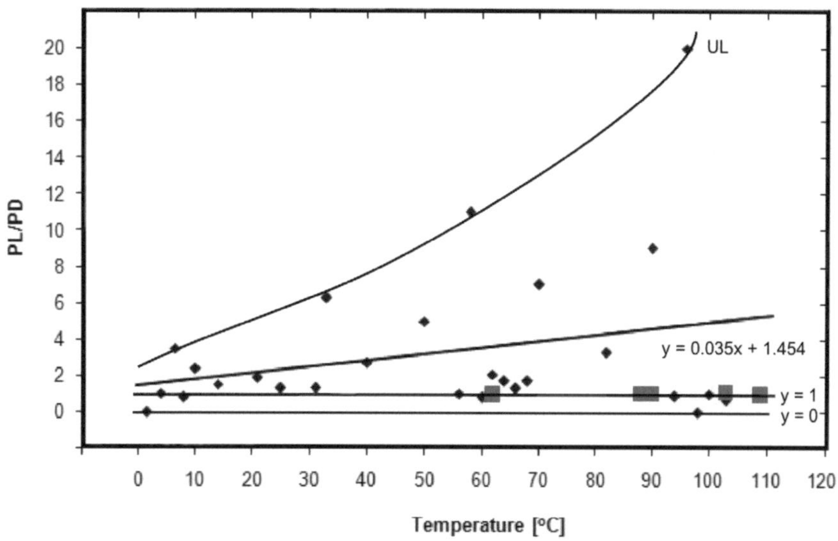

Fig. 12. Effect of temperature on the ratio PL/PD, where PL is the probability of nucleation of L forms, and PD is the probability of creation of D forms. Diamonds-nucleation in open crystallizers, squares- nucleation in closed crystallizers. Curve UL represents the upper limit values of ratio PL/PD.

3.4 Effect of handedness of first nucleated crystals on the handedness of secondary crystals

Appearing of the first crystal or a group of initial crystals and an introduction of a seed into growth medium, in our case into the aqueous solution, initiate secondary nucleation. As a result, secondary and primary heterogeneous or homogeneous nucleation participates in the formation of enantiomers, and various parameters controlling primary and secondary nucleation affect handedness of the nucleated crystals. Figure 13 and Table 2 present the effect of handedness of initial crystals on the formation of sodium bromate enantiomers. Figure 13 shows the effect of seeds on handedness of daughter crystals created by the secondary nucleation processes. It is seen that if the single L-crystal, or a group of L crystals nucleate as the first, mainly or exclusively L-crystals are formed, but if D-crystal is the first mainly D-crystals nucleate (Fig 13 a,b). In the first case peak L, and in the second case peak D is formed, similarly as in the case of sodium chlorate (Szurgot & Szurgot, 1995b). The scatter of ee values is high between -1 and 0, in spite of the fact that high enantiomeric excess is obtained, and probability PD for seeded crystals equals 0.68, whereas PL = 0.68 (Table 2, Fig 13 a,b).

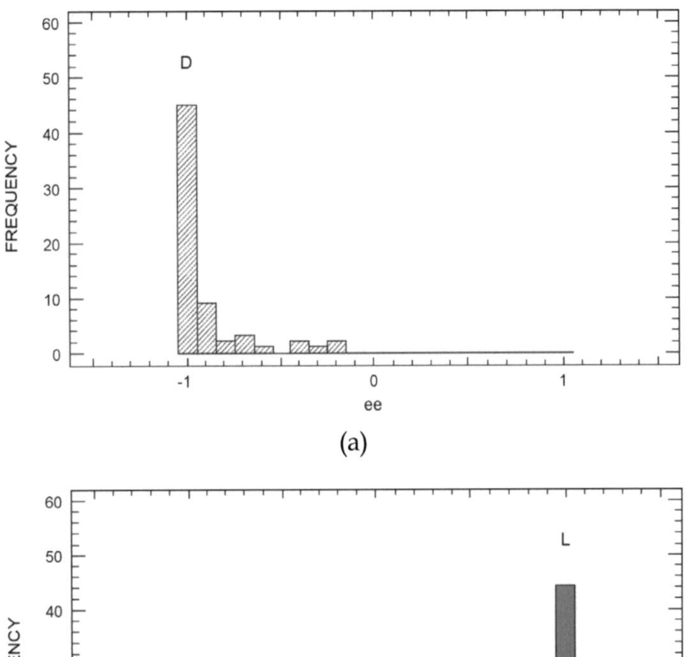

(a)

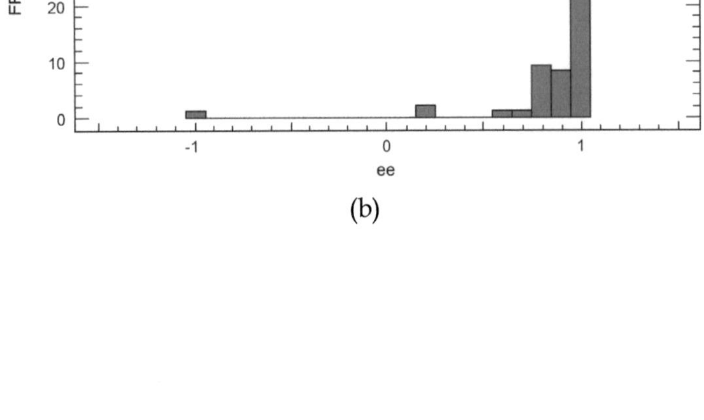

(b)

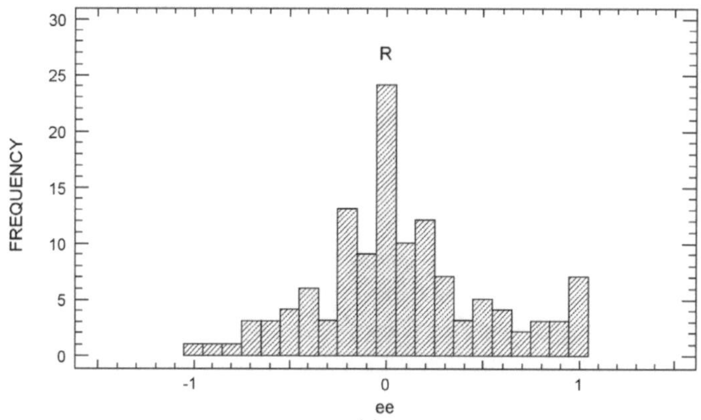

(c)

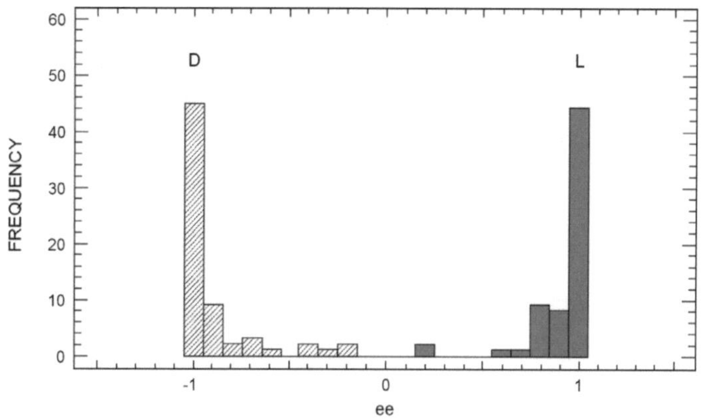

(d)

Fig. 13. Effect of handedness of seed crystals on the formation of $NaBrO_3$ enantiomers in unstirred solution. Distributions of ee values conditioned by secondary nucleation induced by: (a) D forms, (b) L forms, (c) a pair of L and D forms. (d) Combined data for single D forms (dashed), and single L forms (filled). Probability of nucleation and probability ratios: (a) PD = 0.68, PL =0, PR = 0, (b) PD = 0.015, PL =0.68, PR = 0, (c) PD = 0.008, PR = 0.19, PL = 0.06, PL/PD = 7.5, PR/(PL+PD) = 2.8, (d) PD = 0.34, PL =0.34, PR = 0, PL/PD = 1, PR/(PL+PD)= 0. Number of growth runs: (a) 65, (b) 65, (c) 124, (d) 130. Here three independent cases of secondary nucleation induced by single D, single L, and a pair LD of initial crystals are presented (Figs. (a), (b), and (c) respectively), whereas the bimodal distribution presented in Figure (d) reveals the combined data secondary nucleation for single D (Fig (a)), and single L (Fig (b)) initial crystals. Since of 130 experiments presented in Fig (d) 65 represents seeding by a single D enantiomers, and the same number 65 experiments represents seeding by a single D enantiomers this binomial distribution proves that PL=PD.

We can notice that there exists the finite probability of the formation of enantiomers with the opposite handedness with respect to the initial crystals, i.e. L crystals for D initial enantiomers, and D crystals for L parent enantiomers. The probabilities are low but different from zero (PD = 0.015 for L seed, Fig 13b, Table 2). The last results show that during secondary nucleation of sodium bromate enantiomers there exist processes responsible for generation of opposite handedness as it was predicted by Frank in 1953 (Frank, 1953), although the dominant role in unstirred crystallization play processes of chiral autocatalysis. In unstirred crystallization of sodium bromate cloning of initial crystals has 68 % efficiency. The dominant role of handedness of the initial crystals in the formation of enantiomers has been proven.

If a pair LD (or a group of LD pairs) crystals nucleate as first, then central maximum, i.e. peak R appears (Figs 13c, for a single LD pair). Such a situation has been shown earlier for sodium chlorate in which LD pairs of seeds played the role of first crystals (Szurgot, Szurgot, 1995b). Thus, the presence of peaks D, L, and R in distributions of ee values is conditioned by the handedness of first crystals.

Type of seed	Number of crystallization runs and probabilities PL, PR, PD of nucleation				Number of peaks and type of ee distribution
	Total	L	R	D	
Single D	65	0 PL = 0	0 PR = 0	44 PD = 0.68	1 (D) unimodal
Single L	65	44 PL = 0.68	0 PR = 0	1 PD = 0.015	1 (L) unimodal
LD pair	129	7 PL = 0.06	24 PR = 0.19	1 PD = 0.008	2 (R dominant, L) Bimodal

Table 2. Secondary nucleation of $NaBrO_3$ enantiomers induced by chiral seeds in unstirred aqueous solutions. (L - left forms with ee = 1 or $0.95 \leq ee \leq 1$, D - right forms with ee = -1 or $-1 \leq ee \leq -0.95$, R –racemate 50:50 mixture with $ee \approx 0$ or $-0.05 < ee < 0.05$).

The distributions of enantiomeric excess resulted from the effect of handedness of initial crystals enable one to explain experimental ee distributions and predict the possibilities of the formation of new distributions, that have not been discovered yet. Combined results presented in Figures 13a,b,c lead to obtaining various types of ee distributions presented in Figs. 6-11. Figure 13d shows that the bimodal distribution with D and L peaks is the result of nucleation exclusively single D and single L enantiomers as initial. This type of distribution has been obtained in closed, small crystallizers in which supersaturation was relatively low (Figs. 7 and 8), and at 109 °C, i.e. at boiling point when the mother solution was violently moved. The trimodal ee distributions with peaks D, L and R are the result of initial nucleation of single L, single D, and a pair of LD enantiomers. Figures 10 and 11 resulting from the crystallization in open, and in closed crystallizers are just such a type, so we now know that in these crystallization conditions both single L and single D enantiomers, as well as pairs of LD enantiomers are nucleated as the initial crystals. The type of bimodal distribution shown in Figure 9 for nucleation at 58 °C results from the initial nucleation of single L forms and a pair of LD forms.

Values of probabilities PD, PL, and PR revealed in secondary nucleation experiments are in relatively good agreement with the experimentally determined probabilities. For example, values of PD and PL in the bimodal distribution with D and L peaks shown in Figure 7 are equal to about 0.37 (PD = 0.35, PL = 0.39), those in Fig 8 about 0.42 (PD = 0.41, PL =0.43), those in Fig 13d are equal to 0.34 (PD = 0.34, PL = 0.34). The unimodal distribution with peak R presented in Figure 6 gives PR = 0.30, and peak R in Figure 13c gives PR = 0.19, somewhat smaller, but acceptable value.

3.5 Effect of supersaturation on probability of nucleation of enantiomers and racemates

The number of nucleated crystals in one growth run varies markedly between 8 and 273. This means that the supersaturation of the solution at which nuclei are formed also assumes very different values from low to high. In most our experiments supersaturation was not very high. Formation of pure enantiomers and the high enantiomeric excess takes place when the number of nucleated crystals and the supersaturation is relatively low. Spontaneous nucleation at relatively low or medium supersaturation in unstirred solutions (in closed crystallizers, and in small open crystallizers) leads to high enantiomeric excess and to the bimodal and trimodal distributions of ee. In open, large crystallizers, the supersaturations are higher, both enantiomers nucleate simultaneously, ee tends to zero, and monomodal distributions of ee result. Peak R is the one and the only peak in the distributions.

The supersaturation is, however, not the only factor affecting the formation of enantiomers. Kondepudi and co-workers showed in 1995 that the stirring rate is the critical parameter in chiral symmetry breaking in stirred crystallization of sodium chlorate (Kondepudi et al., 1995 a,b). Secondary nucleation studies of sodium chlorate enantiomers in aqueous solution conducted by Denk & Botsaris (Denk & Botsaris, 1972) have shown that apart from supersaturation, stirring rate and impurities, also handedness of seeds plays an important role in the formation of chiral forms during crystallization of sodium chlorate. Our previous results for sodium chlorate (Szurgot & Szurgot, 1995b) and present results for isomorphous sodium bromate confirmed the important role of handedness of initial crystals.

To explain appearing of D, R, and L peaks and various values of ee occurring in the ee distributions the effect of supersaturation on the formation of enantiomers will be analysed. The first crystals appear during the primary nucleation at the supersaturation above the metastable limit, which is here assumed as 10% at 30 °C for sodium bromate. This leads to the drop of supersaturation, as it is was monitored by Kondepudi for sodium chlorate, and confirmed by the analysis of distribution of sizes of sodium bromate crystals. Further acts of nucleation occur at lower supersaturations, in the secondary nucleation region. Denk and Botsaris' secondary nucleation results (Denk & Botsaris, 1972) are our base for explaining of the distributions of sodium bromate enantiomers. The way we applied here is the same as applied earlier for sodium chlorate (Szurgot & Szurgot, 1995b).

Figure 14 presents three possible relations ee and supersaturation: the upper curve for first L crystal, the lower curve for first D crystal, and the central, straight line for two initial, one L, and one D crystal nucleated spontaneously. Dashed lines indicate possible scatter in ee values for each of the dependencies. Four regions of supersaturations are distinguished to describe the formation of sodium bromate enantiomers.

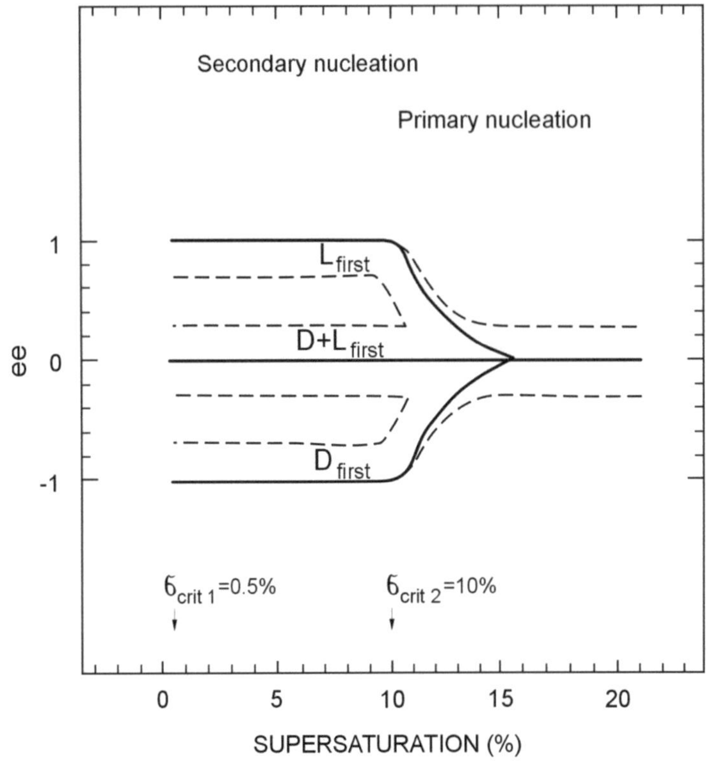

Fig. 14. Expected dependence of the nucleation of D- and L-forms of sodium bromate on supersaturation in an unstirred solution at constant temperature 30 °C.

At very low supersaturations (0-0.5%, region I) no nucleation takes place. At low, and medium supersaturations (0.5-10%, region II) secondary nucleation produces the crystals with the same structure as the first crystals. Strictly speaking a large part of the population between 60% and 100 % of the crystals have the same structure as the parent crystal. This means that for the first D enantiomer we have ee between -1 and -0.7, and for the first L crystal ee values will be between 0.7 and 1 with maximal values ee = -1 and ee = 1, as it has been shown in our experiments with the secondary nucleation (Fig. 13). In the case of pair of LD initial crystals peak R is formed at ee = 0, and in the dashed range of ee values, between -0.3 and +0.3 about 70 % (between 60 and 80 %) of crystals is then nucleated (Figs. 13c, 6).

At still higher supersaturations (above 10 %) primary nucleation begins and exists together with the secondary nucleation or is the only mechanism, forming crystals. In the transitional region (10-15%, region III) ee tends to zero, and any value of ee between -1 and 1 is possible. In the highest supersaturation range (above 15 %, region IV) racemates are only formed, and racemate 50:50 mixture (ee = 0) is the only peak in ee distributions. In any region ee may be equal to 0 if both L and D enantiomers nucleate as first (the central straight).

As Figure 14 summarises the results on the formation of enantiomers in various runs, at various supersaturations, and various first crystals, it reflects various possibilities resulting from the above analysed model. This includes the creation of peaks D, R and L as a consequence of autocatalysis caused by the first crystals but also explains the scatter in ee values. Bimodal distributions of enantiomers with peaks for pure L and D enantiomers (peaks D and L) and absence of the peak R prove that in closed, small crystallizers at low supersaturations only single enantiomers are created, and simultaneous nucleation of a pair of D and L forms is not in these conditions sufficiently competitive process.

Our results show that chiral symmetry breaking takes place in unstirred solution so it is not a forbidden process. Since in unstirred crystallization there exist natural convection, and fluid movement around the growing crystals takes place with the velocities of the order of 0.01-0.1cm/s, the convection-induced transport of the solution from the parent crystal region to the other parts of the crystallizer should operate. In small crystallizers the volume of the solution is small and at low supersaturations when closed crystallizers are used there is enough time to transport the solution from the parent crystal to each place of the crystallizer before a new crystal has nucleated spontaneously. It is enough time because the nucleation rate is low at low supersaturation. Conversely, at high supersaturations, in open crystallizers new spontaneously created crystals of various handedness can be effectively formed during short period of time, and if additionally the crystallizer is large the formation of racemates will be preferred. In stirred crystallization fluid velocities are high (order of 10 cm/s) and process of transport of whiskers formed on the surfaces and broken off the parent crystals is highly probable. Collisions of parent crystals with the stirrer contribute to the cloning of the crystals in stirred solution. Thus bimodal distributions of enantiomers in stirred crystallization are commonly formed. The preferred formation of bimodal distributions of enantiomers in stirred solutions instead of the monomodal racemate distribution can be also caused by much deeper drop in supersaturation in stirred solution after first acts of nucleation in comparison to unstirred crystallization as it was established by Kondepudi and co-workers (Kondepudi et al., 1993) The supersaturation drop occurring in stirred and unstirred solutions leads to the transition from regions IV or from III to region II in which only one type of enantiomer is formed and to region I where no nucleation takes place. In unstirred solution supersaturation drop also contributes to interruption of primary nucleation. This is why bimodal distributions of enantiomers occur in sodium bromate unstirred crystallization. But not only pure enantiomers are formed in unstirred solutions but also this process is highly probable, and expected in small, closed crystallizers. As a result, the effective cloning of crystals in unstirred solutions exist no matter what the microscopic mechanisms of autocatalysis and competition between both chiral forms are.

3.6 Effect of temperature on probability of nucleation of enantiomers and racemates

A large number of nucleation experiments have been conducted to establish the effect of temperature on nucleation of sodium bromate enantiomers. Figures 15 and 16 show the results for crystallization in closed and in open crystallizers in a wide temperature range between -2 and 109 ^{o}C. Does temperature affect the formation of enantiomers? Figures 15 and 16 reveal that in closed crystallizers the probability of the creation of L and D enantiomers is very high, about 0.4 for each of the enantiomers, and probability PR of the formation of racemates is very low (about 0.01). Bimodal distributions in these conditions have much in common. It is seen that in small crystallizers and at low supersaturations temperature does not affect the distribution of enantiomers.

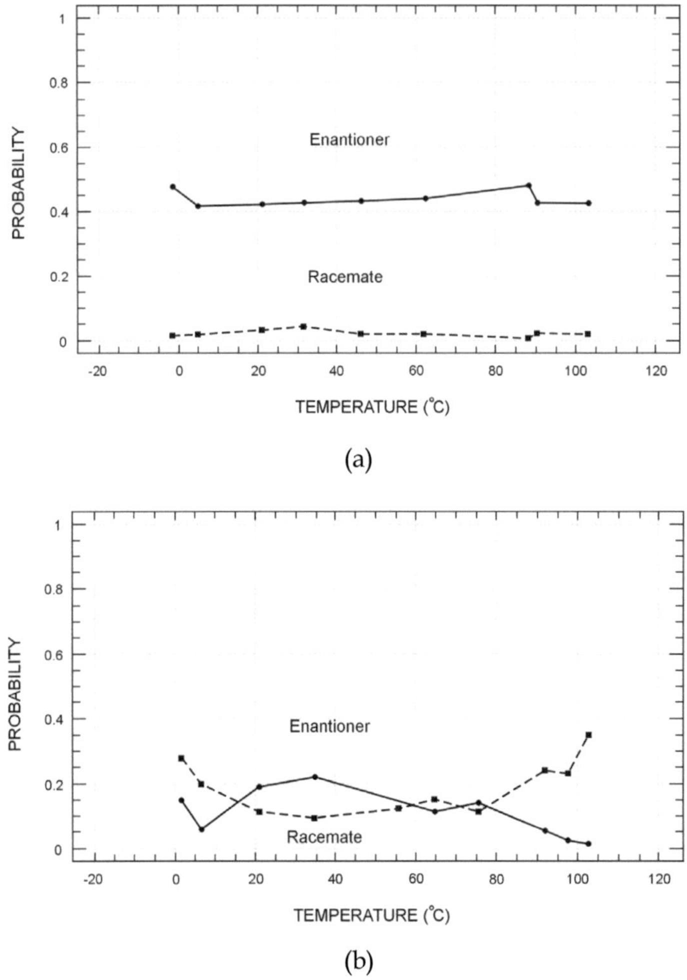

Fig. 15. Effect of temperature on probability of nucleation of pure enantiomers and racemates (50:50 mixture) of sodium bromate in unstirred solution. Growth conditions: (a) low supersaturation (closed, small crystallizers), and (b) high supersaturation (open crystallizers). Probability of nucleation of pure enantiomers is arithmetic average of L and D form.

In open crystallizers temperature affects the formation of enantiomers and racemates (Figures 15 (b), 16 (b)). It is seen that three temperature regions may be considered: low-temperature region (between -2 and +10 °C), medium-temperature region (between +10 and 80 °C), and high temperature region (between 80 and 109 °C). We can see that in open crystallizers, at high supersaturations PR is relatively high (between 0.10 and 0.27) in the whole temperature range, and the probabilities PL and PD of the formation of enantiomers are relatively low (between 0.01 and 0.2) in comparison with nucleation in closed crystallizers. At medium temperatures both probabilities PR and the mean probability of the formation of

enantiomers (PL+PD)/2) are comparable, and does nor vary significantly with the temperature. At low and at high temperatures probability PR of formation of racemates is high, and probabilities of formation of enantiomers are low.
To reveal both effects: temperature and the supersaturation the probabilities of formation of enantiomers and racemates have been drawn separately, one for enantiomers (Fig. 16 (a)) and another for racemates (Fig. 16 (b)). Figs. 15 (a) and 16 (a) illustrate that parity violation in unstirred crystallization occurs at any temperature and this is a common phenomenon, with high values of probabilities for the formation of enantiomers, on an average about 0.4 for each of enantiomers. Such a high probability cannot be neglected in the considerations of chiral symmetry breaking phenomenon.

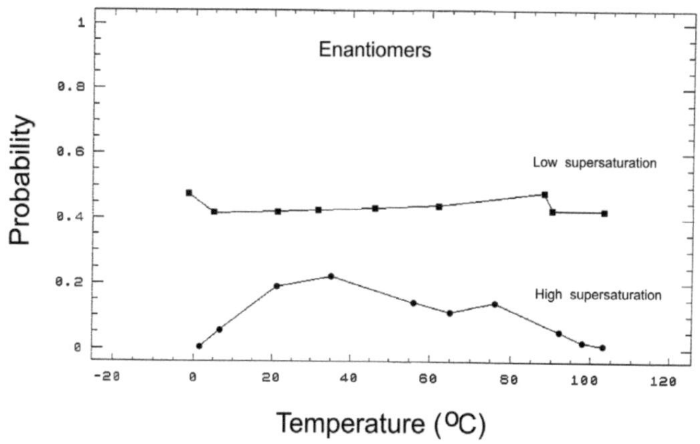

(a)

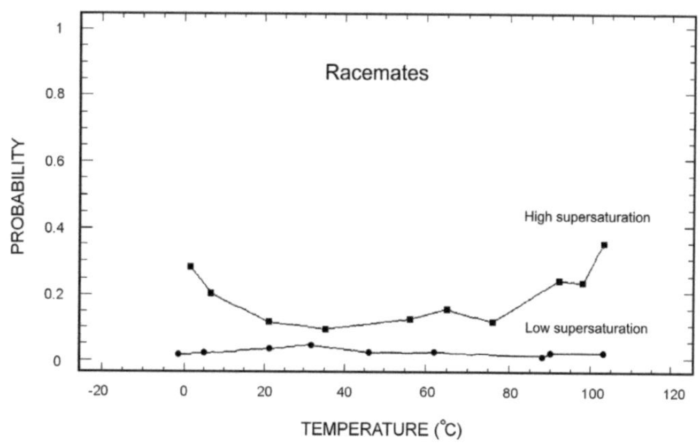

(b)

Fig. 16. Effect of temperature and supersaturation on probability of nucleation of: (a) enantiomers, (b) racemates.

4. Conclusions

First nucleated crystals govern the handedness of secondary crystals and depending on their handedness, nucleation temperature and supersaturation various unimodal, bimodal and trimodal distributions of enantiomeric excess are obtained. Nucleation temperature affects the distribution of enantiomers. Simultaneous seeding of L and D enantiomers leads to a unimodal, broad distribution of enantiomeric excess with maximum at ee = 0, i.e. for a 50:50 racemate mixture. L forms of sodium bromate have a greater probability of spontaneous nucleation than D forms. Unstirred crystallization of sodium bromate leads to a strong breakage of chiral symmetry, formation of pure enantiomers and bimodal distributions of enantiomeric excess, the same type as discovered in stirred solutions.

5. Acknowledgement

I would like to thank Mrs. Jadwiga Szurgot, my wife for her interest in the work, encouragement and stimulating discussion on the results. I would like to express my gratitude to Professor Aleksandar Lazinica, the Book Editor for his valuable comments and suggestions.

6. References

Kondepudi, D.P.; Kaufman, R.J. & Singh N. 1990. Chiral Symmetry Breaking in Sodium Chlorate Crystallization. *Science,* Vol.250, (16 November 1990), pp. 975-976, ISSN 0036-8075

Kondepudi, D.P.; Bullock, K.L.; Digits, J.A.; Hall, J.K.; & Miller, J.M. 1993. Kinetics of Chiral Symmetry Breaking in Crystallization. *Journal of the American Chemical Society,* Vol.115, No.22 (November 1993), pp. 10211-10216, ISSN 0002-7863

Kondepudi, D.P.; Bullock, K.L.; Digits, J.A.; Hall, J.K. & Yarborough, P.D. 1995a. Stirring Rate as a Critical Parameter in Chiral Symmetry Breaking Crystallization. *Journal of the American Chemical Society,* Vol.117, No.1 (January 1995), pp. 401-404, ISSN 0002-7863

Kondepudi, D.P.; Digits, J.A.; Bullock, K.L.; & . 1995b. Studies in Chiral Symmetry Breaking Crystallization I: The Effects of Stirring and Evaporation Rates. *Chirality,* Vol.7, No.2 (March 1995), pp. 62-68, ISSN 1520-636

Kondepudi, D.P. & Sabanayagam, C. 1994. Secondary nucleation that leads to Chiral Symmetry Breaking in Stirred Crystallization. *Chemical Physics Letter,* Vol.217, No.4 (21 January 1994), pp. 364-367, ISSN 0009-2614

McBride, J.M. & Carter R.L. 1991. Spontaneous Resolution by Stirred Crystallization. *Angewandte Chemie,* Vol.30, No.3 (March 1991), pp. 293-295, ISSN 0570-0833

Qian, R.Y. & Botsaris, G.D. 1998. Nuclei Breeding from a Chiral Crystal Seed of $NaClO_3$. *Chemical Engineering Science,* Vol.53, No.9 (1 May 1998), pp. 1745-1756, ISSN 0009-2509

Qian, R.Y. & Botsaris, G.D. 2004. Effect of Seed Preparation on the Chirality of the Secondary Nuclei. *Chemical Engineering Science,* Vol.59, No.14 (July 2004), pp. 2841-2852, ISSN 0009-2509

Veintemillas-Verdaguer, S.; Osuna Esteban S. & Herrero, M.A. 2007. The Effect of Stirring on Sodium Chlorate Crystallization under Symmetry Breaking Conditions. *Journal of Crystal Growth,* Vol.303, No. 2 (15 May 2007), pp. 562-567, ISSN 0022-0248

Viedma, C. 2004. Experimental Evidence of Chiral Symmetry Breaking in Crystallization from Primary Nucleation. *Journal of Crystal Growth,* Vol.261, No.1 (15 January 2004), pp. 118-121, ISSN 0022-0248

Cartwright, J.H.E.; Garcia-Ruiz, J.M.; Piro, O.; Sainz-Diaz, C.I.; & Tuval, I. 2004. Chiral Symmetry Breaking During Crystallization: An Advection-Mediated Nonlinear Autocatalytic Process. *Physical Review Letters,* Vol.93, No.3 (16 July 2004), pp. 035502-1-035502-4, ISSN 0031-9007

Martin, B.; Tharrington, A. & Wu X-I. 1996. Chiral Symmetry Breaking in Crystal Growth: Is Hydrodynamic Convection Relevant? *Physical Review Letters,* Vol.77, No.13 (23 September 1996), pp. 2826-2829, ISSN 0031-9007

Metcalfe, G. & Ottino, J.M. 1994. Autocatalytic Processes in Mixing Flows. *Physical Review Letters,* Vol.72, No.18 (2 May 1994), pp. 2875-2878, ISSN 0031-9007

Bushe, T.; Durand, D.; Kondepudi, D.; Laudadio, J. & Spilker, S. 2000. Chiral Symmetry Breaking in Crystallization: Role of Convection. *Physical Review Letters,* Vol.84, No.19 (8 May 2000), pp. 4405-4408, ISSN 0031-9007

Szurgot, J. & Szurgot, M. 1995a. Enantiomorphism in Sodium Chlorate Crystals. *Crystal Research and Technology,* Vol.30, No.1 (January 1995), pp. 71-79, ISSN 1521-4079

Szurgot, M. & Szurgot, J. 1995b. Chiral Symmetry Breaking in Sodium Chlorate Crystallization from Unstirred Solution. *Crystal Research and Technology,* Vol.30, No.7 (September 1995), pp. 949-956, ISSN 1521-4079

Vogl, O., Qin, M., Bartus, J. & Jaycox G.D. 1995. Chiral Nucleation. *Monatshefte fur Chemie,* Vol.126, No. 1 (January 1995), pp. 67-73, ISSN 0026-9247

Szurgot, M. 1995. Chiral Etch Pits on Sodium Chlorate Crystals. *Crystal Research and Technology,* Vol.30, No.5 (July 1995), pp. 621-628, ISSN 1521-4079

Szurgot, J. & Szurgot, M. 1994. On the Nature of Growth Hillocks on Sodium Chlorate Crystals. *Crystal Research and Technology,* Vol.29, No.6 (June 1994), pp. 829-836, ISSN 1521-4079

Szurgot, M. & Sangwal, K. 1987. Influence of Free Convection on the Formation of Growth Defects in Potassium Bichromate Crystals Grown from Aqueous Solutions. *Crystal Research and Technology,* Vol.22, No.12 (December 1987), pp. 1477-1484, ISSN 1521-4079

Wilcox, W.R. 1983. Influence of Convection on the Growth of Crystals from Solutions. *Journal of Crystal Growth,* Vol.65, No.1-3 (2 December 1983), pp. 133-142, ISSN 0022-0248

Szurgot, M. 2010. Investigations of Impurity Striations in Potassium Bichromate Crystals. *Crystal Research and Technology,* Vol.45, No.4 (June 2010), pp. 347-354, ISSN 1521-4079

Szurgot, M. 1991. Effect of Buoyancy-driven Convection on Surface Morphology and Macromorphology of Potassium Bichromate Crystals. *Crystal Research and Technology,* Vol.26, No.1 (January 1991), pp. 43-52, ISSN 1521-4079

Frank, F.C. 1953. On Spontaneous Asymmetric Synthesis. *Biochimica et Biophysica Acta*, Vol.11, (1953), pp. 459-463, ISSN 0006-3002

Denk, E.G & Botsaris, G.D. 1972. Fundamental Studies in Secondary Nucleation from Solution. *Journal of Crystal Growth*, Vol.13/14, No.14 (May 1972), pp. 493-499, ISSN 0022-0248